Solid State Physics

Solid State Physics

Hilary. D. Brewster

Oxford Book Company
Jaipur ▪ New Delhi

Published By

Oxford Book Company

Regd. Off. "Bony Residency", Gate No. 2
Opp. Tilak Public School, Vishweswariya Nagar,
Gopalpura Road, Jaipur - 302018, India
Ph: 0141 - 2761280, Telefax: 2761381
email - oxfordbookjpr@gmail.com
website - oxfordbookcompany.com

Delhi Office: 102, 1st Floor, Satyam House,
4327/3, Ansari Road, Darya Ganj,
New Delhi - 110002, India
Ph: 011-45652440

Published in India, 2025

ISBN : 9789380179070

Price : ₹ 2250/-

Typeset by Shivangi Computers, Jaipur

Printed and bound in India by Thomson Press Ltd.,
New Delhi

Preface

Solid-state physics, the largest branch of condensed matter physics, is the study of rigid matter, or solids, through methods such as quantum mechanics, crystallography, electromagnetism and metallurgy. Solid-state physics considers how the large-scale properties of solid materials result from their atomic-scale properties. Solid-state physics thus forms the theoretical basis of materials science, as well as having direct applications, for example in the technology of transistors and semiconductors.

The introductory text, written by author is designed to become a landmark in the field of solid state physics through its organized unification of key topics. This book provides necessary information about solid state physics. each topic discussed are explained with selected examples and diagrams. An extraordinarily readable writing style combines with chapter opening principles, study problems and beautiful illustration to make this book an ideal choice for students and teachers of physics.

Hilary. D. Brewster

Contents

Chapter 1

Energy Dispersion Relations in Solids

The transport properties of solids are closely related to the energy dispersion relations $E\ (\vec{k})$ in these materials and in particular to the behavior of $E\ (\vec{k})$ near the Fermi level. Conversely, the analysis of transport measurements provides a great deal of information on $E(\vec{k})$. Although transport measurements donot generally provide the most sensitive tool for studying $E(\vec{k})$, such measurements are fundamental to solid state physics because they can be carried out on nearly all materials and therefore provide a valuable tool for char-acterizing materials. To provide the necessary background for the discussion of transport properties, we give here a brief review of the energy dispersion relations $E\ (\vec{k})$ in solids. In this connection, the two limiting cases of weak and tight binding.

ONE ELECTRON $E\ (\vec{k})$ IN SOLIDS

In the weak binding approximation, we assume that the periodic potential $V\ (\vec{r}) = V\ (\vec{r} + \vec{R}_n)$ is sufficiently weak so that the electrons behave almost as if they were free and the effect of the periodic potential can be handled in perturbation theory. In this formulation $V\ (\vec{r})$ can be an arbitrary periodic potential. The weak binding approximation has achieved some success in describing the valence electrons in metals. For the core elec-trons, however, the potential energy is comparable with the kinetic energy so that core electrons are tightly bound and the weak binding approximation is not applicable. In the weak binding approximation we solve the SchrÄodinger equation in the limit of a very weak periodic potential

$$H\psi = E\psi.$$

Using time{independent perturbation theory we write

$$E(\vec{k}) = E^0(\vec{k}) + E^{(1)}(\vec{k}) + E^{(2)}(\vec{k}) + \cdots$$

and take the unperturbed solution to correspond to $V\ (\vec{r}) = 0$ so that $E^{(0)}(\vec{k})$

is the plane wave solution

$$E^{(0)}(\vec{k}) = \frac{\hbar^2 k^2}{2m}.$$

The corresponding normalized eigenfunctions are the plane wave states

$$\psi_{\vec{k}}^{0}(\vec{r}) = \frac{e^{i\vec{k}\cdot\vec{r}}}{\Omega^{1/2}}$$

in which Ω is the volume of the crystal.

The first order correction to the energy $E^{(1)}(\vec{k})$ is the diagonal matrix element of the perturbation potential taken between the unperturbed states:

$$E^{(1)}(\vec{k}) = \left\langle \psi_{\vec{k}}^{(0)} \,|\, V(\vec{r}) \,|\, \psi_{\vec{k}}^{(0)} \right\rangle = \frac{1}{\Omega}\int_{\Omega} e^{-i\vec{k}\cdot\vec{r}} V(\vec{r}) e^{i\vec{k}\cdot\vec{r}} d^3r$$

$$\frac{1}{\Omega_0}\int_{\Omega_0} V(\vec{r}) d^3r = \overline{V(\vec{r})}$$

where $\overline{V(\vec{r})}$ is independent of $\vec{k}$, and Ω_0 is the volume of the unit cell. Thus, in first order perturbation theory, we merely add a constant energy $\overline{V(\vec{r})}$ to the free particle energy, and that constant term is exactly the mean potential energy seen by the electron, averaged over the unit cell. The terms of interest arise in second order perturbation theory and are

$$E^{(2)}(\vec{k}) = \sum_{\vec{k}'}{}' \frac{\left|\langle \vec{k}' \,|\, V9\vec{r}) \,|\, \vec{k} \rangle\right|^2}{E^{(2)}(\vec{k}) - E^{(0)}(\vec{k}')}$$

where the prime on the summation indicates that $\vec{k}' \neq \vec{k}$.. We next compute the matrix element $\langle \vec{k}' \,|\, V(\vec{r}) \,|\, \vec{k} \rangle$ as follows:

$$\langle \vec{k}' \,|\, V(\vec{r}) \,|\, \vec{k} \rangle = \int_{\Omega} \psi_{\vec{k}'}^{(0)*} V(\vec{r}) \psi_{\vec{k}}^{(0)} d^3r$$

$$= \frac{1}{\Omega}\int_{\Omega} e^{-i(\vec{k}'-\vec{k})\cdot} V(\vec{r}) d^3r$$

$$= \frac{1}{\Omega}\int_{\Omega} e^{i\vec{q}\cdot\vec{r}} V(\vec{r}) d^3r$$

where $\vec{q}$ is the difference wave vector $\vec{q} = \vec{k} - \vec{k}'$ and the integration is over the whole crystal. We now exploit the periodicity of $V(\vec{r})$. Let $\vec{r} =| \vec{r}' + \vec{R}_n$ where $\vec{r}'$ is an arbitrary vector in a unit cell and $\vec{R}_n$ is a lattice vector. Then since $V(\vec{r}) = V(\vec{r})$

$$\langle \vec{k}' \,|\, V(\vec{r}) \,|\, \vec{k} \rangle = \frac{1}{\Omega}\sum_{n}\int_{\Omega_0} e^{i\vec{q}\cdot(\vec{r}+\vec{R}_n)} V(\vec{r}') d^3r'$$

where the sum is over unit cells and the integration is over the volume of one unit cell.

Then

$$\langle \vec{k}' | V(\vec{r}) | \vec{k} \rangle = \frac{1}{\Omega} \sum_n \int_{\Omega_0} e^{i\vec{q}\cdot\vec{R}_n} V(\vec{r}') d^3 r'.$$

Writing the following expressions for the lattice vectors $\vec{R}_n$ and for the wave vector $\vec{q}$

$$\vec{R}_n = \sum_{j=1}^{3} n_j \vec{a}_j$$

$$\vec{q} = \sum_{j=1}^{3} \alpha_j \vec{b}_j$$

where n_j is an integer, then the lattice sum $\sum_n e^{i\vec{q}\cdot\vec{R}_n}$ can be carried out exactly to yield

$$\sum_n e^{i\vec{q}\cdot\vec{R}_n} = \left[\prod_{j=1}^{3} \frac{1 - e^{2\pi i N_j \alpha_j}}{1 - e^{2\pi i \alpha_j}} \right]$$

where $N = N_1 N_2 N_3$ is the total number of unit cells in the crystal and a_j is a real number. This sum fluctuates wildly as $\vec{q}$ varies and is appreciable only if

$$\vec{q} = \sum_{j=1}^{3} m_j \vec{b}_j$$

where m_j is an integer and $\vec{b}_j$ is a primitive vector in reciprocal space, so that $\vec{q}$ must be a reciprocal lattice vector. Hence we have

$$\sum_n e^{i\vec{q}\cdot\vec{R}_n} = N \delta_{\vec{q},\vec{G}}$$

since $\vec{b}_j \cdot \vec{R}_n = 2\pi l_{jn}$ where l_{jn} is an integer.

This discussion shows that the matrix element $\langle \vec{k}' | V(\vec{r}) | \vec{k} \rangle$ is only important when $\vec{q} = \vec{G}$ is a reciproczal lattice vector = $\vec{k} - \vec{k}'$ from which we conclude that the periodic potential $\vec{k} - \vec{k}'$ only connects wave vectors $\vec{k}$ and $\vec{k}'$ separated by a reciprocal lattice vector. We note that this is the same relation that determines the Brillouin zone boundary. The matrix element is then

$$\langle \vec{k}' | V(\vec{r}) | \vec{k} \rangle = \frac{N}{\Omega} \int_{\Omega_0} e^{i\vec{G}\cdot\vec{r}'} V(\vec{r}') d^3 r' \delta_{\vec{k}'-\vec{k},\vec{G}}$$

where

$$\frac{N}{\Omega} = \frac{1}{\Omega_0}$$

and the integration is over the unit cell. We introduce $V_{\vec{G}}$ = Fourier coefficient of $V(\vec{r})$ where

$$V_{\vec{G}} = \frac{1}{\Omega_0} \int_{\Omega_0} e^{i\vec{G}\cdot\vec{r}'} V(\vec{r}') d^3 r'$$

so that

$$\langle \vec{k}' | V(\vec{r}) | \vec{k} \rangle \delta_{\vec{k}-\vec{k}',\vec{G}} V_{\vec{G}}.$$

We can now use this matrix element to calculate the 2nd order change in the energy

$$E^{(2)}(\vec{k}) = \sum_{\vec{G}} \frac{|V_{\vec{G}}|^2}{k^2 - (k')^2} \left(\frac{2m}{\hbar^2} \right) = \frac{2m}{\hbar^2} \sum_{\vec{G}} \frac{(V_{\vec{G}}|^2}{k^2 - (\vec{G}+\vec{k})^2}.$$

We observe that when $k^2 = (\vec{G}+\vec{k})^2$ the denominator vanishes and $E^{(2)}(\vec{k})$ can become very large. This condition is identical with the Laue diffraction condition. Thus, at a Brillouin zone boundary, the weak perturbing potential has a very large effect and therefore non degenerate perturbation theory will not work in this case.

For $\vec{k}$ values near a Brillouin zone boundary, we must then use degenerate perturbation theory. Since the matrix elements coupling the plane wave states $\vec{k}$ and $\vec{k}+\vec{G}$ do not vanish, first order degenerate perturbation theory is sufficient and leads to the determinantal equation

$$\begin{vmatrix} e^{(0)}(\vec{k}) + E^{(1)}(\vec{k}) - E & \langle \vec{E}+\vec{G} | V(\vec{r}) | \vec{k} \rangle \\ \langle \vec{k} | V(\vec{r}) | \vec{k}+\vec{G} \rangle & E^{(0)}(\vec{k}+\vec{G}) + E^{(1)}(\vec{k}+\vec{G}) - E \end{vmatrix} = 0$$

in which

$$E^{(0)}(\vec{k}) = \frac{\hbar^2 k^2}{2m}$$

$$E^{(0)}(\vec{k}+\vec{G}) = \frac{\hbar^2 \vec{k}+\vec{G})^2}{2m}$$

and

$$E^{(1)}(\vec{k}) = \langle \vec{k} | V(\vec{r}) | \vec{k} \rangle = \overline{V(\vec{r})} = V_0$$

$$E^{(1)}(\vec{k}+\vec{G}) = \langle \vec{k}+\vec{G} | V(\vec{r}) | \vec{k}+\vec{G} \rangle = V_0.$$

Solution of this determinantal equation yields:

$$\left[E - V_0 - E^{(0)}(\vec{k}) \right] \left[E - V_0 - E^{(0)}(\vec{k}+\vec{G}) \right] - |V_{\vec{G}}|^2 = 0,$$

or equivalently

$$E^2 - E[2V_0 + E^{(0)}(\vec{k}) + E^{(0)}(\vec{k}+\vec{G})] + [V_0 + E^{(0)}(\vec{k})][V_0 + E^{(0)}(\vec{k}+\vec{G})] - |V_{\vec{G}}|^2 = 0.$$

Solution of the quadratic equation yields

$$E^{\pm} = V_0 + \frac{1}{2}[E^{(0)}(\vec{k}) + E^{(0)}(\vec{k}+\vec{G})] \pm \sqrt{\frac{1}{4}[E^{(0)}(\vec{k}) - E^{(0)}(\vec{k}+\vec{G})]^2 + |V_{\vec{G}}|^2}$$

and we come out with two solutions for the two strongly coupled states. It is of interest to look at these two solutions in two limiting cases:

$$|V_{\vec{G}}| \ll \frac{1}{2}|[E^{(0)}(\vec{k}) - E^{(0)}(\vec{k}+\vec{G})]|$$

In this case we can expand the square root expression in equation for small $|V_{\vec{G}}|$ to obtain:

$$E(\vec{k}) = V_0 + \frac{1}{2}[E^{(0)}(\vec{k}) + E^{(0)}(\vec{k}+\vec{G})]$$

$$\pm \frac{1}{2}[E^{(0)}(\vec{k}) - E^{(0)}(\vec{k}+\vec{G})] \cdot \left[1 + \frac{2|V_G|^2}{[E^{(0)}(\vec{k}) - E^{(0)}(\vec{k}+\vec{G})]^2} + \cdots\right]$$

which simplifies to the two solutions:

$$E^{-}(\vec{k}) = V_0 + E^{(0)}(\vec{k}) + \frac{|V_{\vec{G}}|^2}{e^{(0)}(\vec{k}) - E^{(0)}(\vec{k}+\vec{G})}$$

$$E^{+}(\vec{k}) = V_0 + E^{(0)}(\vec{k}+\vec{G}) + \frac{|V_{\vec{G}}|^2}{E^{(0)}(\vec{k}+\vec{G}) - E^{(0)}(\vec{k})}$$

and we recover the result equation obtained before using non{degenerate perturbation theory. This result in equation is valid far from the Brillouin zone boundary, but near the zone boundary the more complete expression of equation must be used.

case (ii)

$$|V_{\vec{G}}| \gg \frac{1}{2}|[E^{(0)}(\vec{k}) - E^{(0)}(\vec{k}+\vec{G})]|$$

Sufficiently close to the Brillouin zone boundary

$$|E^{(0)}(\vec{k}) - E^{(0)}(\vec{k}+\vec{G})| \ll |V_{\vec{G}}|$$

so that we can expand $E(\vec{k})$ as given by equation to obtain

$$E^{\pm}(\vec{k}) = \frac{1}{2}[E^{(0)}(\vec{k}) + E^{(0)}(\vec{k}+\vec{G})] + V_0 \pm$$

$$\left[|V_{\vec{G}}| + \frac{1}{8}\frac{[E^{(0)}(\vec{k}) - E^{(0)}(\vec{k}+\vec{G})]^2}{|V_{\vec{G}}|} + \cdots\right]$$

$$\cong \frac{1}{2}[E^{(0)}(\vec{k}) + E^{(0)}(\vec{k}+\vec{G})] + V_0 \pm |V_{\vec{G}}|,$$

so that at the Brillouin zone boundary $E^{+}(\vec{k})$ is elevated by $|V_{\vec{G}}|$, while $E^{-}(\vec{k})$ is depressed by $|V_{\vec{G}}|$ and the band gap that is formed is $2|V_{\vec{G}}|$, where

$\vec{G}$ is the reciprocal lattice vector for which $E(\vec{k}_{B,Z.}) = E(\vec{k}_{B,Z.} + \vec{G})$ and

$$V_{\vec{G}} = \frac{1}{\Omega_0} \int_{\Omega_0} e^{i\vec{G}\cdot\vec{r}} V(\vec{r}) d^3r.$$

From this discussion it is clear that every Fourier component of the periodic potential gives rise to a specific band gap. We see further that the band gap represents a range of energy values for which there is no solution to the eigenvalue problem of equation for real k. In the band gap we assign an imaginary value to the wave vector which can be interpreted as a highly damped and non{propagating wave.

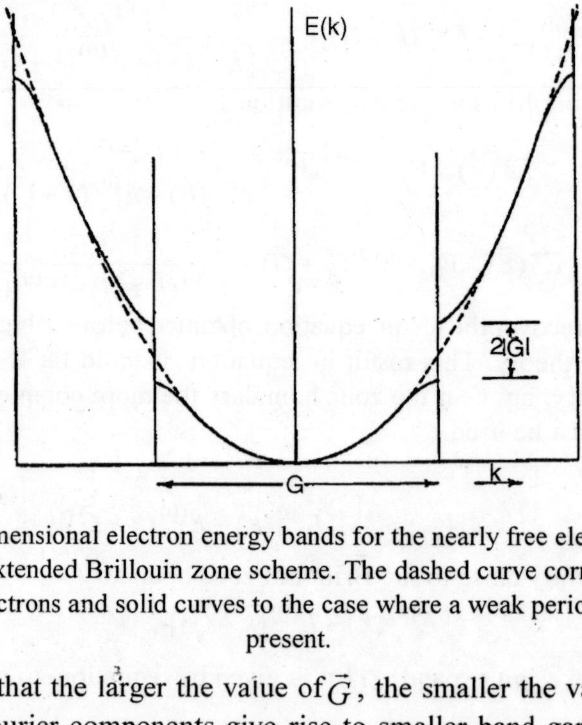

Fig. One dimensional electron energy bands for the nearly free electron model shown in the extended Brillouin zone scheme. The dashed curve corresponds to the case of free electrons and solid curves to the case where a weak periodic potential is present.

We note that the larger the value of $\vec{G}$, the smaller the value of $V_{\vec{G}}$, so that higher Fourier components give rise to smaller band gaps. Near these energy discontinuities, the wave functions become linear combinations of the unperturbed states

$$\psi_{\vec{k}} = \alpha_1 \psi_{\vec{k}}^{(0)} + \beta_1 \psi_{\vec{k}+\vec{G}}^{(0)}$$

$$\psi_{\vec{k}+\vec{G}} = \alpha_2 \psi_{\vec{k}}^{(0)} + \beta_2 \psi_{\vec{k}+\vec{G}}^{(0)}$$

and at the zone boundary itself, instead of traveling waves $e^{i\vec{k}\cdot\vec{r}}$, the wave functions become standing waves cos $\vec{k}\cdot\vec{r}$ and sin $\vec{k}\cdot\vec{r}$. We note that the cos $(\vec{k}\cdot\vec{r})$ solution corresponds to a maximum in the charge density at the lattice sites and therefore corresponds to an energy minimum (the lower level). Likewise, the sin $(\vec{k}\cdot\vec{r})$ solution corresponds to a minimum in the charge density and therefore corresponds to a maximum in the energy, thus forming the upper level.

In constructing $E(\vec{k})$ for the reduced zone scheme we make use of the periodicity of $E(\vec{k})$ in reciprocal space

$$E(\vec{k}+\vec{G}) = E(\vec{k}).$$

The reduced zone scheme more clearly illustrates the formation of energy bands, band gaps E_g and band widths. We now discuss the connection between the $E(\vec{k})$ relations shown above and the trans-port properties of solids, which can be illustrated by considering the case of a semiconductor. An intrinsic semiconductor at temperature $T = 0$ has no carriers so that the Fermi level runs right through the band gap. This would mean that the Fermi level might run between bands (1) and (2), so that band (1) is completely occupied and band (2) is completely empty.

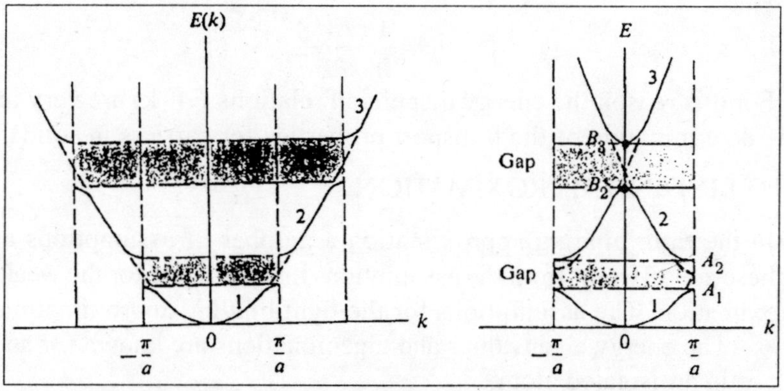

Fig. (a) One dimensional electron energy bands for the nearly free electron model shown in the extended Brillouin zone scheme for the three bands of lowest energy. (b) The same $E(\vec{k})$ as in (a) but now shown on the reduced zone scheme. The shaded areas denote the band gaps and the white areas the band states.

One further property of the semiconductor is that the band gap E_g be small enough so that at some temperature (e.g., room temperature) there will be reasonable numbers of thermally excited carriers, perhaps 10^{15} /cm^3. The doping with donor (electron donating) impurities will raise the Fermi level and doping with accep-tor (electron extracting) impurities will lower the Fermi level. Neglecting for the moment the effect of impurities on the $E(\vec{k})$ relations for the perfectly periodic crystal, let us con-sider what happens when we raise the Fermi level into the bands. If we know the shape of the $E(\vec{k})$ curve, we are in a position to estimate the velocity of the electrons and also the so called effective mass of the electrons. We see that the conduction bands tend to fill up electron states starting at their energy extrema.

Since the energy bands have zero slope about their extrema, we can write $E(\vec{k})$ as a quadratic form in $\vec{k}$. It is convenient to write the proportionality in terms of the quantity called the effective mass m^*

$$E(\vec{k}) = E(0) + \frac{\hbar^2 k^2}{2m^*}$$

so that m^* is defined by

$$\frac{1}{m^*} \equiv \frac{\partial^2 E(\vec{k})}{\hbar^2 \partial k^2}$$

and we can say in some approximate way that an electron in a solid moves as if it were a free electron but with an effective mass m£ rather than a free electron mass. The larger the band curvature, the smaller the effective mass. The mean velocity of the electron is also

found from $E(\vec{k})$, according to the relation

$$\vec{u}_k = \frac{1}{\hbar}\frac{\partial E(\vec{k})}{\partial \vec{k}}.$$

For this reason the energy dispersion relations E (~k) are very important in the determination of the transport properties for carriers in solids.

TIGHT BINDING APPROXIMATION

In the tight binding approximation a number of assumptions are made and these are different from the assumptions that are made for the weak binding approximation. The assumptions for the tight binding approximation are:

- The energy eigenvalues and eigenfunctions are known for an electron in an isolated atom.
- When the atoms are brought together to form a solid they remain sufficiently far apart so that each electron can be assigned to a particular atomic site. This assumption is not valid for valence electrons in metals and for this reason, these valence electrons are best treated by the weak binding approximation.
- The periodic potential is approximated by a superposition of atomic potentials.
- Perturbation theory can be used to treat the difference between the actual potential and the atomic potential.

Thus both the weak and tight binding approximations are based on perturbation theory. For the weak binding approximation the unperturbed state is the free electron plane{wave state while for the tight binding approximation the unperturbed state is the atomic state. In the case of the weak binding approximation, the perturbation Hamiltonian is the weak periodic potential itself, while for the tight binding case, the perturbation is the differ-ence between the periodic potential and the atomic potential around which the electron is localized.

We review here the major features of the tight binding approximation. Let $\phi(\vec{r}-\vec{R}_n)$ represent the atomic wave function for an atom at a lattice position denoted by ~R$_{n}$, which is measured with respect to the origin. The SchrÄodinger equation for an electron in an isolated atom is then:

$$\left[-\frac{\hbar^2}{2m}\nabla^2 + U(\vec{r}-\vec{R}_n) - E^{(0)}\right]\phi(\vec{r}-\vec{R}_n) = 0$$

where $U(\vec{r}-\vec{R}_n)$ is the atomic potential and $E^{(0)}$ is the atomic eigenvalue. We now assume that the atoms are brought together to form the crystal for which $V(\vec{r})$ is the periodic potential, and $\psi(\vec{r})$ and $E(\vec{k})$ are, respectively, the wave function and energy eigenvalue for the electron in the crystal:

$$\left[-\frac{\hbar^2}{2m}\Delta^2 + V(\vec{r}) - E\right]\psi(\vec{r}) = 0.$$

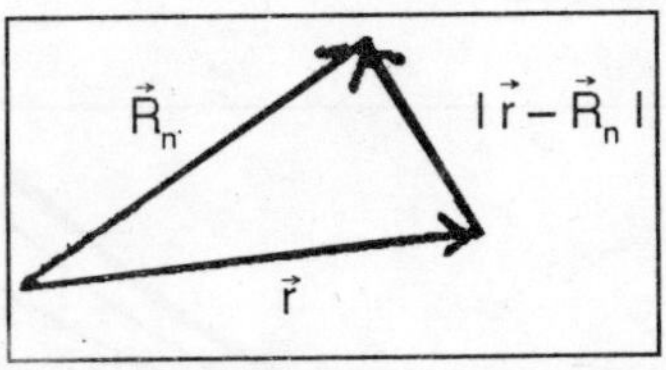

Fig. Definition of the vectors used in the tight binding approximation.

In the tight binding approximation we write $V(\vec{r})$ as a sum of atomic potentials:

$$V(\vec{r}) \simeq \sum_n U(\vec{r}-\vec{R}_n).$$

If the interaction between neighboring atoms is ignored, then each state has a degeneracy of N = number of atoms in the crystal. However, the interaction between the atoms lifts this degeneracy.

The energy eigenvalues $E(\vec{k})$ in the tight binding approximation for a non{degenerate s{state is simply given by

$$E(\vec{k}) = \frac{\langle \vec{k} \mid H \mid \vec{k} \rangle}{\langle \vec{k} \mid \vec{k} \rangle}.$$

The normalization factor in the denominator $\langle \vec{k} \mid \vec{k} \rangle$ is inserted because the wave functions $\psi_{\vec{k}}(\vec{r})$ in the tight binding approximation are usually not normalized. The Hamiltonian in the tight binding approximation is written as

$$H = -\frac{\hbar^2}{2m}\nabla^2 + V(\vec{r}) = \left\{-\frac{\hbar^2}{2m}\nabla^2 + [V(\vec{r}) - U(\vec{r} - \vec{R}_n)] + U(\vec{r} - \vec{R}_n)\right\}$$

$$H = H_0 + H'$$

in which H_0 is the atomic Hamiltonian at site n

$$H_0 = -\frac{\hbar^2}{2m}\nabla^2 + U(\vec{r} - \vec{R}_n)$$

and H' is the difference between the actual periodic potential and the atomic potential at lattice site n

$$H' = V(\vec{r}) - U(\vec{r} - \vec{R}_n).$$

We construct the wave functions for the unperturbed problem as a linear combination of atomic functions $\phi_j(\vec{r} - \vec{R}_n)$ labeled by quantum number j

$$\psi_j(\vec{r}) = \sum_{n=1}^{N} C_{j,n}\phi_j(\vec{r} - \vec{R}_n)$$

and so that $\psi_j(\vec{r})$ is an eigenstate of a Hamiltonian satisfying the periodic potential of the lattice. In this treatment we assume that the tight binding wave-functions ψ_j can be

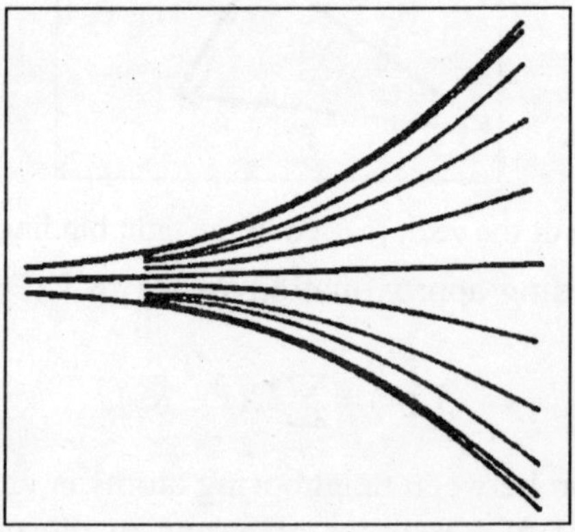

Fig. The relation between atomic states and the broadening due to the presence of neighboring atoms. As the interatomic distance decreases (going to the right in the diagram), the level broadening increases so that a band of levels occurs at atomic separations characteristic of solids.

identified with a single atomic state ψ_j; this approximation must be relaxed in dealing with degenerate levels. According to Bloch's theorem $\psi_j(\vec{r})$ must satisfy the relation:

$$\psi_j(\vec{r} - \vec{R}_m) = e^{i\vec{k}\cdot\vec{R}_m}\psi_j(\vec{r})$$

where $\vec{R}_m$ is an arbitrary lattice vector. This restriction imposes a special form on the coefficients $C_{j,n}$.

Substitution of the expansion in atomic functions $\psi_j(\vec{r})$ from equation into the left side of equation yields:

$$\psi_j(\vec{r}+\vec{R}_n) = \sum_n C_{j,n}\phi_j(\vec{r}-\vec{R}_n+\vec{R}_m)$$
$$= \sum_Q C_{j,Q+m}\phi_j(\vec{r}-\vec{R}_Q)$$
$$= \sum_n C_{j,n+m}\phi(\vec{r}-\vec{R}_n)$$

where we have utilized the substitution $\vec{R}_Q = \vec{R}_n - \vec{R}_m$ and the fact that Q is a dummy index. Now for the right side of the Bloch theorem we have

$$e^{i\vec{k}\cdot\vec{R}_m}\psi_j(\vec{r}) = \sum_n C_{j,n}e^{i\vec{k}\cdot\vec{R}_m}\phi_j(\vec{r}-\vec{R}_n).$$

The coefficients $C_{j,n}$ which relate the actual wave function $\psi_j(\vec{r})$ to the atomic functions $\phi_j(\vec{r}-\vec{R}_n)$ are therefore not arbitrary but must thus satisfy:

$$C_{j,n+m} = e^{i\vec{k}\cdot\vec{R}_m}C_{j,n}$$

which can be accomplished by setting:

$$C_{j,n} = \xi_j e^{i\vec{k}\cdot\vec{R}_n}$$

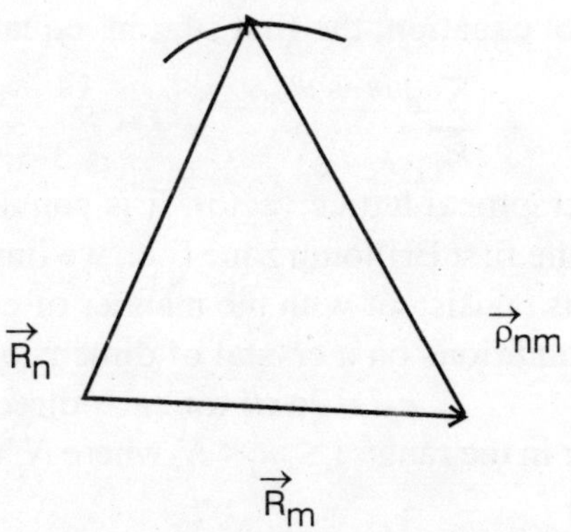

Fig. Definition of $\vec{\rho}_{nm}$ denoting the distance between atoms at $\vec{R}_m$ and $\vec{R}_n$.

where the new coefficient ξ_j is independent of n. We therefore obtain:

$$\psi_{j,\vec{k}}(\vec{r}) = \xi_j \sum_n e^{i\vec{k}\cdot\vec{R}_n}\phi_j(\vec{r}-\vec{R}_n)$$

where j is an index labeling the particular atomic state of degeneracy N and $\vec{k}$ is the quantum number for the translation operator and labels the Bloch state $\psi_{j,\vec{k}}(\vec{r})$.

For simplicity, we will limit the present discussion of the tight binding approximation to

s{bands (non degenerate atomic states) and therefore we can suppress the j index on the wave functions. (The treatment for p-bands is similar to what we will do here, but more complicated because of the degeneracy of the atomic states.) To find matrix elements of the Hamiltonian we write

$$\langle \vec{k}' | H | \vec{k} \rangle = |\xi|^2 \sum_{n,m} e^{i(\vec{k}\cdot\vec{R}_n)} \int_{\Omega} \phi^*(\vec{r} - \vec{R}_m) H \phi(\vec{r} - \vec{R}_n) d^3r$$

in which the integration is carried out throughout the volume of the crystal. Since II is a function which is periodic in the lattice, the only significant distance is

$$(\vec{R}_n - \vec{R}_m) = \vec{\rho}_{nm}$$

We then write the integral in equation as:

$$\langle \vec{k}' | H | \vec{k} \rangle = |\xi|^2 \sum_{\vec{R}_m} e^{i(\vec{k}\cdot\vec{k}')\cdot\vec{R}_m} \sum_{\vec{\rho}_{nm}} e^{ik\vec{\rho}_{nm}} H_{mn}(\vec{\rho}_{nm})$$

where we have written the matrix element H_{mn} $(\vec{\rho}_{nm})$ as

$$H_{mn}(\vec{\rho}_{nm}) = \int_{\Omega} \phi^*(\vec{r} - \vec{R}_m) H \phi(\vec{r} - \vec{R}_m - \vec{\rho}_{nm} d^3r = \int_{\Omega} \phi^*(\vec{r}' - \vec{\rho}_{nm}) d^3r'$$

We note here that the integral in equation depends only on $\vec{\rho}_{nm}$ and not on $\vec{R}_m$. According to equation, the first sum in equation is

$$\sum_{\vec{R}_m} e^{i(\vec{k}-\vec{k}')\cdot R_m} = \delta_{\vec{k}',\vec{k}+\vec{G}} N$$

where $\vec{G}$ is a reciprocal lattice vector. It is convenient to restrict the $\vec{k}$ vectors to lie within the first Brillouin zone (i.e., we limit ourselves to reduced wave vectors). This is consistent with the manner of counting states with the periodic boundary conditions on a crystal of dimension d on a side

$$k_i d = 2\pi m i \text{ for each direction } i$$

where m_i is an integer in the range $1 \leq m_i < N_i$ where $N_i \approx N^1 = 3$. From equation we have

$$k_i = \frac{2\pi m_i}{d}.$$

The maximum value that a particular m_i can assume is N_i and the maximum value for k_i is $2\pi = a$ at the Brillouin zone boundary since $N_i/d = 1/a$. With this restriction, $\vec{k}$ and $\vec{k}'$ must both lie within the 1st B.Z. and thus cannot differ by any reciprocal lattice vector other than $\vec{G} = 0$. We thus obtain the following form for the matrix ment of H (and also for the matrix elements H_0 and H'):

$$\langle \vec{k}' | H | \vec{k} \rangle = |\xi|^2 N \delta_{\vec{k},\vec{k}'} \sum_{\vec{\rho}_{nm}} e^{i\vec{k}\cdot\vec{\rho}_{nm}} H_{mn}(\vec{\rho}_{nm})$$

yielding the result

$$E(\vec{k}) = \frac{\langle \vec{k} | H | \vec{k} \rangle}{\langle \vec{k} | \vec{k} \rangle} = \frac{\sum_{\vec{\rho}_{nm}} e^{i\vec{k}\cdot\vec{\rho}_{nm}} H_{mn}(\vec{\rho}_{nm})}{\sum_{\vec{\rho}_{nm}} e^{i\vec{k}\cdot\vec{\rho}_{nm}} S_{mn}(\vec{\rho}_{nm})}$$

in which

$$\langle \vec{k}' | \vec{k} \rangle = |\xi|^2 \delta_{\vec{k},\vec{k}'} N \sum_{\vec{\rho}_{nm}} e^{i\vec{k}\cdot\vec{\rho}_{nm}} S_{mn}(\vec{\rho}_{nm})$$

where the matrix element $S_{mn}(\vec{\rho}_{nm})$ measures the overlap of atomic functions on different sites

$$S_{mn}(\vec{\rho}_{nm}) = \int_{\Omega} \phi^*(\vec{r})\phi(\vec{r})\phi(\vec{r} - \vec{\rho}_{nm}) d^3r.$$

The overlap integral $S_{mn}(\vec{\rho}_{nm})$ will be nearly 1 when $\vec{\rho}_{nm} = 0$ and will fall off rapidly as $\vec{\rho}_{nm}$ increases, which exemplifies the spirit of the tight binding approximation. By selecting $\vec{k}$ vectors to lie within the first Brillouin zone, the orthogonality condition on the $\psi_{\vec{k}}(\vec{r})$ is automa-tically satisfied. Writing $H = H_0 + H'$ yields:

$$H_{mn} = \int_{\Omega} \phi^*(\vec{r} - \vec{R}_n) \left[-\frac{\hbar^2}{2m}\nabla^2 + U(\vec{r} - \vec{R}_n) \right] \phi(\vec{r} - \vec{R}_n) d^3r$$

$$\int_{\Omega} \phi^*(\vec{r} - \vec{R}_m)[V(\vec{r}) - U(\vec{r} - \vec{R}_n)]\phi(\vec{r} - \vec{R}_n) d^3r$$

or

$$H_{mn} = E^{(0)} S_{mn}(\vec{\rho}_{nm}) + H'_{mn}(\vec{\rho}_{nm})$$

which results in the general expression for the tight binding approximation:

$$E(\vec{k}) = E^{(0)} + \frac{\sum_{\vec{\rho}_{nm}} e^{i\vec{k}\cdot\vec{\rho}_{nm}} H'_{mn}(\vec{\rho}_{nm})}{\sum_{\vec{\rho}_{nm}} e^{i\vec{k}\cdot\vec{\rho}_{nm}} S_{mn}(\vec{\rho}_{nm})}.$$

In the spirit of the tight binding approximation, the second term in equation is assumed to be small, which is a good approximation if the overlap of the atomic wave functions is small. We classify the sum over $\vec{\rho}_{nm}$ according to the distance between site m and site n: (i) zero distance, (ii) the nearest neighbour distance, (iii) the next nearest neighbour distance, etc.

$$\sum_{\vec{\rho}_{nm}} e^{i\vec{k}\cdot\vec{\rho}_{nm}} H'_{mn}(\vec{\rho}_{nm}) = H'_{nn}(0) + \sum_{\vec{\rho}_1} e^{i\vec{k}\cdot\vec{\rho}_{nm}} H'_{mn}(\vec{\rho}_{nm}) + \cdots$$

The zeroth neighbour term $H'_{nn}(0)$ in equation results in a constant additive energy, independent of $\vec{k}$. The sum over nearest neighbour distances $\vec{\rho}_1$ gives rise to a $\vec{k}$-dependent perturbation, and hence is of particular interest in calculating the band structure. The terms $H'_{nn}(0)$ and the sum over the nearest neighbour terms in equation are of comparable magnitude, as can be seen by the following argument. In the integral

$$H'_{nn}(0) = \int \phi^*(\vec{r}-\vec{R}_n)[V - U(\vec{r}-\vec{R}_n)]\phi(\vec{r}-\vec{R}_n)d^3r$$

we note that $|\phi(\vec{r}-\vec{R}_n)|^2$ has an appreciable amplitude only in the vicinity of the site $\vec{R}_n$. But at site $\vec{R}_n$, the potential energy term $[V - U(\vec{r}-\vec{R}_n)] = H'$ is a small term, so that $H'_{nn}(0)$ represents the product of a small term times a large term. On the other hand, the integral $H'_{mn}(\vec{\rho}_{nm})$ taken over nearest neighbour distances has a factor $[V - U(\vec{r}-\vec{R}_n)]$ which is large near the mth site; however, in this case the wave functions $\phi^*(\vec{r}-\vec{R}_n)$ and $\phi(\vec{r}-\vec{R}_n)$ are on different atomic sites and have only a small overlap on nearest neighbour sites. Therefore $H'_{mn}(\vec{\rho}_{nm})$ over nearest neighbour sites also results in the product of a large quantity times a small quantity.

In treating the denominator in the perturbation term of equation, we must sum

$$\sum_{\vec{\rho}_{nm}} e^{i\vec{k}\cdot\vec{\rho}_{nm}} S_{mn}(\vec{\rho}_{nm}) = S_{nn}(0) + \sum_{\vec{\rho}_1} e^{i\vec{k}\cdot\vec{\rho}_{nm}} S_{mn}(\vec{\rho}_{nm}) + \cdots$$

In this case the leading term $S_{nn}(0)$ is approximately unity and the overlap integral $S_{mn}(\vec{\rho}_{nm})$ over nearest neighbour sites is small, and can be neglected to lowest order in comparison with unity. The nearest neighbour term in equation is of comparable magnitude to the next nearest neighbour terms arising from $H_{mn}(\vec{\rho}_{nm})$ in equation .

We will here make several explicit evaluations of $E(\vec{k})$ in the tight{binding limit to show how this method incorporates the crystal symmetry. For illustrative purposes we will give results for the simple cubic lattice (SC), the body centered cubic (BCC) and face centered cubic lattice (FCC). We shall assume here that the overlap of atomic potentials on neighboring sites is sufficiently weak so that only nearest neighbour terms need be considered in the sum on H^{0mn} and only the leading term in the sum of S_{mn}.

For the simple cubic structure there are 6 terms in the nearest neighbour sum on H'_{mn} with $\vec{\rho}_1$ vectors given by:

$$\vec{\rho}_1 = a(\pm1,0,0), a(0,\pm1,0), a(0,0,\pm1).$$

By symmetry $H'_{mn}(\vec{\rho}_1)$ is the same for all of the $\vec{\rho}_1$ vectors so that

$$E(\vec{k}) = E^{(0)} + H'_{nn}(0) + 2H'_{mn}(\vec{\rho}_1)[\cos k_x a + \cos k_y a + \cos k_z a] + \cdots$$

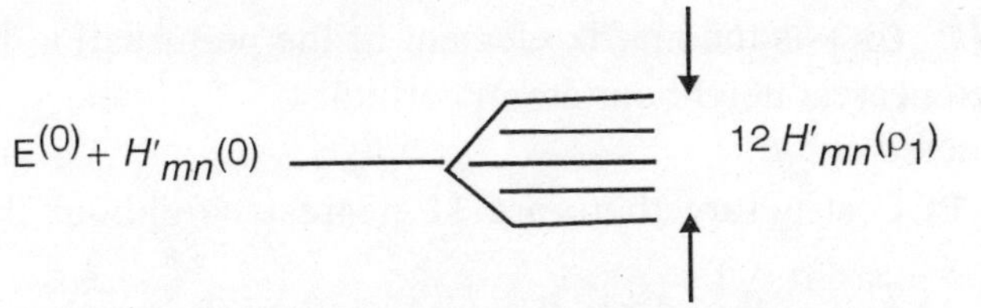

Fig. The relation between the atomic levels and the broadened level in the tight binding approximation.

where $\vec{\rho}_1$ = nearest neighbour separation and k_x, k_y, k_z are components of wave vector $\vec{k}$ in the first Brillouin zone.

This dispersion relation $E(\vec{k})$ clearly satisfies three properties which characterize energy eigenvalues in typical periodic structures:

1. Periodicity in $\vec{k}$ space under translation by a reciprocal lattice vector $\vec{k} \to \vec{k} + \vec{G}$,
2. $E(\vec{k})$ is an even function of $\vec{k}$ (i.e., $E(k) = E(-k)$)
3. $\partial E/\partial k = 0$ at the Brillouin zone boundary

In the above expression for $E(\vec{k})$, the maximum value for the term in brackets is ± 3. Therefore for a simple cubic lattice in the tight binding approximation we obtain a bandwidth of $12\,H'_{mn}(\rho_1)$ from nearest neighbour interactions. Because of the different locations of the nearest neighbour atoms in the case of the BCC and FCC lattices, the expression for $E(\vec{k})$ will be different for the various cubic lattices. Thus the form of the tight binding approximation explicitly takes account of the crystal structure. These results are summarized below.

simple cubic

$E(\vec{k}) = \text{const} + 2H'_{mn}(\vec{\rho}_1)\,[\cos k_x a + \cos k_y a + \cos k_z a] + ...$

body centered cubic

The eight $\vec{\rho}_1$ vectors for the nearest neighbour distances in the BCC structure are ($\pm\, a/2$, $\pm\, a/2$, $\pm\, a/2$) so that there are 8 exponential terms which combine in pairs such as:

$$\left[\exp\frac{ik_x a}{2}\exp\frac{ik_y a}{2}\exp\frac{ik_z a}{2} + \exp\frac{-ik_x a}{2}\exp\frac{ik_y a}{2}\exp\frac{ik_z a}{2}\right]$$

to yield

$$2\cos\left(\frac{k_x a}{2}\right)\exp\frac{ik_y a}{2}\exp\frac{ik_z a}{2}.$$

We thus obtain for the BCC structure:

$$E(\vec{k}) = \text{const} + 8H'_{mn}(\vec{\rho}_1)\cos\left(\frac{k_x a}{2}\right)\cos\left(\frac{k_y a}{2}\right)\cos\left(\frac{k_z a}{2}\right) + \cdots$$

where $H'_{mn}(\vec{\rho}_1)$ is the matrix element of the perturbation Hamilton-ian taken between nearest neighbour atomic orbitals.

face centered cubic

For the FCC structure there are 12 nearest neighbour distances $\vec{\rho}_1$: $\left(0,\pm\frac{a}{2},\pm\frac{a}{2}\right),\left(\pm\frac{a}{2},\pm\frac{a}{2},0\right),\ \left(\pm\frac{a}{2},0,\pm\frac{a}{2}\right)$, so that the twelve exponential terms combine in groups of 4 to yield:

$$\exp\frac{ik_xa}{2}\exp\frac{ik_ya}{2}+\exp\frac{ik_xa}{2}\exp\frac{-ik_ya}{2}+\exp\frac{-ik_xa}{2}\exp\frac{ik_ya}{2}$$

$$+\exp\frac{-ik_xa}{2}\exp\frac{-ik_ya}{2}=4\cos\left(\frac{k_xa}{2}\right)\cos\left(\frac{k_ya}{2}\right)$$

thus resulting in the energy dispersion relation

$$E(\vec{k})=const+4H'_{mn}(\vec{\rho}_1)$$

$$\left[\cos\left(\frac{k_ya}{2}\right)+\cos\left(\frac{k_za}{2}\right)+\cos\left(\frac{k_xa}{2}\right)\cos\left(\frac{k_za}{2}\right)+\cos\left(\frac{k_xa}{2}\right)\cos\left(\frac{k_ya}{2}\right)\right]+\cdots$$

We note that $E(\vec{k})$ for the FCC is different from that for the SC or BCC structures. The tight-binding approximation has symmetry considerations built into its formulation through the symmetrical arrangement of the atoms in the lattice. The situation is quite different in the weak binding approximation where symmetry enters into the form of $V(\vec{r})$ and determines which Fourier components $V_{\vec{G}}$ will be important in creating band gaps.

Weak and Tight Binding Approximations

We will now make some general statements about bandwidths and forbidden band gaps which follow from either the tight binding or weak binding approximations. With increasing energy, the bandwidth tends to increase. On the tight{binding picture, the higher atomic states are less closely bound to the nucleus, and the resulting increased overlap of the wave functions results in a larger value for $H'_{mn}(\vec{\rho}_1)$ in the case of the higher atomic states: that is, for silicon, which has 4 valence electrons in the $n = 3$ shell, the overlap integral $H'_{mn}(\vec{\rho}_1)$ will be smaller than for which is isoelectronic to silicon but has instead 4 valence electrons in the $n = 4$ atomic shell. On the weak{binding picture, the same result follows, since for higher energies, the electrons are more nearly free; therefore, there are more allowed energy ranges available, or equivalently, the energy range of the forbidden states is smaller. Also in the weak{binding approximation the band gap of $2|V_{\vec{G}}|$ tends to decrease as $\vec{G}$ increases, because of the oscillatory character of $e^{i\vec{G}\cdot\vec{r}}$ in

$$V_{\vec{G}} = \frac{1}{\Omega_0} \int_{\Omega_0} e^{-i\vec{G}\cdot\vec{r}} V(\vec{r}) d^3 r.$$

From the point of view of the tight{binding approximation, the increasing bandwidth with increasing energy is also equivalent to a decrease in the forbidden band gap. At the same time, the atomic states at higher energies become more closely spaced, so that the increased bandwidth eventually results in band overlaps. When band overlaps occur, the tight-binding approximation as given above must be generalized to treat coupled or interacting bands using degenerate perturbation theory.

Tight Binding Approximation with 2 Atoms/Unit Cell

We present here a simple example of the tight binding approximation for a simplified version of polyacetylene which has two carbon atoms (with their appended hydrogens) per unit cell.

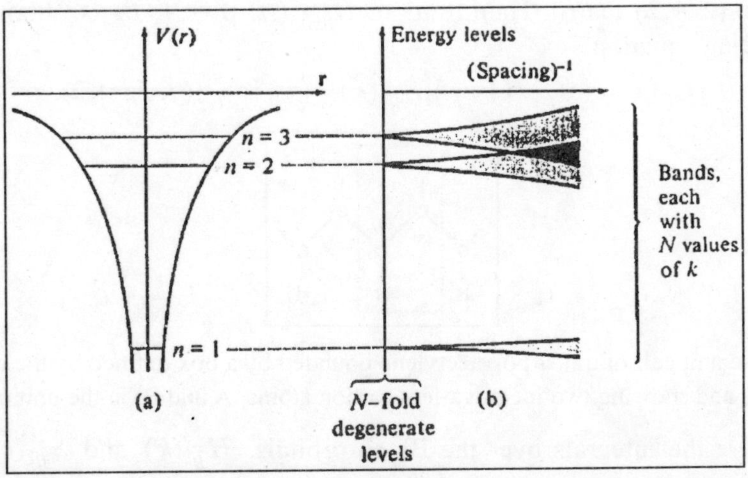

Fig. Schematic diagram of the increased bandwidth and decreased band gap in the tight binding approximation as the interatomic separation decreases.

The box defined by the dotted lines, the unit cell for trans-polyacetylene $(CH)_x$. This unit cell of an infinite one-dimensional chain contains two inequivalent carbon atoms, A and B. There is one π-electron per carbon atom, thus giving rise to two π-energy bands in the first Brillouin zone. These two bands are called bonding π-bands for the valence band, and anti-bonding π-bands for the conduction band.

The lattice unit vector and the reciprocal lattice unit vector of this one-dimensional polyacetylene chain are given by $\vec{a}_1 = (a,0,0)$ and $\vec{b}_1 = (2\pi / a, 0, 0),$, respectively. The Brillouin zone is the line segment $-\pi / a$

$< k < \pi / a$. The Bloch orbitals consisting of A and B atoms are given by

$$y_j(r) = \frac{1}{\sqrt{N}} \sum_{R_\alpha} e^{ikR_\alpha} \phi_j(r - R_\alpha), (\alpha = A, B)$$

where the summation is taken over the atom site coordinate R_α for the A or B carbon atoms in the solid.

To solve for the energy eigenvalues and wavefunctions we need to solve the general equation:

$$H\psi = ES\psi$$

where H is the $n \times n$ tight binding matrix Hamiltonian for the n coupled bands ($n = 2$ in the case of polyacetylene) and S is the corresponding $n \times n$ overlap integral matrix. To obtain a solution to this matrix equation, we require that the determinant $|H - ES|$ vanish.

This approach is easily generalized to periodic structures with more than 2 atoms per unit cell.

The (2 × 2) matrix Hamiltonian, $H_{\alpha\beta}$, (α, $\beta = A$, B) is obtained by substituting equation

$$H_{jj'}(\vec{k}) = \langle \psi_j | H | \psi_{j'} \rangle, S_{jj'}(\vec{k}) = \langle \psi_j | \psi_{j'} \rangle (j, j' = 1, 2),$$

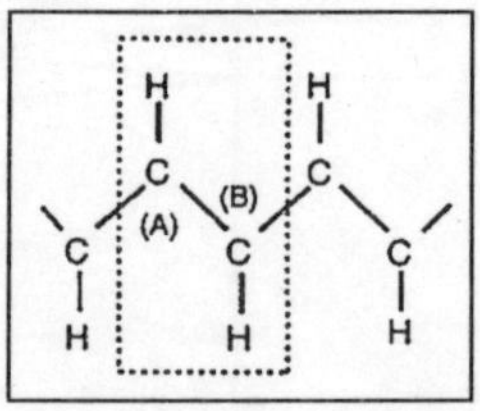

Fig. The unit cell of trans- polyacetylene bounded by a box defined by the dotted lines, and showing two inequiva-lent carbon atoms, A and B, in the unit cell.

where the integrals over the Bloch orbitals, $H_{jj'}(\vec{k})$ and $S_{jj'}(\vec{k})$,, are called transfer integral matrices and overlap integral matrices, respectively. When $\alpha = \beta =$ A, we obtain the diagonal matrix element

$$H_{AA}(r) = \frac{1}{N} \sum_{R,R'} e^{ik(R-R')} \langle \phi_A(r - R') | H | \phi_A(r - R) \rangle$$

$$\frac{1}{N} \sum_{R,R'} E_{2p} + \frac{1}{N} \sum_{R'=R\pm a} e^{\pm ika} \langle \phi_A(r - R') | H | \phi_A(r - R) \rangle$$

+(terms equal to or more distant than $R' = R \pm 2a$).

$= E_{2p}$ + (terms equal to or more distant than $R' = R \pm a$).

The main contribution to the matrix element H_{AA} comes from $R' = R$, and this gives the orbital energy of the $2p$ level, E_{2p}. We note that E_{2p} is not simply the atomic energy value for the free atom, because the Hamiltonian H also includes a crystal potential contribution. The next order contribution to H_{AA} in equation comes from terms in $R' = R \pm a$, which are here neglected for simplicity. Similarly, H_{BB} also gives E_{2p} to the same order of approximation.

Next let us consider the off-diagonal matrix element H_{AB} (r) which explicitly couples the A unit to the B unit. The largest contribution to H_{AB} (r) arises when atoms A and B are nearest neighbors. Thus in the summation over R', we only consider the terms with $R' = R \pm a/2$ as a first approximation and neglect more distant terms to obtain

$$H_{AB}(r) = \frac{1}{N}\sum_{R}\left\{e^{-ika/2}\langle \phi_A(r-R)|H|\phi_B(r-R-a/2)\rangle\right.$$

$$\left. + e^{-ika/2}\langle \phi_A(r-R)|H|\phi_B(r-R-a/2)\rangle\right\}$$

$$= 2t\cos(ka/2)$$

where t is the transfer integral appearing in equation and is denoted by

$$t = \langle \phi_A(r-R)|H|\phi_B(r-R \pm a/2)\rangle .$$

Here we have assumed that all the π bonding orbitals are of equal length (1.5°A bonds). In the real $(CH)_x$ compound, bond alternation occurs, in which the bonding between adjacent carbon atoms alternates between single bonds (1.7Å) and double bonds (1.3°A). With this bond alternation, the two matrix elements between atomic wavefunctions in equation are not equal. Although the distortion of the lattice lowers the total energy, the electronic energy always decreases more than the lattice energy in a one-dimensional material. This distortion deforms the lattice by a process called the Peierls instability. This instability arises for example when a distortion is introduced into a system containing a previously degenerate system with 2 equivalent atoms per unit cell. The distortion making the atoms inequivalent increases the unit cell by a factor of 2 and decreases the reciprocal lattice by a factor of 2. If the energy band was formally half filled, a band gap is introduced by the Peierls instability at the Fermi level, which lowers the total energy of the system. It is stressed that t has a negative value. The matrix element $H_{BA}(r)$ is obtained from H_{AB} (r) through the Hermitian conjugation relation $H_{BA} = H^*_{AB}$, but since H_{AB} is real, we obtai

$$H_{BA} = H_{AB}.$$

The overlap matrix S_{ij} can be calculated by a similar method as was used for H_{ij}, except that the intra-atomic integral S_{ij} yields a unit matrix in the limit of large in-teratomic distances, if we assume that the atomic wavefunction is normalized so that $S_{AA} = S_{BB} = 1$. It is assumed that for polyacetylene the S_{AA} and S_{BB} matrix elements are still approxi-mately unity. For the off-diagonal matrix element for polyacetylene we have $S_{AB} = S_{BA} = 2s\cos(ka/2)$, where s is an overlap integral between the nearest A and B atoms,

$$s = \langle \phi_A(r-R) | \phi_B(r - R \pm a/2) \rangle.$$

The secular equation for the 2_{pz} orbital of CH_x is obtained by setting he determinant of

$$\begin{vmatrix} E_{2p} - E & 2(t-sE)\cos(ka/2) \\ 2(t-sE)\cos(ka/2) & E_{2p} - E \end{vmatrix} = (E_{2p} - E)^2 - 4(t - sE)^2 \cos^2(ka/2) = 0$$

yielding the eigenvalues of the energy dispersion relations of equation

$$E_{\pm}(\vec{k}) = \frac{E_{2p} \pm 2t\cos(ka/2)}{1 \pm 2s\cos(ka/2)}, \left(-\frac{\pi}{a} < k < \frac{\pi}{a}\right)$$

in which the + sign is associated with the bonding π-band and the – sign is associated with the antibonding π*-band. Here it is noted that by setting E_{2p} to zero (thereby defining the origin of the energy), the levels E_+ and E_- are degenerate at $ka = \pm \pi$. Figure is constructed for $t < 0$ and $s > 0$. Since there are two π electrons per unit cell, each with a different spin orientation, both electrons occupy the bonding π energy band. The effect of the inter-atomic bonding is to lower the total energy below E_{2p}.

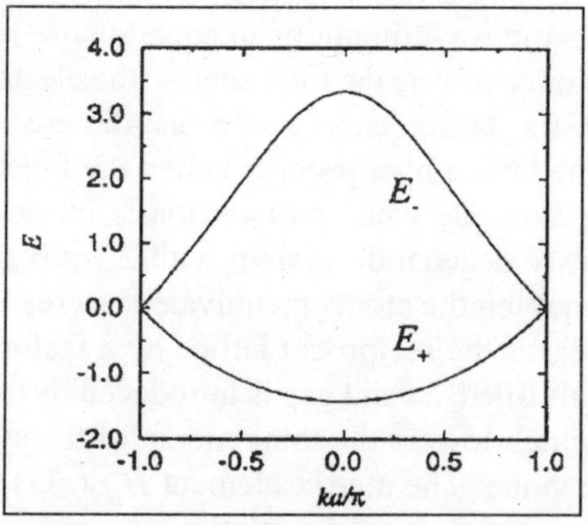

Fig. The energy dispersion relation E§(~k) for polyacetylene [(CH)$_x$], given by values for the parameters t = ¡1 and s = 0:2. Curves E$_+$(~k) and E$_¡$(~k) are called bonding ¼ and antibonding ¼$^£$ energy bands, respectively, and the energy is plotted in units of t.

ENERGY BANDS IN SOLIDS

We present here some examples of energy bands which are representative of metals, semi-conductors and insulators and point out some of the characteristic features in each case. A schematic way between insulators (a), metals (b), semimetals (c), a thermally excited semiconductor (d) for which at $T = 0$ all states in the valence band are occupied and all states in the conduction band are unoccupied, assuming no impurities or crystal defects. Finally, we see a p-doped semiconductor which is deficient in electrons, not having sufficient electrons to fill the valence band completely as in (d).

The electron dispersion relations for an insulator (a), while (c) shows dispersion relations for a metal. A semimetal if the number of electrons equals the number of holes, but a metal otherwise.

METALS

Alkali Metals-e.g., Sodium

For the alkali metals the valence electrons are nearly free and the weak binding approxima-tion describes these electrons quite well. The Fermi surface is nearly spherical and the band gaps are small. The crystal structure for the alkali metals is body centered cubic (BCC) and the $E(\vec{k})$ diagram is drawn starting with the bottom of the half-filled conduction band.

For example, the $E(\vec{k})$ for sodium begins at ~ –0.6 Rydberg and represents the 3s conduction band. The filled valence bands lie much lower in energy.

For the case of sodium, the 3s conduction band is very nearly free electron like and the $E(\vec{k})$ relations are closely isotropic. Thus the $E(\vec{k})$ relations along the $\Delta(100)$, $\Sigma(110)$ and $\Lambda(111)$ directions are essentially coincident and can be so plotted. For these metals, the Fermi level is determined so that the 3s band is exactly half-occupied, since the Brillouin zone is large enough to accommodate 2 electrons per unit cell. Thus the radius of the Fermi surface k_F satisfies the relation where $V_{B.Z.}$ and a are, respectively, the volume of the Brillouin zone and the lattice constant.

$$\frac{4}{3}\pi k_F^3 = \frac{1}{2}V_{B.Z.} = \frac{1}{2}(2)\left(\frac{2\pi}{a}\right)^3 \text{.or} \frac{k_F a}{2\pi} \sim 0.63,$$

For the alkali metals, the effective mass m^* is nearly equal to the free electron mass m and the Fermi surface is nearly spherical and never comes close to the Brillouin zone boundary. The zone boundary for the Σ, Λ and Δ directions are indicated in the $E(\vec{k})$ by vertical lines. For the alkali metals the band gaps are very small compared to the band widths and the $E(\vec{k})$ relations are parabolic $(E = \hbar^2 k^2 / 2m^*)$ almost up to the Brillouin zone boundaries. By comparing $E(\vec{k})$ for Na with the BCC empty lattice bands for which the potential $V(r) = 0$, we can see the effect of the very weak periodic potential in partially lifting the band degeneracy at the various high symmetry points in the Brillouin zone. The threshold for optical transitions corresponds to photons having sufficient energy to take an electron from an occupied state at k_F to an unoccupied state at k_F since the wave vector for photons is very small compared with k_F and wave vector conservation (i.e., crystal momentum conservation) is required for optical transitions. The threshold for optical transitions is indicated by $\hbar\omega$. Because of the low density of initial and final states for a given energy separation, we would expect optical interband transitions for alkali metals to be very weak and this is in agreement with experimental observations for all the alkali metals. The notation a.u. stands for atomic units and expresses lattice constants in units of Bohr radii. The electron energy is given in Rydbergs where 1 Rydberg = 13.6 eV, the ionization energy of a hydrogen atom.

NOBLE METALS

The noble metals are copper, silver and gold and they crystallize in a face centered cubic (FCC) structure; the usual notation for the high symmetry points in the FCC Brillouin zone. As in the case of the alkali metals, the noble metals have one valence electron/atom and therefore one electron per primitive unit cell. However, the free electron picture does not work so well for the noble metals.

In the case of copper, the bands near the Fermi level are derived from the 4*s* and 3*d* atomic levels. The so-called 4*s* and 3d bands accommodate a total of 12 electrons, while the number of available electrons is 11. Therefore the Fermi level must cross these bands. Consequently copper is metallic. We see that the 3*d* bands are relatively flat and show little dependence on wave vector $\vec{k}$. We can trace the 3d bands by starting at $\vec{k} = 0$ with the $\Gamma_{25'}$ and Γ_{12} levels. On the other hand, the 4s band has a strong *k*-dependence and large curvature.

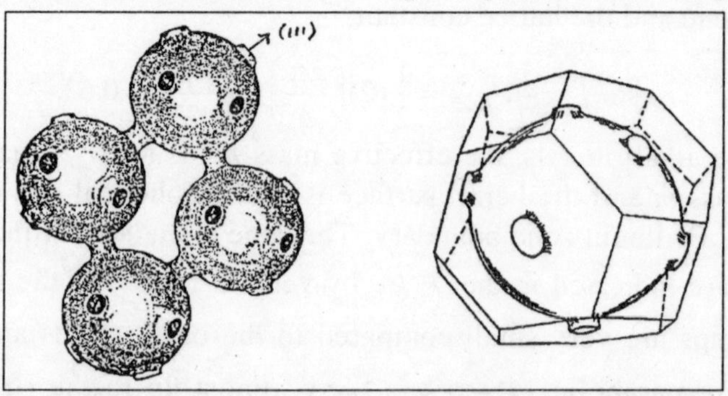

Fig. (a) The copper Fermi surface in the extended zone scheme. (b) A sketch of the Fermi surface of copper inscribed within the FCC Brillouin zone.

This band can be traced by starting at $\vec{k} = 0$ with the Γ_1 level. About halfway between Γ and *X*, the 4*s* level approa-ches the 3*d* levels and mixing or hybridization occurs. As we further approach the *X*-point, we can again pick up the 4*s* band (beyond where the interaction with the 3*d* bands occurs) because of its high curvature. This 4*s* band eventually crosses the Fermi level before reaching the Brillouin Zone boundary at the *X* point. A similar mixing or hybridizat-ion between 4*s* and 3*d* bands occurs in going from Γ to *L*, except that in this case the 4*s* band reaches the Brillouin Zone boundary before crossing the Fermi level.

Of particular significance for the transport properties of copper is the band gap that opens up at the *L*-point. In this case, the band gap is between the $L_{2'}$ level below the Fermi level E_F and the L_1 level above E_F. Since this bandgap is comparable with the typical bandwidths in copper, we cannot expect the Fermi surface to be free electron like. By looking at the energy bands $E(\vec{k})$ along the major high symmetry directions such as the (100), (110) and (111) directions, we can readily trace the origin of the copper Fermi surface.

Here we see basically a spherical Fermi surface with necks pulled out in the (111) directions and making contact with the Brillouin zone boundary through these necks, thereby linking the Fermi surface in one zone to that in the next zone in the extended zone scheme. In the (100) direction, the cross section of the Fermi surface is nearly circular, indicative of the nearly parabolic E (~k) relation of the 4s band at the Fermi level in going from Γ to *X*. In contrast, in going from Γ to *L*, the 4s band never crosses the Fermi level. Instead the 4s level is depressed from the free electron parabolic curve as the Brillouin zone boundary is reached, thereby producing a higher density of states. Thus, near the zone boundary, more electrons can be accommodated per unit energy range, or to say this another way, there will be increasingly more $\vec{k}$ vectors with approximately the same energy.

This causes the constant energy surfaces to be pulled out in the direction of the Brillouin zone boundary. This "pulling out" effect follows both from the weak binding and tight binding approximations and the effect is more pronounced as the strength of the periodic potential (or $V_{\vec{G}}$) increases. If the periodic potential is sufficiently strong so that the resulting bandgap at the zone boundary straddles the Fermi level, as occurs at the *L*-point in copper, the Fermi surface makes contact with the Brillouin zone boundary. The resulting Fermi surfaces are called open surfaces because the Fermi surfaces between neighboring Brillouin zones are connected.

The electrons associated with the necks are contained in the electron pocket shown in the $E(\vec{k})$ diagram away from the *L*-point in the *LW* direction which is ? to the {111} direction. The copper Fermi surface bounds electron states. Hole pockets are formed in copper in the extended zone and constitute the unoccupied space between the electron surfaces. Direct evidence for hole pockets is provided by Fermi surface measurements.

From the $E(\vec{k})$ diagram for copper we see that the threshold for optical interband transitions occurs for photon energies sufficient to take an electron at constant $\vec{k}$ {vector from a filled 3d level to an unoccupied state above the Fermi level.

Such interband transitions can be made near the *L*-point in the Brillouin zone. Because of the high density of initial states in the *d*-band, these transitions will be quite intense. The occurrence of these interband transitions at ~ 2 *eV* gives rise to a large absorption of electromagnetic energy in this photon energy region. The reddish colour of copper metal is thus due to a higher reflectivity for photons (below the threshold for interband transitions) than for photons (above this threshold).

POLYVALENT METALS

The simplest example of a polyvalent metal is aluminum with 3 electrons/atom and having a $3s^2 3p$ electronic configuration for the valence electrons. (As far as the number of electrons/atom is concerned, two electrons/atom completely fill a non-degenerate band_one for spin up, the other for spin down.) Because of the partial filling of the $3s^2$ $3p^6$ bands, aluminum is a metal.

Aluminum crystallizes in the FCC structure so we can use the same notation as for the Brillouin zone. The energy bands for aluminum are very free electron like. This follows from the small magnitudes of the band gaps relative to the band widths on the energy band.

The lowest valence band is the 3s band which can be traced by starting at zero energy at the Γ point ($\vec{k}$ = 0) and going out to X_4 at the *X*-point, to W_3 at the *W*-point, to L'_2 at the *L*-point and back to Γ_1 at the Γ point ($\vec{k}$ = 0). Since this band always lies below the Fermi level, it is completely filled, containing 2 electrons. The third valence electron partially occupies the second and third *p*-bands (which are more accurately described as hybridized 3*p*-bands with some admixture of the 3*s* bands with which they interact).

The second band is partly filled; the occupied states extend from the Brillouin zone boundary inward toward the centre of the zone; this can be seen in going from the *X* point to Γ, on the curve labeled Δ_1.

Since the second band states near the centre of the Brillouin zone remain unoccupied, the volume enclosed by the Fermi surface in the second band is a hole pocket. Because $E(\vec{k})$ for the second band in the vicinity of E_F is free electron like, the masses for the holes are approximately equal to the free electron mass.

The 3rd zone electron pockets are small and are found around the $K-$ and W–points.

These pockets are $\vec{k}$ space volumes that enclose electron states, and because of the large curvature of $E(\vec{k})$, these electrons have relatively small masses. This diagram gives no evidence for any 4th zone pieces of Fermi surface, and for this reason we can conclude that all the electrons are either in the second band or in the third band. The total electron concentration is sufficient to exactly fill a half of a Brillouin zone:

$$V_{e,2}+V_{e,3}=\frac{V_{BZ}}{2}.$$

With regard to the second zone, it is partially filled with electrons and the rest of the zone is empty (since holes correspond to the un filled states):

$$V_{h,2}+V_{e,2}=V_{BZ},$$

with the volume that is empty slightly exceeding the volume that is occupied. Therefore we focus attention on the more dominant second zone holes. Substitution cf equation into equation then yields for the second zone holes and the third zone electrons

$$V_{h,2}-V_{e,3}=\frac{V_{BZ}}{2}$$

where the subscripts *e*, *h* on the volumes in $\vec{k}$ space refer to electrons and holes and the Brillouin zone (B.Z.) index is given for each of the carrier pockets. Because of the small masses and high mobility of the 3rd zone electrons, they play a more important role in the transport properties of aluminum than would be expected from their small numbers.

From the $E(\vec{k})$ we see that at the same $\vec{k}$ -points (near the $K-$ and $W-$ points in the Brillouin zone) there are occupied 3s levels and unoccupied $3p$ levels separated by ~ 1eV. From this we conclude that optical interband transitions should be observable in the 1eV photon energy range. Such interband transitions are in fact observed experimentally and are responsible for the departures from nearly perfect reflectivity of aluminum mirrors in the vicinity of 1 eV.

SEMICONDUCTORS

Assume that we have a semiconductor at $T = 0\ K$ with no impurities. The Fermi level will then lie within a band gap. Under these conditions, there are no carriers, and no Fermi surface. We now illustrate the energy band structure for several representative semiconductors in the limit of T = 0K and no impurities. Semiconductors having no impurities or defects are called intrinsic semiconductors.

PBTE

The energy bands for PbTe. This direct gap semiconductor is chosen initially for illustrative purposes because the energy bands in the valence and conduction bands that are of particular interest are non-degenerate.

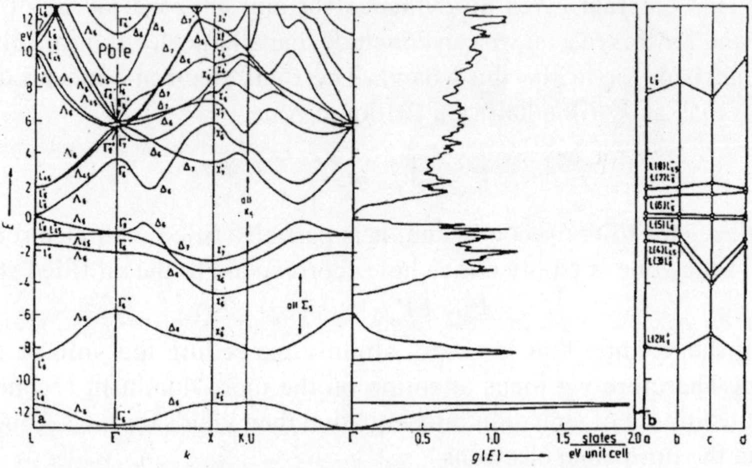

Fig. (a) Energy band structure and density of states for PbTe obtained from an empirical pseudopotential calculation. (b) Theoretical values for the L point bands calculated by different models (labeled a, b, c, d on the x-axis).

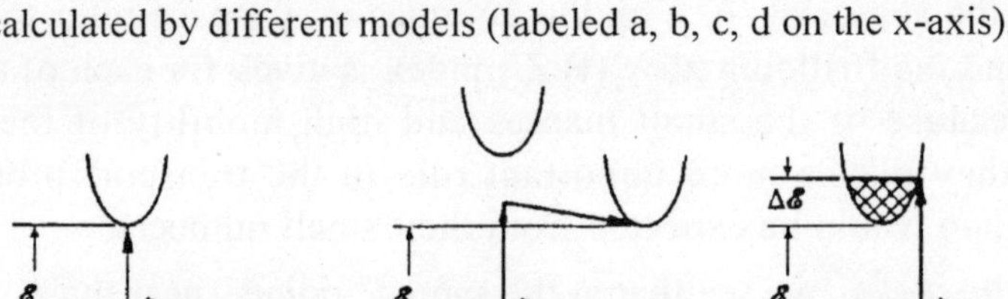

Fig. Optical absorption processes for (a) a direct band gap semiconductor, (b) an indirect band gap semiconductor, and (c) a direct band gap semiconductor with the conduction band filled to the level shown.

Therefore, the energy states in PbTe near E_F are simpler to understand than for the more common semiconductors silicon and , and many of the III-V and II-VI compound semiconductors.

The position of E_F for the idealized conditions of the intrinsic (no carriers at $T = 0$) semiconductor PbTe. From a diagram like this, we can obtain a great deal of information which could be useful for making semiconductor devices. For example, we can calculate effective masses from the band curvatures, and electron velocities from the slopes of the $E(\vec{k})$ dispersion relations.

Suppose we add impurities (e.g., donor impurities) to PbTe. The donor impurities will raise the Fermi level and an electron pocket will eventually be formed in the L^{-}_{6} conduction band about the *L*-point. This electron pocket will have an ellipsoidal Fermi surface because the band curvature is different as we move away from the *L* point in the $L\Gamma$ direction as compared with the band curvature as we move away from *L* on the Brillouin zone boundary containing the *L* point (e.g., L*W* direction). $E(\vec{k})$ from *L* to Γ corresponding to the (111) direction. Since the effective masses

$$\frac{1}{m_{ij}^{*}} = \frac{1}{\hbar^2}\frac{\partial^2 E(\vec{k})}{\partial k_i \partial k_j}$$

for both the valence and conduction bands in the longitudinal $L\Gamma$ direction are heavier than in the *LK* and *LW* directions, the ellipsoids of revolution describing the carrier pockets are prolate for both holes and electrons. The *L* and Σ point room temperature band gaps are 0.311 eV and 0.360 eV, respectively. For the electrons, the effective mass parameters are $m_{\perp}$ = 0:053me and mk = 0:620me. The experimental hole effective masses at the *L* point are $m_{\perp} = 0{:}0246m_e$ and $m_{\|} = 0{:}236m_e$ and at the Σ point the hole effective mass values are $m_{\perp} = 0{:}124m_e$ and $m_{\|} = 1{:}24m_e$.

Thus for the L-point carrier pockets, the semi-major axis of the constant energy surface along L¡ will be longer than along LK. From the E(~k) diagram for PbTe one would expect that hole carriers could be thermally excited to a second band at the Σ point, which is indicated on the *E* (~k) diagram. At room temperature, these Σ point hole carriers contribute significantly to the transport properties.

Because of the small gap (0.311 eV) in PbTe at the L-point, the threshold for interband transitions will occur at infrared frequencies. PbTe crystals can be prepared either p-type or n-type, but never perfectly stoichiometrically (i.e., intrinsic PbTe has not been prepared).

Therefore, at room temperature the Fermi level E_F often lies in either the valence or conduction band for actual PbTe crystals. Since optical transitions conserve wavevector, the interband transitions will occur at k_F and at a higher photon energy than the direct band gap. This increase in the threshold energy for interband transitions in degenerate semiconductors (where E_F lies within either the valence or conduction bands) is called the Burstein shift.

We will next look at the $E(\vec{k})$ relations for:

1. The group IV semiconductors which crystallize in the diamond structure and

2. The closely related III-V compound semiconductors which crystallize in the zincblende structure. These semiconductors have degenerate valence bands at $\vec{k} = 0$

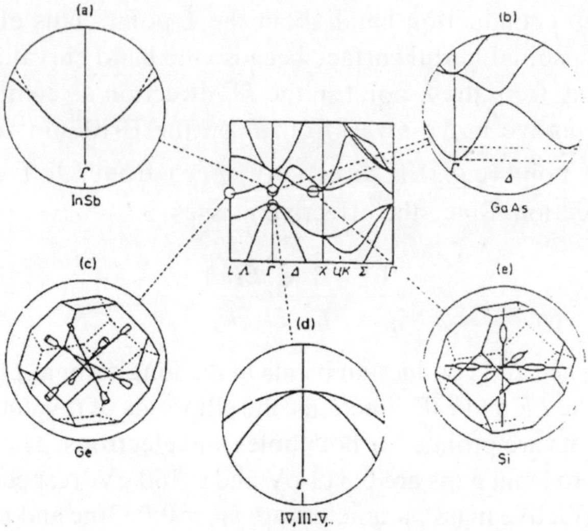

Fig. Important details of the band structure of typical group IV and III-V semiconductors.

and for this reason have more complicated $E(\vec{k})$ relations for carriers than is the case for the lead salts. The $E(\vec{k})$ diagram for . Ge is a semiconductor with a bandgap occurring between the top of the valence band at Γ_{25}', and the bottom of the lowest conduction band at L_1.

Since the valence and conduction band extrema occur at different points in the Brillouin zone, Ge is an indirect gap semiconductor. Using the same arguments for the Fermi surface of PbTe, we see that the constant energy surfaces for electrons in are ellipsoids of revolution. As for the case of PbTe, the ellipsoids of revolution are elongated along ΓL which is the heavy mass direction in this case. Since the multiplicity of L-points is 8, we have 8 half-ellipsoids of this kind within the first Brillouin zone, just as for the case of PbTe. By translation of these half-ellipsoids by a reciprocal lattice vector we can form 4 full-ellipsoids. The $E(\vec{k})$ diagram for further shows that the next highest conduction band above the L point minimum is at the Γ–point ($\vec{k} = 0$) and after that along the ΓX axis at a point commonly labeled as a Δ–point. Because of the degeneracy of the highest valence band, the Fermi surface for holes in is more complicated than for electrons. The lowest direct band gap in is at $\vec{k} = 0$ between the $\Gamma_{25'}$ valence band and the $\Gamma_{2'}$ conduction band.

From the $E(\vec{k})$ diagram we note that the electron effective mass for the $\Gamma_{2'}$ conduction band is very small because of the high curvature of the $\Gamma_2 0$ band about $\vec{k} = 0$, and this effective mass is isotropic so that the constant energy surfaces are spheres.

The optical properties for show a very weak optical absorption for photon energies corresponding to the indirect gap. Since the valence and conduction band extrema occur at a different $\vec{k}$-point in the Brillouin zone, the indirect gap excitation requires a phonon to conserve crystal momentum. Hence the threshold for this indirect

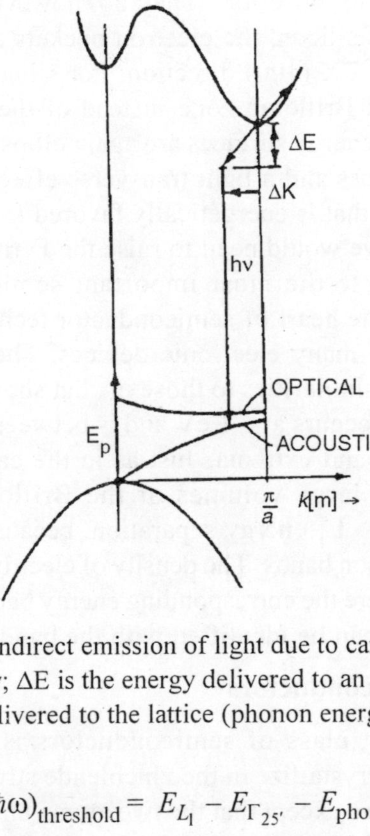

Fig. Illustration of the indirect emission of light due to carriers and phonons in Ge. [h° is the photon energy; ΔE is the energy delivered to an electron; E_p is the energy delivered to the lattice (phonon energy)].

transition is

$$(\hbar\omega)_{\text{threshold}} = E_{L_1} - E_{\Gamma_{25'}} - E_{\text{phonon}}.$$

The optical absorption for increases rapidly above the photon energy corresponding to the direct band gap $E_{\Gamma_{2'}} - E_{\Gamma_{25'}}$, because of the higher probability for the direct optical excitation process.

However, the absorption here remains low compared with the absorption at yet higher photon energies because of the low density of states for the Γ–

point transition, as seen from the $E(\vec{k})$ diagram. Very high optical absorption, however, occurs for photon energies corresponding to the energy separation between the $L_{3'}$ and L_1 bands which is approximately the same for a large range of $\vec{k}$ values, thereby giving rise to a very large joint density of states (the number of states with constant energy separation per unit energy range). A large joint density of states arising from the tracking of conduction and valence bands is found for , silicon and the III-V compound semiconductors, and for this reason these materials tend to have high dielectric constants.

Silicon

From the energy band diagram for silicon, we see that the energy bands of Si are quite similar to those for . They do, however, differ in detail. For example, in the case of silicon, the electron pockets are formed around a Δ point located along the ΓX (100) direction. For silicon there are 6 electron pockets within the first Brillouin zone instead of the 8 half-pockets which occur in . The constant energy surfaces are again ellipsoids of revolution with a heavy longitudinal mass and a light transverse effective mass. The second type of electron pocket that is energetically favored is about the L_1 point, but to fill electrons there, we would need to raise the Fermi energy by ~ 1 eV.

Silicon is of course the most important semiconductor for device applications and is at the heart of semiconductor technology for transistors, integrated circuits, and many electronic devices. The optical properties of silicon also have many similarities to those in , but show differences in detail. For Si, the indirect gap occurs at ~ 1 eV and is between the $\Gamma_{25'}$ valence band and the Δ conduction band extrema. Just as in the case for , strong optical absorption occurs for large volumes of the Brillouin zone at energies comparable to the $L_{3'} \rightarrow L_1$ energy separation, because of the "tracking" of the valence and conduction bands. The density of electron states for Si covering a wide energy range where the corresponding energy band. Most of the features in the density of states can be identified with the band model.

III-V Compound Semiconductors

Another important class of semiconductors is the III-V compound semiconductors which crystallize in the zincblende structure; this structure is like the diamond structure except that the two atoms/unit cell are of a different chemical species. The III-V compounds also have many practical applications, such as semiconductor lasers for fast electronics, GaAs in light emitting diodes, and InSb for infrared detectors.

The $E(\vec{k})$ diagram for GaAs is shown and we see that the electronic levels are very similar to those of Si and Ge. One exception is that the lowest conduction band for GaAs is at $\vec{k} = 0$ so that both valence and conduction band extrema are at $\vec{k} = 0$. Thus GaAs is a direct gap semiconductor, and for this reason GaAs shows a stronger and more sharply defined optical absorption threshold than Si or Ge.

Here we see that the lowest conduction band for GaAs has high curvature and therefore a small effective mass. This mass is isotropic so that the constant energy surface for electrons in GaAs is a sphere and there is just one such sphere in the Brillouin zone. The next lowest conduction band is at a Δ point and a significant carrier density can be excited into this Δ point pocket at high temperatures.

The constant energy surface for electrons in the direct gap semiconductor InSb is likewise a sphere, because InSb is also a direct gap semiconductor. InSb differs from GaAs in having a very small band gap (~ 0.2 eV), occurring in the infrared. Both direct and indirect band gap materials are found in the III-V compound semiconductor family. Except for optical phenomena close to the band gap, these compound semiconductors all exhibit very similar optical properties which are associated with the band-tracking phenomena.

"Zero Gap" Semiconductors

It is also possible to have "zero gap" semiconductors. An example of such a material is which also crystallizes in the diamond structure. The energy band model for without spin-orbit interaction. On this diagram the zero gap occurs between the $\Gamma_{25'}$ valence band and the Γ_2' conduction band, and the Fermi level runs right through this degeneracy point between these bands. Spin-orbit interaction (to be discussed later in this course) is very important for in the region of the $\vec{k} = 0$ band degeneracy and a detailed diagram of the energy bands near $\vec{k} = 0$ and including the effect of spin-orbit interaction. In the effective mass for the conduction band is much lighter than for the valence band as can be seen by the band curvatures.

Optical transitions labeled *B* occur in the far infrared from the upper valence band to the conduction band. In the near infrared, interband transitions labeled

A are induced from the Γ_7^- valence band to the Γ_8^+ conduction band. We note that is classified as a zero gap semiconductor rather than a semimetal because there are no band overlaps in a zero-gap semiconductor anywhere in the Brillouin zone.

Because of the zero band gap in grey tin, impurities play a major role in determining the position of the Fermi level. is normally prepared n-type which means that there are some electrons always present in the conduction band (for example a typical electron concentration would be 10^{15} /cm^3 which amounts to less than 1 carrier/10^7 atoms).

Molecular Semiconductors – Fullerenes

Other examples of semiconductors are molecular solids such as C_{60}. For the case of solid C_{60}, a C_{60} molecule, which crystallizes in a FCC structure with four C_{60} molecules per conventional simple cubic unit cell. A small distortion of the bonds, lengthening the C-C bond lengths on the single bonds to 1.46°A and shortening the double bonds to 1.40°A, stabilizes a band gap of ~1.5eV. In this semiconductor the energy bandwidths are very small compared with the band gaps, so that this material can be considered as an organic molecular semiconductor. The transport properties of C_{60} differ markedly from those for conventional group IV or III-V semiconductors.

SEMIMETALS

Another type of material that commonly occurs in nature is the semimetal. Semimetals have exactly the correct number of electrons to completely fill an integral number of Brillouin zones. Nevertheless, in a semimetal the highest occupied Brillouin zone is not filled up completely, since some of the electrons find lower energy states in "higher" zones.

For semimetals the number of electrons that spill over into a higher Brillouin zone is exactly equal to the number of holes that are left behind. This is illustrated schematically in where a two-dimensional Brillouin zone and a circular Fermi surface of equal area is inscribed. Here we can easily see the electrons in the second zone at the zone edges and the holes at the zone corners that are left behind in the first zone. Translation by a reciprocal lattice vector brings two pieces of the electron surface together to form a surface in the shape of a lens, and the 4 pieces at the zone corners form a rosette shaped hole pocket. Typical examples of semimetals are bismuth and graphite. For these semimetals the carrier density is on the order of one carrier/10^6 atoms.

The carrier density of a semimetal is thus not very different from that which occurs in doped semiconductors, but the behavior of the conductivity $\sigma(T)$ as a function of temperature is very different.

For intrinsic semiconductors, the carriers which are excited thermally contribute significantly to conduction. Consequently, the conductivity tends to rise rapidly with increasing temperature. For a semimetal, the carrier concentration does not change significantly with temperature because the carrier density is determined by the band overlap. Since the electron scattering by lattice vibrations increases with increasing temperature, the conductivity of semimetals tends to fall as the temperature increases.

A schematic diagram of the energy bands of the semimetal bismuth. Electron and hole carriers exist in equal numbers but at different locations in the Brillouin zone. For Bi, electrons are at the L-point, and holes at the T - point. The crystal structure for Bi can be understood from the NaCl structure by considering a very small displacement of the Na FCC structure relative to the Cl FCC structure along one of the body diagonals and an elongation of that body diagonal relative to the other 3 body diagonals. The special f111g direction corresponds to $\Gamma - T$ in the Brillouin zone while the other three $\{111\}$ directions are labeled as $\Gamma - L$.

Instead of a band gap between valence and conduction bands (as occurs for semiconductors), semimetals are characterized by a band overlap in the millivolt range. In bismuth, a small band gap also occurs at the L-point between the conduction band and a lower filled valence band. Because the coupling between these L-point valence and conduction bands is strong, some of the effective mass components for the electrons in bismuth are anomalously small. As far as the optical properties of bismuth are concerned, bismuth behaves much like a metal with a high reflectivity at low frequencies due to strong free carrier absorption.

INSULATORS

The electronic structure of insulators is similar to that of semiconductors, in that both insulators and semiconductors have a band gap separating the valence and conduction bands. However, in the case of insulators, the band gap is so large that thermal energies are not sufficient to excite a significant number of carriers.

The simplest insulator is a solid formed of rare gas atoms. An example of a rare gas insulator is solid argon which crystallizes in the FCC structure with one Ar atom/primitive unit cell. With an atomic configuration $3s^2\ 3p^6$, argon has filled 3s and 3p bands which are easily identified in the energy band diagram. These occupied bands have very narrow band widths compared to their band gaps and are therefore well described by the tight binding approximation. This higher energy states forming the conduction bands (the hybridized 4s and 3d bands) show more dispersion than the more tightly bound valence bands. The band diagram shows argon to have a direct band gap at the Γ point of about 1 Rydberg or 13.6 eV. Although the 4s and 3d bands have similar energies, identification with the atomic levels can easily be made near $k = 0$ where the lower lying 4s-band has considerably more band curvature than the 3d levels which are easily identified because of their degeneracies [the so called three-fold t_g ($\Gamma_{25'}$) and the two-fold e_g (Γ_{12}) crystal field levels for *d*-bands in a cubic crystal].

Another example of an insulator formed from a closed shell configuration is found. Here the closed shell configuration results from charge transfer, as occurs in all ionic crystals.

For example in the ionic crystal LiF (or in other alkali halide compounds) the valence band is identified with the filled anion orbitals (fluorine p-orbitals in this case) and at much higher energy the empty cation conduction band levels will lie (lithium s-orbitals in this case). Because of the wide band gap separation in the alkali halides between the valence and conduction bands, such materials are transparent at optical frequencies.

Insulating behavior can also occur for wide bandgap semiconductors with covalent bonding, such as diamond, ZnS and GaP. The $E(\vec{k})$ diagrams for these materials are very similar to the dispersion relations for typical III-V semiconducting compounds and the group IV semiconductors silicon and ; the main difference, however, is the large band gap separating valence and conduction bands.

Even in insulators there is a finite electrical conductivity. For these materials the band electronic transport processes become less important relative to charge hopping from one atom to another by over-coming a potential barrier. Ionic conduction can also occur in insulating ionic crystals. From a practical point of view, one of the most important applications of insulators is the control of electrical breakdown phenomena.

The principal experimental methods for studying the electronic energy bands depend on the nature of the solid. For insulators, the optical properties are the most important, while for semiconductors both optical and transport studies are important. For metals, optical properties are less important and Fermi surface studies become more important.

In the case of insulators, electrical conductivity can arise through the motion of lattice ions as they move from one lattice vacancy to another, or from one interstitial site to another. Ionic conduction therefore occurs through the presence of lattice defects, and is promoted in materials with open crystal structures. In ionic crystals there are relatively few mobile electrons or holes even at high temperature so that conduction in these materials is predominantly due to the motions of ions.

Ionic conductivity (σ_{ionic}) is proportional both to the density of lattice defects (vacancies and interstitials) and to the diffusion rate, so that we can write

$$\sigma_{\text{ionic}} \sim e^{-(E+E_0)/k_BT}$$

where E_0 is the activation energy forv ionic motion and E is the energy for formation of a defect (a vacancy, a vacancy pair, or an interstitial). Being an activated process, ionic conduction is enhanced at elevated temperatures. Since defects in ionic crystals can be observed visibly as the migration of colour through the crystal, ionic conductivity can be distinguished from electronic conductivity by comparing the transport of charge with the transport of mass, as can, for example, be measured by the material plated out on electrodes in contact with the ionic crystal.

In this course, we will spend a good deal of time studying optical and transport properties of solids. In connection with topics on magnetism, we will also study Fermi surface measurements which are closely connected to issues relevant to transport properties. Since Fermi surface studies, for the most part, are resonance experiments and involve the use of a magnetic field, it is pedagogically more convenient to discuss these topics in Part III of the course, devoted to Magnetism. We have presented this review of the electronic energy bands of solids because the $E(\vec{k})$ relations are closely connected with a large number of common measurements in the laboratory, and because a knowledge of the $E(\vec{k})$ relations forms the basis for many device applications.

WHAT IS CONDENSED MATTER PHYSICS?

LENGTH, TIME, ENERGY SCALES

We will be concerned with:

- ω, T $\ll$ 1eV
- $|x_i - x_j|, \dfrac{1}{q} \gg 1\text{Å}$

as compared to energies in the MeV for nuclear matter, and GeV or even T eV, in particle physics.

The properties of matter at these scales is determined by the behaviour of collections of many ($\sim 10^{23}$) atoms.

In general, we will be concerned with scales much smaller than those at which gravity becomes very important, which is the domain of astrophysics and cosmology.

MICROSCOPIC EQUATIONS VS. STATES OF MATTER, PHASE TRANSITIONS, CRITICAL POINTS

Systems containing many particles exhibit properties which are special to such systems. Many of these properties are fairly insensitive to the details at length scales shorter than 1Å and energy scales higher than 1eV – which are quite adequately described by the equations of non-relativistic quantum mechanics. Such properties are emergent. For example, precisely the same microscopic equations of motion – Newton's equations – can describe two different systems of 10^{23} H_2O molecules.

$$m\frac{d^2\vec{x}_i}{dt^2} = -\sum_{j \neq i} \vec{\nabla}_i V(\vec{x}_i - \vec{x}_j)$$

Or, perhaps, the Schrodinger equation:

$$-\frac{\hbar^2}{2m}\sum_i \nabla_i^2 + \sum_{i,j} V(\vec{x}_i - \vec{x}_j)\,\psi(\vec{x}_1,\ldots,\vec{x}_N) = E\psi(\vec{x}_1,\ldots,\vec{x}_N)$$

However, one of these systems might be water and the other ice, in which case the properties of the two systems are completely different, and the similarity between their microscopic descriptions is of no practical consequence. As this example shows, many-particle systems exhibit various phases – such as ice and water – which are not, for the most part, usefully described by the microscopic equations. Instead, new low-energy, long-wavelength physics emerges as a result of the interactions among large numbers of particles. Different phases are separated by phase transitions, at which the low-energy, long-wavelength description becomes non-analytic and exhibits singularities. In the above example, this occurs at the freezing point of water, where its entropy jumps discontinuously.

BROKEN SYMMETRIES

As we will see, different phases of matter are distinguished on the basis of symmetry. The microscopic equations are often highly symmetrical – for instance, Newton's laws are translationally and rotationally invariant – but a given phase may exhibit much less symmetry. Water exhibits the full translational and rotational symmetry of of Newton's laws; ice, however, is only invariant under the discrete translational and rotational group of its crystalline lattice. We say that the translational and rotational symmetries of the microscopic equations have been spontaneously broken.

EXPERIMENTAL PROBES: X-RAY SCATTERING, NEUTRON SCATTERING, NMR, THERMODYNAMIC, TRANSPORT

There are various experimental probes which can allow an experimentalist to determine in what phase a system is and to determine its quantitative properties:

- *Scattering:* send neutrons or X-rays into the system with prescribed energy, momentum and measure the energy, momentum of the outgoing neutrons or X-rays.
- *NMR:* apply a static magnetic field, B, and measure the absorption and emission by the system of magnetic radiation at frequencies of the order of $\omega_c = geB/m$. Essentially the scattering of magnetic radiation at low frequency by a system in a uniform B field.

- *Thermodynamics:* measure the response of macroscopic variables such as the energy and volume to variations of the temperature, pressure, etc.
- *Transport:* set up a potential or thermal gradient, $\nabla\varphi$, ∇T and measure the electrical or heat current $\vec{j}, \vec{j}_Q$. The gradients $\nabla\varphi$, ∇T can be held constant or made to oscillate at finite frequency.

THE SOLID STATE: METALS, INSULATORS, MAGNETS, SUPERCONDUCTORS

In the solid state, translational and rotational symmetries are broken by the arrangement of the positive ions. It is precisely as a result of these broken symmetries that solids are solid, i.e. that they are rigid. It is energetically favorable to break the symmetry in the same way in different parts of the system. Hence, the system resists attempts to create regions where the residual translational and rotational symmetry groups are different from those in the bulk of the system.

The broken symmetry can be detected using X-ray or neutron scattering: the X-rays or neutrons are scattered by the ions; if the ions form a lattice, the X-rays or neutrons are scattered coherently, forming a diffraction pattern with peaks. In a crystalline solid, discrete subgroups of the translational and rotational groups are preserved. For instance, in a cubic lattice, rotations by $\pi/2$ about any of the crystal axes are symmetries of the lattice (as well as all rotations generated by products of these). Translations by one lattice spacing along a crystal axis generate the discrete group of translations.

In this course, we will be focussing on crystalline solids. Some examples of noncrystalline solids, such as plastics and glasses will be discussed below. Crystalline solids fall into three general categories: metals, insulators, and superconductors. In addition, all three of these phases can be further subdivided into various magnetic phases. Metals are characterized by a non-zero conductivity at $T = 0$. Insulators have vanishing conductivity at $T = 0$. Superconductors have inπ finite conductivity for

$T < T_c$ and, furthermore, exhibit the Meissner effect: they expel magnetic field s.

In a magnetic material, the electron spins can order, thereby breaking the spinrotational invariance. In a ferromagnet, all of the spins line up in the same direction, thereby breaking the spin-rotational invariance to the subgroup of rotations about this direction while preserving the discrete translational symmetry of the lattice. (This can occur in a metal, an insulator, or a superconductor.) In an antiferromagnet, neighboring spins are oppositely directed, thereby breaking spin-rotational invariance to the subgroup of rotations about the preferred direction and breaking the lattice translational symmetry to the subgroup of translations by an even number of lattice sites. Recently, new states of matter – the fractional quantum Hall states – have been discovered in effectively two-dimensional systems in a strong magnetic field at very low T. Tomorrow's experiments will undoubtedly uncover new phases of matter.

OTHER PHASES: LIQUID CRYSTALS, QUASICRYSTALS, POLYMERS, GLASSES

The liquid – with full translational and rotational symmetry – and the solid – which only preserves a discrete subgroup – are but two examples of possible realizations of translational symmetry. In a liquid crystalline phase, translational and rotational symmetry is broken to a combination of discrete and continuous subgroups.

For instance, a nematic liquid crystal is made up of elementary units which are line segments. In the nematic phase, these line segments point, on average, in the same direction, but their positional distribution is as in a liquid. Hence, a nematic phase breaks rotational invariance to the subgroup of rotations about the preferred direction and preserves the full translational invariance. Nematics are used in LCD displays.

In a smectic phase, on the other hand, the line segments arrange themselves into layers, thereby partially breaking the translational symmetry so that discrete translations perpendicular to the layers and continuous translations along the layers remain unbroken. In the smectic-A phase, the preferred orientational direction is the same as the direction perpendicular to the layers; in the smectic-C phase, these directions are different.

In a hexatic phase, a two-dimensional system has broken orientational order, but unbroken translational order; locally, it looks like a triangular lattice. A quasicrystal has rotational symmetry which is broken to a 5-fold discrete subgroup. Translational order is completely broken (locally, it has discrete translational order). Polymers are extremely long molecules. They can exist in solution or a chemical reaction can take place which cross-links them, thereby forming a gel. A glass is a rigid, 'random' arrangement of atoms. Glasses are somewhat like 'snapshots' of liquids, and are probably non-equilibrium phases, in a sense.

Chapter 2

Effective Mass Theory and Transport Phenomena

WAVEPACKETS IN CRYSTALS AND OF ELECTRONS IN SOLIDS

In a crystal lattice, the electronic motion which is induced by an applied field is conveniently described by a wavepacket composed of eigenstates of the unperturbed crystal. These eigenstates are Bloch functions

$$\psi_{nk}(\vec{r}) = e^{i\vec{k}\cdot\vec{r}} u_{nk}(\vec{r})$$

and are associated with band n. These wavepackets are solutions of the time-dependent Schroodinger equation

$$H_0 \psi_n(\vec{r},t) = i\hbar \frac{\partial \psi_n(\vec{r},t)}{\partial t}$$

where the time independent part of the Hamiltonian can be written as

$$H_0 = \frac{p^2}{2m} + V(\vec{r})$$

where $V(\vec{r}) = V(\vec{r} + \vec{R}_n)$ is the periodic potential. The wave packets $\psi_n(\vec{r},t)$ can be written in terms of the Bloch states $\psi_{nk}(\vec{r})$ as

$$\psi_n(\vec{r},t) = \sum_k A_{n,k}(t)\psi_{nk}(\vec{r}) = \int d^3k A_{n,k}(t)\psi_{nk}(\vec{r})$$

where we have replaced the sum by an integration over the Brillouin zone, since permissible $\vec{k}$ values for a macroscopic solid are very closely spaced. If the Hamiltonian H_0 is time-independent as is often the case, we can write

$$A_{n,k}(t) = A_{n,k} e^{-i\omega n(\vec{k})t}$$

where

$$\hbar\omega_n(\vec{k}) = E_n(\vec{k})$$

and thereby obtain

$$\psi_n(\vec{r},t) = \int d^3k A_{n,k} u_{nk}(\vec{r}) e^{i[\vec{k}\cdot\vec{r}-\omega_n(\vec{k})t]}$$

We can localize the wavepacket in $\vec{k}$-space by requiring that the coefficients $A_{n,k}$ be large only in a confinedregion of ~k-space centered at $\vec{k} = \vec{k}_0$. If we now expand the band energy in a Taylor series around $\vec{k} = \vec{k}_0$ we obtain:

$$E_n(\vec{k}) = E_n(\vec{k}_0) + (\vec{k}-\vec{k}_0)\cdot\left.\frac{\partial E_n(\vec{k})}{\partial\vec{k}}\right|_{\vec{k}=\vec{k}_0} + \cdots,$$

where we have written $\vec{k}$ as

$$\vec{k} = \vec{k}_0 + (\vec{k}-\vec{k}_0).$$

Since $|\vec{k}-\vec{k}_0|$ is assumed to be small compared with Brillouin zone dimensions, we are justified in retaining only the first two terms of the Taylor expansion in equation given above. Substitution into equation for the wave packet yields:

$$\psi_n(\vec{r},t) \simeq e^{i(\vec{k}_0\cdot\vec{r}-w_n(\vec{k}_0)t)} \int d^3k A_{n,k} u_{nk}(\vec{r}) e^{i(\vec{k}-\vec{k}_0)\cdot\left[\vec{r}-\frac{\partial w_n(\vec{k})}{\partial\vec{k}}t\right]}$$

where

$$\hbar\omega_n(\vec{k}_0) = E_n(\vec{k}_0)$$

and

$$\hbar\frac{\partial\omega_n(\vec{k})}{\partial\vec{k}} = \frac{\partial E_n(\vec{k})}{\partial\vec{k}}$$

and the derivative $\partial\omega_n(\vec{k})/\partial\vec{k}$ which appears in the phase factor of equation is evaluated at $\vec{k} = \vec{k}_0$. Except for the periodic function $u_{nk}(\vec{r})$ the above expression is in the standard form for a wavepacket moving with "" $\vec{v}_g$

$$\vec{v}_g = \frac{\partial\omega_n(\vec{k})}{\partial\vec{k}}$$

so that

$$\vec{v}_g = \frac{1}{\hbar}\frac{\partial E_n(\vec{k})}{\partial\vec{k}}$$

while the phase velocity

$$\frac{\vec{v}_p = \omega_n(\vec{k})}{\vec{k} = \frac{\partial E_n(\vec{k})}{\vec{k}}}.$$

In the limit of free electrons the becomes

$$\vec{v}_g = \frac{\vec{p}}{m} = \frac{\hbar\vec{k}}{m}.$$

This result also follows from the above discussion using

$$E_n(\vec{k}) = \frac{\hbar^2 k^2}{2m}$$

$$\frac{\partial E_n(\vec{k})}{\hbar\partial\vec{k}} = \frac{\hbar\vec{k}}{m}.$$

We shall show later that the electron wavepacket moves through the crystal very much like a free electron provided that the wavepacket remains localized in k space during the time interval of interest in the particular problem under consideration. Because of the uncertainty principle, the localization of a wavepacket in reciprocal space implies a delocalization of the wavepacket in real space.

We use wavepackets to describe electronic states in a solid when the crystal is perturbed in some way (e.g., by an applied electric or magnetic field). We make frequent applications of wavepackets to transport theory (e.g., electrical conductivity). In many practical applications of transport theory, use is made of the Effective-Mass Theorem, which is the most important result of transport theory.

We note that the above discussion for the wavepacket is given in terms of the perfect crystal. In our discussion of the Effective-Mass Theorem we will see that these wavepackets are also of use in describing situations where the Hamiltonian which enters Schroodinger's equation contains both the unperturbed Hamiltonian of the perfect crystal H_0 and the perturbation Hamiltonian H' arising from an external perturbation. Common perturbations are applied electric or magnetic field s, or a lattice defect or an impurity atom.

THE EFFECTIVE MASS THEOREM

We shall now present the Effective Mass theorem, which is central to the consideration of the electrical and optical properties of solids. An elementary proof of the theorem will be given here for a simple but important case, namely the non-degenerate band which can be identified with the corresponding atomic state. The theorem will be discussed from a more advanced point of view which considers also the case of degenerate bands in the following courses in the physics of solids sequence.

For many practical situations we find a solid in the presence of some perturbing field (e.g., an externally applied electric field, or the perturbation created by an impurity atom or a crystal defect).

The perturbation may be either time-dependent or time-independent. We will show here that under many common circumstances this perturbation can be treated in the effective mass approximation whereby the periodic potential is replaced by an effective Hamiltonian based on the $E(\vec{k})$ relations for the perfect crystal. To derive the effective mass theorem, we start with the time-dependent Schroodinger equation

$$(H_0 + H')\psi_n(\vec{r},t) = i\hbar\frac{\partial\psi_n(\vec{r},t)}{\partial t}$$

We then substitute the expansion for the wave packet

$$\psi_n(\vec{r},t) = \int d^3kA_{nk}(t)e^{i\vec{k}\cdot\vec{r}}u_{nk}(\vec{r})$$

into Schroodinger's equation and make use of the Bloch solution

$$H_0e^{i\vec{k}\cdot\vec{r}}u_{nk}(\vec{r}) = E_n(\vec{k})e^{i\vec{k}\cdot\vec{r}}u_{nk}(\vec{r})$$

to obtain:

$$(H_0 + H')\psi_n(\vec{r},t) = \int d^3k[E_n(\vec{k}) + H']A_{nk}(t)e^{i\vec{k}\cdot\vec{r}}u_{nk}(\vec{r}) = i\hbar(\partial\psi_n(\vec{r},t)/\partial t$$

$$= i\hbar\int d^3k\dot{A}_{nk}(t)e^{i\vec{k}\cdot\vec{r}}u_{nk}(\vec{r}).$$

It follows from Bloch's theorem that $E_n(\vec{k})$ is a periodic function in the reciprocal lattice. We can therefore expand $E_n(\vec{k})$ in a Fourier series in the direct lattice

$$E_n(\vec{k}) = \sum_{\vec{R}_l} E_{nl}e^{i\vec{k}\cdot\vec{R}_l}$$

where the $\vec{R}_l$ are lattice vectors. Now consider the differential operator $E_n(-i\vec{\nabla})$ formed by replacing $\vec{k}$ by $-i\vec{\nabla}$

$$E_n(-i\vec{\nabla}) = \sum_{\vec{R}_l} E_{nl}e^{\vec{R}_l\cdot\vec{\nabla}}$$

Consider the effect of $E_n(-i\vec{\nabla})$ on an arbitrary function $f(\vec{r})$. Since $e^{\vec{R}_l\cdot\vec{\nabla}}$ can be expanded in a Taylor series, we obtain

$$e^{\vec{R}_l\cdot\vec{\nabla}}f(\vec{r}) = \left[1 + \vec{R}_l\cdot\vec{\nabla} + \frac{1}{2}(\vec{R}_l\cdot\vec{\nabla})(\vec{R}_l\cdot\vec{\nabla}) + \cdots\right]f(\vec{r})$$

$$= f(\vec{r}) + \vec{R}_l \cdot \vec{\nabla} f(\vec{r}) + \frac{1}{2!} R_{l,\alpha} R_{l,\beta} \frac{\partial^2}{\partial r_\alpha \partial r_\beta} f(\vec{r}) + \cdots$$

$$= f(\vec{r} + \vec{R}_l).$$

Thus the effect of $E_n(-i\vec{\nabla})$ on a Bloch state is

$$E_n(-i\vec{\nabla})\psi_{nk}(\vec{r}) = \sum_{\vec{R}_l} E_{nl}\psi_{nk}(\vec{r} + \vec{R}_l)$$

$$= \sum_{\vec{R}_l} E_{nl} e^{i\vec{k}\cdot\vec{R}_l} e^{i\vec{k}\cdot\vec{r}} u_{nk}(\vec{r})$$

$$= E_n(\vec{k})\psi_{nk}(\vec{r})$$

since from Bloch's theorem

$$\psi_{nk}(\vec{r} + \vec{R}_l) = e^{i\vec{k}\cdot\vec{R}_l}\left[e^{i\vec{k}\cdot\vec{r}} u_{nk}(\vec{r})\right].$$

Substitution of

$$E_n(-i\vec{\nabla})\psi_{nk}(\vec{r}) = E_n(\vec{k})\psi_{nk}(\vec{r})$$

from equation into SchrÄodinger's equation yields:

$$\int d^3k\left[E_n(-i\vec{\nabla}) + H'\right] A_{nk}(t) e^{i\vec{k}\cdot\vec{r}} u_{nk}(\vec{r}) = \left[E_n(-i\vec{\nabla}) + H'\right]$$

$$\int d^3k A_{nk}(t) e^{i\vec{k}\cdot\vec{r}} u_{nk}(\vec{r})$$

so that

$$\left[E_n(-i\vec{\nabla}) + H'\right]\psi_n(\vec{r},t) = i\hbar\frac{\partial\psi_n(\vec{r},t)}{\partial t}.$$

This result is called the effective mass theorem. We observe that the original crystal Hamiltonian $p^2/2m + V(\vec{r})$ does not appear in this equation. It has instead been replaced by an effective Hamiltonian which is an operator formed from the solution $E(\vec{k})$ for the perfect crystal in which we replace $\vec{k}$ by $-i\vec{\nabla}$.

For example, for the free electron ($V(\vec{r}) \equiv 0$)

$$E_n(-i\vec{\nabla}) \to -\frac{\hbar^2\nabla^2}{2m}.$$

In applying the effective mass theorem, we assume that $E(\vec{k})$ is known either from the results of a theoretical calculation or from the analysis of experimental results.

What is important here is that once $E(\vec{k})$ is known, the effect of various perturbations on the ideal crystal can be treated in terms of the solution to the energy levels of the perfect crystal without recourse to consideration of the full Hamiltonian.

In practical cases, the solution to the effective mass equation is much easier to carry out than the solution to the original Schroodinger equation.

According to the above discussion, we have assumed that $E(\vec{k})$ is specified throughout the Brillouin zone. For many practical applications, the region of $\vec{k}$-space which is of importance is confined to a small portion of the Brillouin zone. In such cases it is only necessary to specify $E(\vec{k})$ in a local region (or regions) and to localize our wavepacket solutions to these local regions of $\vec{k}$-space.

Suppose that we localize the wavepacket around $\vec{k} = \vec{k}_0$, and correspondingly expand our Bloch functions around $\vec{k}_0$,

$$\psi_{nk}(\vec{r}) = e^{i\vec{k}\cdot\vec{r}} u_{nk}(\vec{r}) \simeq e^{i\vec{k}\cdot\vec{r}} u_{nk_0}(\vec{r}) = e^{i(\vec{k}-\vec{k}_0)\cdot\vec{r}} \psi_{nk_0}(\vec{r})$$

where we have noted that $u_{nk}(\vec{r}) \simeq u_{nk_0(\vec{r})}$ has only a weak dependence on $\vec{k}$. Then our wavepacket can be written as

$$\psi_{nk}(\vec{r},t) = \int d^3k A_{nk}(t) e^{i(\vec{k}-\vec{k}_0)\cdot\vec{r}} \psi_{nk_0}(\vec{r}) = F(\vec{r},t)\psi_{nk_0}(\vec{r})$$

where $F(\vec{r},t)$ is called the amplitude or envelope function and is defined by

$$F(\vec{r},t) = \int d^3k\, A_{nk}(t) e^{i(\vec{k}-\vec{k}_0)\cdot\vec{r}}.$$

Since the time dependent Fourier coefficients $A_{nk}(t)$ are assumed here to be large only near $\vec{k} = \vec{k}_0$, then $F(\vec{r},t)$ will be a slowly varying function of $\vec{r}$, because in this case

$$e^{i(\vec{k}-\vec{k}_0)\vec{r}} \simeq 1 + i(\vec{k} - \vec{k}_0)\cdot\vec{r} + \cdots$$

It can be shown that the envelope function also satisfies the effective mass equation

$$\left[E_n(-i\vec{\nabla}) + H'\right]F(\vec{r},t) = i\hbar\frac{\partial F(\vec{r},t)}{\partial t}$$

where we now replace $\vec{k} - \vec{k}_0$ in $E_n(\vec{k})$ by $-i\vec{\nabla}$. This form of the effective mass equation is useful for treating the problem of donor and acceptor impurity states in semiconductors, and $\vec{k}_0$ is taken as the band extremum.

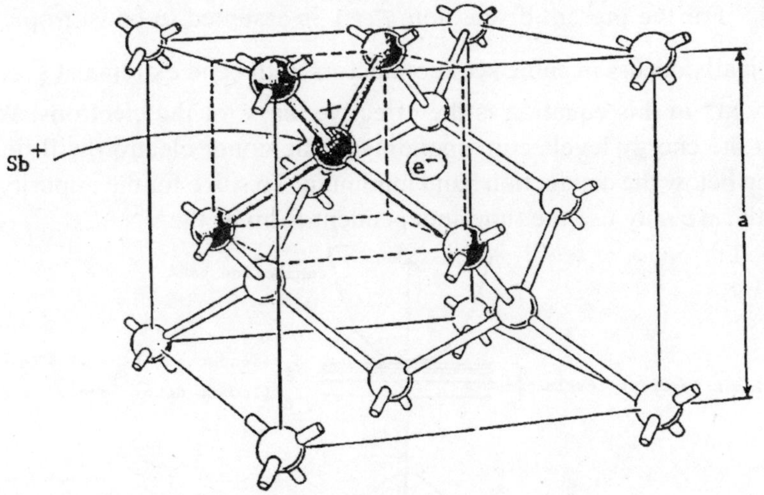

Fig. Crystal structure of diamond, showing the tetrahedral bond arrangement with an Sb+ ion on one of the lattice sites and a free donor electron available for conduction.

Application of the Effective Mass Theorem to Donor Impurity Levels in a Semiconductor

Suppose that we add an impurity from column V in the Periodic Table to a semiconductor such as silicon or , which are both members of column IV of the periodic table. This impurity atom will have one more electron than is needed to satisfy the valency requirements for the tetrahedral bonds which the or silicon atoms form with their 4 valence electrons.

This extra electron from the impurity atom will be free to wander through the lattice, subject of course to the coulomb attraction of the ion core which will have one unit of positive charge. We will consider here the case where we add just a small number of these impurity atoms so that we may focus our attention on a single, isolated substitutional impurity atom in an otherwise perfect lattice. In the course of this discussion we will define more carefully what the limits on the impurity concentration must be so that the treatment given here is applicable.

Let us also assume that the conduction band of the host semiconductor in the vicinity of the band "minimum" at $\vec{k}_0$ has the simple analytic form $E_c(\vec{k}) \simeq E_c(\vec{k}_0) + \dfrac{\hbar^2 (k - k_0)^2}{2m^*}$

We can consider this expression for the conduction band level $E_c(\vec{k})$ as a special case of the Taylor expansion of $E_c(\vec{k})$ about an energy band minimum at $\vec{k} = \vec{k}_0$. For the present discussion, $E(\vec{k})$ is assumed to be isotropic in $\vec{k}$; this typically occurs in cubic semiconductors with band extrema at $\vec{k} = 0$. The quantity m* in this equation is the effective mass for the electrons. We will see that the energy levels corresponding to the donor electron will lie in the band gap below the conduction band minimum. To solve for the impurity levels explicitly, we may use the time-independent form of the

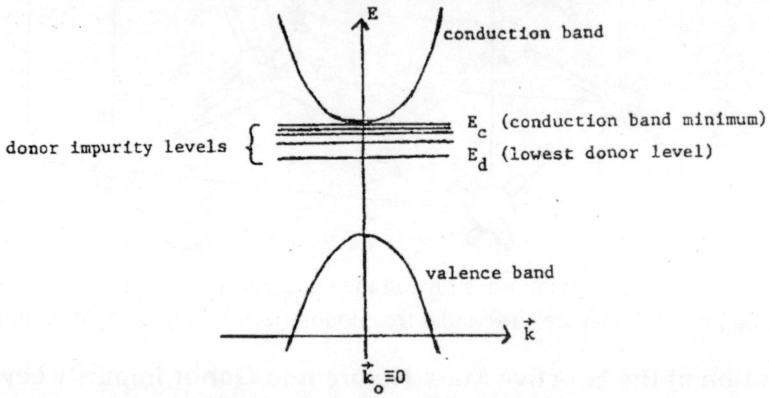

Fig. Schematic Band Diagram Showing Donor Levels in a Semiconductor.

effective mass theorem derived from equation

$$\left[E_n(-i\vec{\nabla}) + H'\right]F(\vec{r}) = (E - E_c)F(\vec{r}) .$$

Equation is applicable to the impurity problem in a semiconductor provided that the amplitude function $F(\vec{r})$ is sufficiently slowly varying over a unit cell.

In the course of this discussion, we will see that the donor electron in a column IV (or III-V or II-VI compound semiconductor) will wander over many lattice sites and therefore this approximation on $F(\vec{r})$ will be justified.

For a singly ionized donor impurity (such as arsenic in), the perturbing potential H' can be represented as a Coulomb potential

$$H' = -\frac{e^2}{\varepsilon r}$$

where ε is an average dielectric constant of the crystal medium which the donor electron sees as it wanders through the crystal. Experimental data on donor impurity states indicate that ε is very closely equal to the low frequency limit of the electronic dielectric constant $\varepsilon_1(\omega)|_{\omega=0}$, which we will discuss extensively in treating the optical properties of solids (Part II of this course). The above discussion involving an isotropic $E(\vec{k})$ is appropriate for semiconductors with conduction band minima at $\vec{k}_0 = 0$. The Effective Mass equation for the unperturbed crystal is

$$E_n(-i\vec{\nabla}) = -\frac{\hbar^2}{2m^*}\nabla^2$$

in which we have replaced $\vec{k}$ by $-i\vec{\nabla}$.

The donor impurity problem in the effective mass approximation thus becomes $\left[-\frac{\hbar^2}{2m^*}\nabla^2 - \frac{e^2}{\varepsilon r}\right]F(\vec{r})$
$= (E - E_c)F(\vec{r})$

where all energies are measured with respect to the bottom of the conduction band E_c. If we replace m^* by m and e^2/by e^2, we immediately recognize this equation as Schroodinger's equation for a hydrogen atom under the identification of the energy eigenvalues with

$$E_n = \frac{e^2}{2n^2 a_0} = \frac{me^4}{2n^2\hbar^2}$$

where a_0 is the Bohr radius $a_0 = \hbar^2/me^2$. This identification immediately allows us to write E_l for the donor energy levels as

$$E_l = E_{l=1} = E_c - \frac{m^* e^4}{2\varepsilon^2\hbar^2}.$$

where $l = 1, 2, 3,...$ is an integer denoting the donor level quantum numbers and we identify the bottom of the conduction band E_c as the ionization energy for this effective hydrogenic problem.

Physically, this means that the donor levels correspond to bound (localized) states while the band states above E_c correspond to delocalized nearly-free electron-like states. The lowest or "ground-state" donor energy level is then written as

$$E_d = E_{l=1} = E_c - \frac{m^* e^4}{2\varepsilon^2\hbar^2}.$$

It is convenient to identify the "effective" first Bohr radius for the donor level as

$$a_0^* = \frac{\varepsilon\hbar^2}{m^* e^2}$$

and to recognize that the wave function for the ground state donor level will be of the form

$$F(\vec{r}) = Ce^{-r/a_0^*}$$

where C is the normalization constant. Thus the solutions to equation for a semiconductor are hydrogenic energy levels with the substitutions $m \rightarrow m^*$, $e^2 \rightarrow (e^2/\varepsilon)$ and the ionization energy usually taken as the zero of energy for the hydrogen atom now becomes E_c, the conduction band extremum.

For a semiconductor like we have a very large dielectric constant, ε; 16. The value for the effective mass is somewhat more difficult to specify in since the constant energy surfaces for are located about the L-points in the Brillouin zone and are ellipsoids of revolution. Since the constant energy surfaces for such semiconductors are non-spherical, the effective mass tensor is anisotropic. However we will write down an average effective mass value m^*/m; 0.12 (Kittel ISSP) so that we can estimate pertinent magnitudes for the donor levels in a typical semiconductor. With these values for ε and m^* we obtain:

$$E_c - E_d \simeq 0.007\text{eV}$$

and the effective Bohr radius

$$a_0^* \simeq 70\text{Å}.$$

These values are to be compared with the ionization energy of 13.6 eV for the hydrogen atom and with the hydrogenic Bohr orbit of $a_0 = \hbar^2/me^2 =$ 0.5Å. Thus we see that a_0^* is indeed large enough to satisfy the requirement that $F(\vec{r})$ be slowly varying over a unit cell. On the other hand, if a_0^* were to be comparable to a lattice dimension, then $F(\vec{r})$ could not be considered as a slowly varying function of $\vec{r}$ and generalizations of the above treatment would have to be made. Such generalizations involve:

1. Treating $E(\vec{k})$ for a wider region of $\vec{k}$-space, and
2. Relaxing the condition that impurity levels are to be associated with a single band. From the uncertainty principle, the localization in momentum space for the impurity state requires a delocalization in real space; and likewise, the converse is true, that a localized impurity in real space corresponds to a delocalized description in $\vec{k}$-space. Thus "shallow" hydrogenic donor levels can be attributed to a specific band at a specific energy extremum at $\vec{k}_0$ in the Brillouin zone. On the other hand, "deep" donor levels are not hydrogenic and have a more complicated energy level structure. Deep donor levels cannot be readily associated with a specific band or a specific $\vec{k}$ point in the Brillouin zone.

In dealing with this impurity problem, it is very tempting to discuss the donor levels in silicon and . For example in silicon where the conduction band extrema are at the Δ point, the effective mass theorem requires us to replace $E(-i\vec{\nabla})$ by $E(-i\vec{\nabla}) \to -\frac{\hbar^2}{2m_l^*}\frac{\partial^2}{\partial x^2} - \frac{\hbar^2}{2m_l^*}\left(\frac{\partial^2}{\partial y^2} + \frac{\partial^2}{\partial z^2}\right)$

and the resulting Schroodinger equation can no longer be solved exactly. Although this is a very interesting problem from a practical point of view, it is too difficult a problem for us to solve in detail for illustrative purposes in an introductory course.

QUASI-CLASSICAL ELECTRON DYNAMICS

According to the "Correspondence Principle" of Quantum Mechanics, wavepacket solutions of SchrÄodinger's equation follow the trajectories of classical particles and satisfy Newton's laws. One can give a Correspondence Principle argument for the form which is assumed by the velocity and acceleration of a wavepacket. According to the Correspondence Principle, the connection between the classical Hamiltonian and the quantum mechanical Hamiltonian is made by the identification of $\vec{p} \to (\hbar / i)\vec{\nabla}$. Thus

$$E_n(-i\vec{\nabla}) + H'(\vec{r}) \leftrightarrow E_n(\vec{p}/\hbar) + H'(\vec{r}) = H_{\text{classical}}(\vec{p},\vec{r}) .$$

In classical mechanics, Hamilton's equations give the velocity according to: $\dot{\vec{r}} = \frac{\partial H}{\partial \vec{p}} = \nabla_p H = \frac{\partial E(\vec{k})}{\hbar \partial \vec{k}}$

in agreement with the for a wavepacket given by equation. Hamilton's equation for the acceleration is given by:

$$\dot{\vec{p}} = -\frac{\partial H}{\partial \vec{r}} = -\frac{\partial H'(\vec{r})}{\partial \vec{r}} .$$

For example, in the case of an applied electric field $\vec{E}$ the perturbation Hamiltonian is

$$H'(\vec{r}) = -e\vec{r}\cdot\vec{E}$$

so that

$$\dot{\vec{p}} = \hbar\dot{\vec{k}} = e\vec{E} .$$

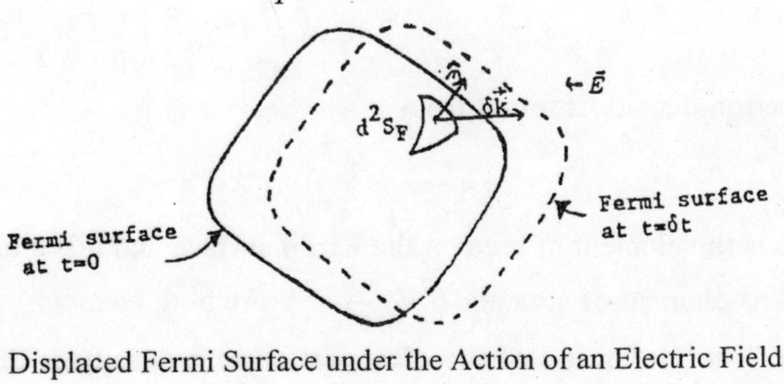

Fig. Displaced Fermi Surface under the Action of an Electric Field $\vec{E}$.

In this equation $e\vec{E}$ is the classical Coulomb force on an electric charge due to an applied field $\vec{E}$. It can be shown that in the presence of a magnetic field $\vec{B}$, the acceleration theorem follows the Lorentz force equation

$$\dot{\vec{p}} = \hbar\dot{\vec{k}} = e[\vec{E} + (1/c)\vec{v} \times \vec{B}]$$

where

$$\vec{v} = \frac{\partial E(\vec{k})}{\hbar \partial \vec{k}}.$$

In the crystal, the crystal momentum ℏ~k for the wavepacket plays the role of the momentum for a classical particle.

Quasi-Classical Theory of Electrical Conductivity – Ohm's Law

We will now apply the idea of the quasi-classical electron dynamics in a solid to the problem of the electrical conductivity for a metal with an arbitrary Fermi surface and band structure. The electron is treated here as a wavepacket with momentum $\hbar\vec{k}$ moving in an external electric field $\vec{E}$ in compliance with Newton's laws. Because of the acceleration theorem, we can think of the electric field as creating a "displacement" of the electron distribution in $\vec{k}$-space. We remember that the Fermi surface encloses the region of occupied states within the Brillouin zone. The effect of the electric field is to change the wave vector $\vec{k}$ of an electron by

$$\delta\vec{k} = \frac{e}{\hbar}\vec{E}\delta t$$

(where we note that the charge on the electron e is a negative number). We picture the displacement $\delta\vec{k}$ of equation by the displacement of the Fermi surface. From this diagram we see that the incremental volume of $\vec{k}$-space $\delta^3 V_{\vec{k}}$ which is "swept out" in the time dt due to the presence of the field $\vec{E}$ is

$$\delta^3 V_{\vec{k}} = d^2 S_F \hat{n} \cdot \delta\vec{k} = d^2 S_F \hat{n} \cdot \left(\frac{e}{\hbar}\vec{E}\delta t\right)$$

and the electron density is found from

$$n = \frac{2}{(2\pi)^3} \int_{E \le E_F} d^3k$$

where d^2S_F is the element of area on the Fermi surface and $\hat{n}$ is a unit vector normal to this element of area and $\delta^3 V_{\vec{k}} \to d^3k$ are both elements of volume in $\vec{k}$-space. The Definition of the electrical current density is the current flowing through a unit area in real space and is given by the product of the (number of electrons per unit volume) with the (charge per electron) and with the () so that the current density $\delta\vec{j}$ created by applying the electric field $\vec{E}$ for a time interval δt is given by

$$\delta\vec{j} = \int [2/(2\pi)^3]\cdot[\delta^3 V_{\vec{k}}]\cdot[e]\cdot[\vec{v}_g]$$

where $\vec{v}_g$ is the for electron wavepacket and $2/(2\pi)^3$ is the density of electronic states in $\vec{k}$-space (including the spin degeneracy of two) because we can put 2 electrons in each phase space state. Substitution for $\delta^3 V_{\vec{k}}$ in equation yields the instantaneous rate of change of the current density averaged over the Fermi surface

$$\frac{\partial\vec{j}}{\partial t} = \frac{e^2}{4\pi^3\hbar}\oint \vec{v}_g \hat{n}\cdot\vec{E}d^2S_F = \frac{e^2}{4\pi^3\hbar}\oint\oint \vec{v}_g\left(\frac{\vec{v}_g\cdot\vec{E}}{|v_g|}\right)(d^2S_F)$$

since the given by equation is directed normal to the Fermi surface. In a real solid, the electrons will not be accelerated inde finitely, but will eventually collide with an impurity, or a lattice defect or a lattice vibration (phonon).

These collisions will serve to maintain the displacement of the Fermi surface at some steady state value, depending on τ, the average time between collisions. We can introduce this relaxation time through the expression

$$n(t) = n(0)e^{-t/\tau}$$

where $n(t)$ is the number of electrons that have not made a collision at time t, assuming that the last collision had been made at time $t = 0$. The relaxation time is the average collision time

$$\langle t\rangle = \frac{1}{\tau}\int_0^\infty te^{-t/\tau}dt = \tau.$$

If in equation , we set $\langle\delta t\rangle = \tau$ and write the average current density as $\vec{j} = \langle\delta\vec{j}\rangle$, then we

$$\vec{j} = \frac{e^2\tau}{4\pi^3\hbar}\int \vec{v}_g\frac{\vec{v}_g\cdot\vec{E}}{|v_g|}(d^2S_F).$$

We define the conductivity tensor as $\vec{j} = \overleftrightarrow{\sigma}\cdot\vec{E}$, so that equation provides an explicit expression for the tensor $\overleftrightarrow{\sigma}$.

$$\overleftrightarrow{\sigma} = \frac{e^2\tau}{4\pi^3\hbar}\int\frac{\vec{v}_g\vec{v}_g}{|v_g|}(d^2S_F).$$

In the free electron limit $\overleftrightarrow{\sigma}$ becomes a scalar (isotropic conduction) and is given by the Drude formula which we derive below from equation. Using the equations for the free electron limit

$$E = \hbar^2 k^2 / 2m$$

$$E_F = \hbar^2 k_F^2 / 2m$$

$$\vec{v}_g = \hbar \vec{k}_F / m .$$

We then obtain

$$\vec{v}_g \vec{v}_g \rightarrow v_x^2 = v_y^2 = v_z^2 = v^2 / 3$$

$$\int d^2 S_F = 4\pi k_F^2 ,$$

so that the number of electrons/unit volume can be written as:

$$n = \frac{1}{4\pi^3} \frac{4\pi}{3} k_F^3 .$$

Therefore $\vec{j} = \dfrac{e^2 \tau}{4\pi^3 \hbar} \left(\dfrac{\hbar k_F}{m} \right) \dfrac{1}{3} \vec{E} (4\pi k_F^2) = \dfrac{ne^2 \tau}{m} \vec{E}.$

Thus the free electron limit gives Ohm's law in the familiar form

$$\sigma = \frac{ne^2 \tau}{m} = ne\mu ,$$

and showing that the electrical conductivity depends on both the carrier density n and the carrier mobility μ.

A slightly modified form of Ohm's law is also applicable to conduction in a material for which the energy dispersion relations are simple parabolic and m has been replaced by the effective mass $m^*, E(\vec{k}) = \hbar^2 / k^2 / 2m^*$. In this case σ is given by

$$\sigma = ne^2 \tau / m^*$$

where the effective mass is found from the band curvature $1/m^* = \partial^2 E / \hbar^2 \partial k^2$. The generalization of Ohm's law can also be made to deal with solids for which the effective mass tensor is anisotropic and this will be discussed later in this course.

TRANSPORT PHENOMENA

In this section we study some of the transport properties for metals and semiconductors. An intrinsic semiconductor at $T = 0$ has no carriers and therefore there is no transport of carriers under the influence of external field

s. However at finite temperatures there are thermally generated carriers. Impurities also can serve to generate carriers and transport properties. For insulators, there is very little charge transport and in this case the defects and the ions themselves can participate in charge transport under the influence of external applied field s. Metals make use of the Fermi-Dirac distribution function but are otherwise similar to semiconductors, for which the Maxwell-Boltzmann distribution function is usually applicable.

At finite fields, the electrical conductivity will depend on the product of the carrier density and the carrier mobility. For a one carrier system, the gives the carrier density and the magnetoresistance gives the mobility, the key parameters governing the transport properties of a semiconductor. From the standpoint of device applications, the carrier density and the carrier mobility are the parameters of greatest importance.

To the extent that electrons can be considered as particles, the electrical conductivity, the electronic contribution to the thermal conductivity and the magnetoresistance are all found by solving the Boltzmann equation. For the case of ultra-small dimensions, where the wave aspects of the electron must be considered (called mesoscopic physics), more sophisticated approaches to the transport properties must be considered. To review the standard procedures for classical electrons, we briefly review the Boltzmann equation and its solution in the next section.

THE BOLTZMANN EQUATION

The Boltzmann transport equation is a statement that in the steady state, there is no net change in the distribution function $f(\vec{r},\vec{k},t)$ which determines the probability of finding an electron at position $\vec{r}$, crystal momentum $\vec{k}$ and time t. Therefore we get a zero sum for the changes in $f(\vec{r},\vec{k},t)$ due to the 3 processes of diffusion, the effect of forces and field s, and collisions:

$$\left.\frac{\partial f(\vec{r},\vec{k},t)}{\partial t}\right|_{\text{diffusion}} + \left.\frac{\partial f(\vec{r},\vec{k},t)}{\partial t}\right|_{\text{fields}} + \left.\frac{\partial f(\vec{r},\vec{k},t)}{\partial t}\right|_{\text{collisions}} = 0$$

It is customary to substitute the following differential form for the diffusion process

$$\left.\frac{\partial f(\vec{r},\vec{k},t)}{\partial t}\right|_{\text{diffusion}} = -\vec{v}(\vec{k}).\frac{\partial f(\vec{r},\vec{k},t)}{\partial \vec{r}}$$

which expresses the continuity equation in real space in the absence of forces, field s and collisions. For the forces and field s we write correspondingly

$$\left.\frac{\partial f(\vec{r},\vec{k},t)}{\partial t}\right|_{\text{fields}} = -\frac{\partial \vec{k}}{\partial t}\cdot\frac{\partial f(\vec{r},\vec{k},t)}{\partial \vec{k}}$$

to obtain the Boltzmann equation:

$$\frac{\partial f(\vec{r},\vec{k},t)}{\partial t}+\vec{v}(\vec{k})\cdot\frac{\partial f(\vec{r},\vec{k},t)}{\partial \vec{r}}+\frac{\partial \vec{k}}{\partial t}\cdot\frac{\partial f(\vec{r},\vec{k},t)}{\partial \vec{k}}=\left.\frac{\partial f(\vec{r},\vec{k},t)}{\partial t}\right|_{\text{collisions}}$$

which includes derivatives for all the variables of the distribution function on the left hand side of the equation and the collision terms appear on the right hand side of equation The first term in equation gives the explicit time dependence of the distribution function and is needed for the solution of ac driving forces or for impulse perturbations. Boltzmann's equation is usually solved using two approximations:

1. The perturbation due to external field s and forces is assumed to be small so that the distribution function can be linearized and written as:

   $$f(\vec{r},\vec{k}) = f_0(E)+f_1(\vec{r},\vec{k})$$

   where $f_0(E)$ is the equilibrium distribution function (the Fermi function) which depends only on the energy E, while $f_1(\vec{r},\vec{k})$ is the perturbation term giving the departure from equilibrium.
2. The collision term in the Boltzmann equation is written in the relaxation time approximation so that the system returns to equilibrium uniformly:

   $$\left.\frac{\partial f}{\partial t}\right|_{\text{collisions}} = -\frac{(f-f_0)}{\tau}=-\frac{f_1}{\tau}$$

   where τ denotes the relaxation time and in general is a function of crystal momentum, i.e., $\tau=\tau(\vec{k})$. The physical interpretation of the relaxation time is the time associated with the rate of return to the equilibrium distribution when the external field s or thermal gradients are switched off. Solution to equation when the field s are switched off at $t=0$ leads to

   $$\frac{\partial f}{\partial t}=-\frac{(f-f_0)}{\tau}$$

   which has solutions

   $$f(t)= f_0\left[f(0)-f_0\right]e^{-t/\tau}$$

where f_0 is the equilibrium distribution and $f(0)$ is the distribution function at time $t=0$. The relaxation in equation follows a Poisson distribution indicating that collisions relax the distribution function exponentially to f_0 with a time constant τ..

With these approximations, the Boltzmann equation is solved to find the distribution function which in turn determines the number density and current density. The current density $\vec{j}(\vec{r},t)$ is given by

$$\vec{j}(\vec{r},t) = \frac{e}{4\pi^3}\int \vec{v}(\vec{k})f(\vec{r},\vec{k},t)d^3k$$

in which the crystal momentum $\hbar\vec{k}$ plays the role of the momentum $\vec{p}$ in specifying a volume in phase space. Every element of size h (Planck's constant) in phase space can accommodate one spin ↑ and one spin ↓ electron. The carrier density $n(\vec{r},t)$ is thus simply given by integration of the distribution function over k-space

$$n(\vec{r},t) = \frac{1}{4\pi^3}\int f(\vec{r},\vec{k},t)d^3k$$

where d^3k is an element of 3D wavevector space. The velocity of a carrier with crystal momentum $\hbar\vec{k}$ is related to the $E(\vec{k})$ dispersion expression by

$$\vec{v}(\vec{k}) = \frac{1}{\hbar}\frac{\partial E(\vec{k})}{\partial \vec{k}}$$

and $f_0(E)$ is the Fermi distribution function

$$f_0(E) = \frac{1}{1+e^{(E-E_F)/k_BT}}$$

which defines the equilibrium state in which E_F is the Fermi energy and k_B is the Boltzmann constant.

ELECTRICAL CONDUCTIVITY

To calculate the static electrical conductivity, we consider an applied electric field $\vec{\sigma}$ which for convenience we will take along the x-direction. We will assume for the present that there is no magnetic field and that there are no thermal gradients present. The electrical

$$\vec{j} = \ddot{\sigma}\cdot\vec{E}$$

conductivity is expressed in terms of the conductivity tensor $\ddot{\sigma}$ which is evaluated explicitly from the relation

$$\vec{j} = \ddot{\sigma}\cdot\vec{E}$$

from solution of equation using $\vec{v}(\vec{k})$ from equation and the distribution function $f(\vec{r},\vec{k},t)$ from solution of the Boltzmann equation represented by equation The first term in equation vanishes since the dc applied field $\vec{E}$ has no time dependence.

For the second term in the Boltzmann equation equation

$$\vec{v}(\vec{k})\ \partial f(\vec{r},\vec{k},t)/\partial\vec{r}$$

we note that

$$\frac{\partial f}{\partial \vec{r}} \simeq \frac{\partial f_0}{\partial \vec{r}} = \frac{\partial f_0}{\partial T}\frac{\partial T}{\partial \vec{r}}.$$

Since there are no thermal gradients present in the simplest calculation of the electrical conductivity given in this section, this term does not contribute to equation For the third term in equation which we write as $\dot{\vec{k}} \cdot \dfrac{\partial f(\vec{r},\vec{k},t)}{\partial \vec{k}} = \sum_{\alpha} \dot{k}_{\alpha} \dfrac{\partial f(\vec{r},\vec{k},t)}{\partial k_{\alpha}}$

where the right hand side shows the summ^tion over the vector components, we do get a contribution, since the equations of motion ($F = ma$) give

$$\hbar\dot{\vec{k}} = e\vec{E}$$

and

$$\frac{\partial f(\vec{r},\vec{k},t)}{\partial \vec{k}} = \frac{\partial (f_0 + f_1)}{\partial \vec{k}} = \frac{\partial f_0}{\partial E}\frac{\partial E}{\partial \vec{k}} + \frac{\partial f_1}{\partial \vec{k}}.$$

In considering the linearized Boltzmann equation, we retain only the leading terms in the perturbing electric field, so that $(\partial f_1/\partial\vec{k})$ can be neglected and only the term $(\partial f_0/\partial E)\hbar\vec{v}(\vec{k})$ need be retained. We thus obtain the linearized Boltzmann equation for the case on an applied static electric field and no thermal gradients: $\dot{\vec{k}} \cdot \dfrac{\partial f(\vec{r},\vec{k},t)}{\partial \vec{k}} = \dfrac{\phi}{\tau}\dfrac{\partial f_0}{\partial E} = -\dfrac{f_1}{\tau}$

where it is convenient to write:

$$f_1 = -\phi\left(\frac{\partial f_0}{\partial E}\right)$$

in order to show the $(\partial f_0/\partial E)$ dependence explicitly. Substitution of Equiation yields $\left[\dfrac{e\vec{E}}{\hbar}\left(\dfrac{\partial f_0}{\partial E}\right)\right] \cdot [\hbar\vec{v}(\vec{k})] = \dfrac{\phi(\vec{k})}{\tau}\left(\dfrac{\partial f_0}{\partial E}\right)$

so that

$$\phi(\vec{k}) = e\tau\vec{E} \cdot \vec{v}(\vec{k}).$$

Thus we can relate $\phi(\vec{k})$ to $f_1(\vec{k})$ by

$$f_1(\vec{k}) = -\phi(\vec{k})\frac{\partial f_0(E)}{\partial E} = -e\tau\vec{E} \cdot \vec{v}(\vec{k})\frac{\partial f_0(E)}{\partial E}.$$

The current density is then found from the distribution function $f(\vec{k})$ by calculation of the average value of $\langle ne\vec{v}\rangle$ over all *k*-space

$$\vec{j} = \frac{1}{4\pi^3}\int e\vec{v}(\vec{k})f(\vec{k})d^3k = \frac{1}{4\pi^3}\int e\vec{v}(\vec{k})f_1(\vec{k})d^3k$$

since

$$\int e\vec{v}(\vec{k})f_0(\vec{k})d^3k = 0.$$

Equation 4.24 states that no net current flows in the absence of an applied electric field, another statement of the equilibrium condition. Substitution for $f_1(\vec{k})$ given equation for $\vec{j}$ yields

$$\vec{j} = -\frac{e^2\vec{E}}{4\pi^3}\cdot\int \tau\vec{v}\vec{v}\frac{\partial f_0}{\partial E}d^3k$$

where in general $\tau = \tau(\vec{k})$ and $\vec{v}$ is given by equation. A comparison of Equations thus yields the desired result for the conductivity tensor $\ddot{\sigma}$

$$\ddot{\sigma} = -\frac{e^2}{4\pi^3}\int \tau\vec{v}\vec{v}\frac{\partial f_0}{\partial E}d^3k$$

where $\ddot{\sigma}$ is a symmetric second rank tensor $(\sigma_{ij} = \sigma_{ji})$. The evaluation of the integral in over all k-space depends on the $E(\vec{k})$ relations through the $\vec{v}\vec{v}$ terms and the temperature dependence comes through the $\partial f_0/\partial E$ term.

ELECTRICAL CONDUCTIVITY OF METALS

To exploit the energy dependence of $(\partial f_0/\partial E)$ in applying equation to metals, it is more convenient to evaluate $\ddot{\sigma}$ if we replace $\int d^3k$ with an integral over the constant energy surfaces

$$\int d^3k = \int d^2S dk_\perp \equiv \int d^2S dE/|\partial E/\partial\vec{k}|$$

Thus equation is written

$$\ddot{\sigma} = -\frac{e^2}{4\pi^3}\int \frac{\tau\vec{v}\vec{v}}{|\partial E/\partial\vec{k}|}\frac{\partial f_0}{\partial E}d^2S\,dE.$$

From the Fermi-Dirac distribution function f_0 (*E*), we see that the derivative $(-\partial f_0/\partial E)$ can approximately be replaced by a d-function so that equation can be written as $\ddot{\sigma} = -\frac{e^2}{4\pi^3\hbar}\int_{\text{Fermi surface}} \tau\vec{v}\vec{v}\frac{d^2S}{v}$

Fermi surface

For a cubic crystal, $[v_x v_x] = v^2/3$ and thus $\overleftrightarrow{\sigma}$ has only diagonal components σ that are all equal to each other:

$$\sigma = \frac{e^2}{4\pi^3\hbar}\int_{\text{Fermi surface}} \tau v \frac{d^2S}{3} = \frac{ne^2\tau}{m^*}$$

since

$$n = (1/4\pi^3)(4\pi/3)k_F^3$$

Fermi surface

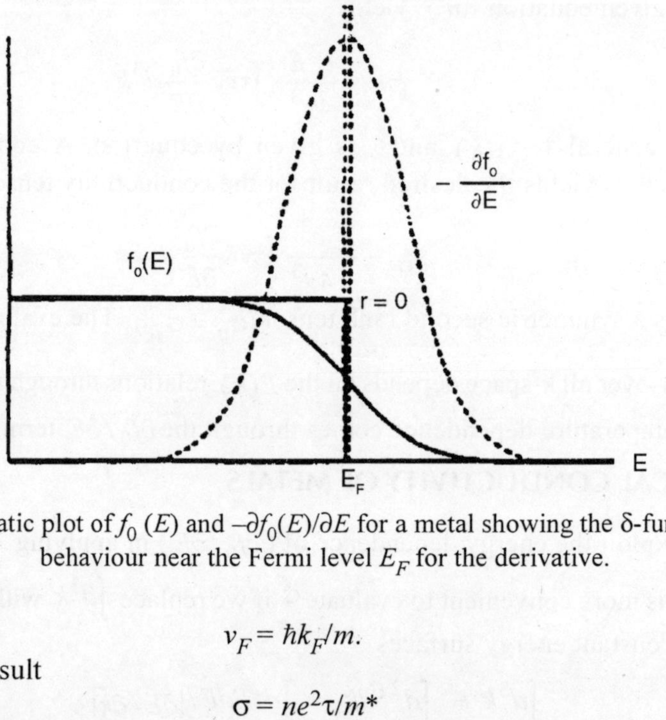

Fig. Schematic plot of f_0 (E) and $-\partial f_0(E)/\partial E$ for a metal showing the δ-function like behaviour near the Fermi level E_F for the derivative.

and

$$v_F = \hbar k_F/m.$$

The result

$$\sigma = ne^2\tau/m^*$$

is called the Drude formula for the dc electrical conductivity.

ELECTRICAL CONDUCTIVITY OF SEMICONDUCTORS

We show in this section that the simple Drude model $\sigma = ne^2\tau = m^*$ can also be recovered for a semiconductor from the general relation given by equation , using a simple parabolic band model and a constant relaxation time. When a more complete theory is used, departures from the simple Drude model will result.

In deriving the Drude model for a semiconductor we make three approximations:

- Approximation #1

In the case of electron states in intrinsic semiconductors having no donor or acceptor impurities, we have the condition $(E - E_F) \gg k_BT$ since E_F is in the band gap and E is the energy of an electron in the conduction band.

Thus the first approximation is equivalent to writing

$$f_0(E) = \frac{1}{1+\exp[E-E_F)/k_BT]} \simeq \exp[E-E_F)/k_BT]$$

which is equivalent to using the Maxwell-Boltzmann distribution in place of the full Fermi-Dirac distribution. Since E is usually measured with respect to the bottom of the conduction band, E_F is a negative energy and it is therefore convenient to write $f_0(E)$ as $f_0(E) \simeq e^{-|E_F|/k_BT_e - E/k_BT}$

so that the derivative of the Fermi function becomes

$$\frac{\partial f_0(E)}{\partial E} = -\frac{e^{-|E_F|/k_BT}}{k_BT} e^{-E/k_BT}$$

- Approximation #2

For simplicity we assume a constant relaxation time τ independent of $\vec{k}$ and E. This approximation is made for simplicity and is not valid for specific cases. Some common scattering mechanisms yield an energy dependent relaxation time such as acoustic deformation potential scattering or ionized impurity scattering where $r = -1/2$ and $r = +3/2$ respectively in the relation $\tau = \tau_0(E/k_BT)^r$.

- Approximation #3

To illustrate the explicit evaluation of the integral in equation , we consider the simplest case, assuming an isotropic, parabolic band $E = \hbar^2k^2/2m^*$ for the evaluation of $\vec{v} = \partial E/\hbar\partial\vec{k}$ about the conduction band extremum.

Using this third approximation we can write

$$\vec{v}\vec{v} = \frac{1}{3}v^2\vec{1}$$

$$k^2 = 2m^*E/\hbar^2$$

$$2kdk = 2m^*dE/\hbar^2$$

$$v^2 = 2E/m^*$$

$$v = \hbar k/m^*$$

where $\vec{1}$ is the unit second rank tensor. We next convert equation to an integration over energy and write $d^3k = 4\pi k^2 dk = 4\pi\sqrt{2}(m^*/\hbar^2)^{3/2}\sqrt{E}\,dE$

so that equation becomes $\sigma = \frac{e^2\tau}{4\pi^3}\left(\frac{8\sqrt{2\pi}\sqrt{m^*}}{3\hbar^3 k_B T}\right)e^{-|E_F/k_BT}\int_0^\infty E^{3/2}dEe^{-E/k_BT}$
in which the integral over energy E is extended to ∞ because there is negligible contribution for large E and because the de finite integral

$$\int_0^\infty x^p dx e^{-x} = \Gamma(p+1)$$

can be evaluated exactly, $\Gamma(p)$ being the Γ function which has the property

$$\Gamma(p+1) = p\Gamma(p)$$

$$\Gamma(1/2) = \sqrt{\pi}\,.$$

Substitution into equation thus yields

$$\sigma = \frac{2e^2\tau}{m^*}\left(\frac{m^* k_B T}{2\pi\hbar^2}\right)^{3/2} e^{-|E_F|/k_BT}$$

which gives the temperature dependence of σ. Now the carrier density calculated same approximations becomes

$$n = (4\pi^3)^{-1}\int f_0(E)d^3k$$

$$= (4\pi^3)^{-1}e^{-|E_F|/k_BT}\int e^{-E/k_BT}\,4\pi k^2 dk$$

$$= (\sqrt{2}/\pi^2)(m^*/\hbar^2)^{3/2}e^{-|E_F|/k_BT}\int_0^\infty \sqrt{E}\,dEe^{-E/k_BT}$$

where

$$\int_0^\infty \sqrt{E}\,dEe^{-E/k_BT} = \frac{\sqrt{\pi}}{2}(k_BT)^{3/2}$$

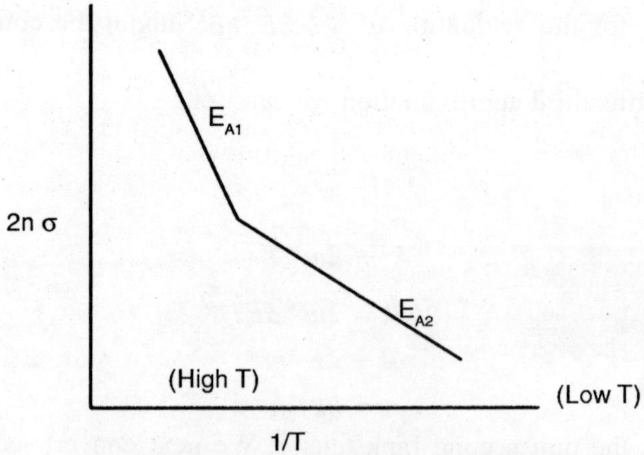

Fig. Schematic Diagram of an Arrhenius Plot of ln s vs 1 = T Showing two Carrier Types with Different Activation Energies.

which gives the final result for the temperature dependence of the carrier density

$$n = 2\left(\frac{m^* k_B T}{2\pi\hbar^2}\right)^{3/2} e^{-|E_F|/k_B T}$$

so that by substitution into equation , the Drude formula is recovered

$$\sigma = \frac{ne^2\tau}{m^*}$$

for a semiconductor with constant τ and isotropic, parabolic dispersion relations.

To find σ for a semiconductor with more than one spherical carrier pocket, the conductivities per carrier pocket are added

$$\sigma = \sum_i \sigma_i$$

where i is the carrier pocket index. We use these simple formulae to make rough estimates for the carrier density and conductivity of semiconductors. The $E(\vec{k})$ relation must be considered, as well as an energy dependent τ and use of the complete Fermi function.

The electrical conductivity and carrier density of a semiconductor with one carrier type exhibits an exponential temperature dependence so that the slope of ln σ vs $1/T$ yields an activation energy. The plot of ln σ vs $1/T$ is called an "Arrhenius plot". If a plot of ln σ vs $1/T$ exhibits one temperature range with activation energy E_{A1} and a second temperature range with activation energy E_{A2}, then two carrier behaviour is suggested. Also in such cases, the activation energies can be extracted from an Arrhenius plot.

Ellipsoidal Carrier Pockets

The conductivity results given above for a spherical Fermi surface can easily be generalized to an ellipsoidal Fermi surface which is commonly found in degenerate semiconductors. Semiconductors are degenerate at $T = 0$ when the Fermi level is in the valence or conduction band rather than in the energy band gap.

For an ellipsoidal Fermi surface, we write

$$E(\vec{k}) = \frac{\hbar^2 k_x^2}{2m_{xx}} + \frac{\hbar^2 k_y^2}{2m_{yy}} + \frac{\hbar^2 k_z^2}{2m_{zz}}$$

where the effective mass components m_{xx}, m_{yy} and m_{zz} are appropriate tures in the x, y, z directions, respectively. Substitution of

$$k'_\alpha = k_\alpha \sqrt{m_0 / m_\alpha}$$

for $\alpha = x, y, z$ brings equation into spherical form

$$E(\vec{k}') = \frac{\hbar^2 k'^2}{2m_0}$$

where $k'^2 = k'^2_x + k'_{2y} + k'_{2z}$. For the volume element d^3k in equation we

$$d^3k = \sqrt{m_{xx} m_{yy} m_{zz} / m_0^2}\; d^3k'$$

and the carrier density associated with a single carrier pocket becomes

$$n_i = 2\sqrt{m_{xx} m_{yy} m_{zz} / m_0^2} \left(\frac{k_B T}{2\pi\hbar^2} \right)^{3/2} e^{-|E_F|/k_B T}.$$

For an ellipsoidal constant energy surface, the directions of the electric field, electron velocity and electron acceleration will in general be different. Let (x, y, z) be the coordinate system for the major axes of the constant energy ellipsoid and (X, Y, Z) be the laboratory coordinate system. Then in the laboratory system the current density $\vec{j}$ and electric field $\vec{E}$ are related by

$$\begin{pmatrix} j_X \\ j_Y \\ j_Z \end{pmatrix} = \begin{pmatrix} \sigma_{XX} & \alpha_{XY} & \sigma_{XZ} \\ \sigma_{YX} & \sigma_{YY} & \sigma_{YZ} \\ \sigma_{ZX} & \sigma_{ZY} & \sigma_{ZZ} \end{pmatrix} \begin{pmatrix} E_X \\ E_Y \\ E_Z \end{pmatrix}$$

As an example, suppose that the electric field is applied in the XY plane along the X axis at an angle θ with respect to the x axis of the constant energy ellipsoid. The conductivity tensor is easily written in the xyz crystal coordinate system where the xyz axes are along the principal axes of the ellipsoid:

$$\begin{pmatrix} j_x \\ j_y \\ j_z \end{pmatrix} = ne^2\tau \begin{pmatrix} 1/m_{xx} & 0 & 0 \\ 0 & 1/m_{yy} & 0 \\ 0 & 0 & 1/m_{zz} \end{pmatrix} \begin{pmatrix} E\cos\theta \\ E\sin\theta \\ 0 \end{pmatrix}$$

A coordinate transformation from the crystal axes to the laboratory frame allows us to relate $\ddot{\sigma}_{\text{crystal}}$ which we have written easily by equation to $\ddot{\sigma}_{\text{Lab}}$ which we measure by equation. In general

$$\ddot{\sigma}_{\text{Lab}} = R\ddot{\sigma}_{\text{crystal}} R^{-1}$$

where

$$R = \begin{pmatrix} \cos\theta & \sin\theta & 0 \\ -\sin\theta & \cos\theta & 0 \\ 0 & 0 & 1 \end{pmatrix}$$

and

$$R^{-1} = \begin{pmatrix} \cos\theta & -\sin\theta & 0 \\ \sin\theta & \cos\theta & 0 \\ 0 & 0 & 1 \end{pmatrix}$$

so that the conductivity tensor $\ddot{\sigma}_{\text{Lab}}$ in the lab frame becomes:

$$\ddot{\sigma}_{\text{Lab}} = ne^2\tau \begin{pmatrix} \cos^2/m_{xx} + \sin^2\theta/m_{yy} & \cos\theta\sin\theta(1/m_{yy} - 1/m_{xx}) & 0 \\ \cos\theta\sin\theta(1/m_{yy} - 1/m_{xx}) & \sin^2\theta/m_{xx} + \cos^2\theta/m_{yy} & 0 \\ 0 & 0 & 1/m_{zz} \end{pmatrix}$$

Semiconductors with ellipsoidal Fermi surfaces usually have several such surfaces located in crystallographically equivalent locations. In the case of cubic symmetry, the sum of the

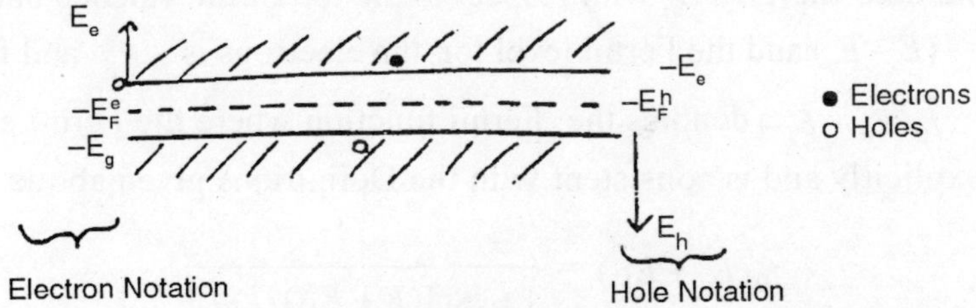

Fig. Schematic Diagram of the Band gap in a Semiconductor showing the Symmetry of Electrons and Holes.

conductivity components results in an isotropic conductivity even though the contribution from each ellipsoid is anisotropic. Thus measurement of the electrical conductivity provides no information on the anisotropy of the Fermi surfaces of cubic materials. However, measurement of the magnetoresistance does provide such information, since the application of a magnetic field gives special importance to the magnetic field direction, thereby lowering the effective crystal symmetry.

ELECTRONS AND HOLES IN INTRINSIC SEMICONDUCTORS

In the absence of doping, carriers are generated by thermal or optical excitations. Thus at T = 0, all valence band states are occupied and all conduction band states are empty. Thus, for each electron that is excited into the conduction band, a hole is left behind. For intrinsic semiconductors, conduction is by both holes and electrons. The Fermi level is thus determined by the condition that the number of electrons is equal to the number of holes. Writing g_v (E_h) and $g_c(E_e)$ as the density of hole and electron states, respectively, we obtain

$$n_h = \int_0^\infty g_v(E_h)\hat{f}_0(E_h + E_F^h)dE_h = \int_0^\infty g_c(E_e)\hat{f}_0(E_e + E_F^e)dE_e = n_e$$

where the notation we have used. Here the energy gap E_g is written as

$$E_F^e + E_F^h = E_g.$$

The condition $n_e = n_h$ for intrinsic semiconductors is used to determine the position of the Fermi levels for electrons and holes within the band gap. If the band curvatures of the valence and conduction bands are the same, then their effective masses are the same magnitude and E_F lies at midgap. We derive in this section the general result for the placement of E_F when $m_e^* \neq m_h^*$.

On the basis of this interpretation, the holes obey Fermi statistics as do the electrons, only we must measure their energies downward, while electron energies are measured upwards. This approach clearly builds on the symmetry relation between electrons and holes. It is convenient to measure electron energies E_e with respect to the bottom of the conduction band E_c so that $E_e = E - E_c$ and hole energies E_h with respect to the top of the valence band E_v so that $E_h = -(E - E_v)$ and the Fermi level for the electrons is $-E_F^e$ and for holes is $-E_F^h$. $\hat{f}_0(E_e + E_F^e)$ denotes the Fermi function where the Fermi energy is written explicitly and is consistent with the Definitions given above.

$$\hat{f}_0(E_e + E_F^e) = \frac{1}{1 + \exp[(E + E_F^e)/k_BT]}$$

In an intrinsic semiconductor, the magnitudes of the energies E_F^e and E_F^h are both much greater than thermal energies, i.e., $|E_F^e| \ll k_BT$ and $|E_F^h| \gg k_BT$, so that the distribution functions can be approximated by the Boltzmann form

$$\hat{f}_0(E_e + E_F^e) \simeq e^{-(E_e + E_F^e)/k_BT}$$

$$\hat{f}_0(E_h + E_F^h) \simeq e^{-(E_h + E_F^h)/k_BT}.$$

If m_e and m_h are respectively the electron and hole effective masses and if we write the dispersion relations around the valence and conduction band extrema as

$$Ee = \hbar^2 k^2/(2m_e)$$

$$E_h = \hbar^2 k^2/(2m_h)$$

then the density of states for electrons at the bottom of the conduction band and for holes at the top of the valence band can be written in their respective nearly free electron forms

$$g_c(E_e) = \frac{1}{2\pi^2}\left(2m_e/\hbar^2\right)^{3/2} E_e^{1/2}$$

$$g_v(E_h) = \frac{1}{2\pi^2}\left(2m_h/\hbar^2\right)^{3/2} E_h^{1/2}.$$

These expressions follow from

$$n = \frac{1}{2\pi^3}\frac{4\pi}{3}k^3$$

and substitution of k via the simple parabolic relation

$$E = \frac{\hbar^2 k^2}{2m^*}$$

so that

$$n = \frac{1}{3\pi^2}\left(\frac{2m^* E}{\hbar^2}\right)^{3/2}$$

and

$$g(E) = \frac{dn}{dE} = \frac{1}{2\pi^2}\left(\frac{2m^*}{\hbar^2}\right)^{3/2} E^{1/2}.$$

Substitution of this density of states expression into equation results in a carrier density

$$n_e = 2\left(\frac{m_e k_B T}{2\pi\hbar^2}\right)^{3/2} e^{-E_F^e/k_B T}.$$

Likewise for holes we obtain

$$n_h = 2\left(\frac{m_h k_B T}{2\pi\hbar^2}\right)^{3/2} e^{-E_F^h/k_B T}$$

Thus the famous product rule is obtained

$$n_e n_h = 4\left(\frac{k_B T}{2\pi\hbar^2}\right)^3 (m_e m_h)^{3/2} e^{-E_g/k_B T}$$

where $E_g = E_F^e + E_F^h$. But for an intrinsic semiconductor $n_e = n_h$. Thus by taking the square root of the above expression, we obtain both n_e and n_h

$$n_e = n_h = 2\left(\frac{k_B T}{2\pi\hbar^2}\right)^{3/2} (m_e m_h)^{3/4} e^{-E_g/2k_B T}$$

Comparison with the expressions given in Eqs. 4.69 and 4.70 for n_e and n_h allows us to solve for the Fermi levels E_F^e and E_F^h

$$n_e = 2\left(\frac{m_e k_B T}{2\pi\hbar^2}\right)^{3/2} e^{-E_F^e/k_B T} = 2\left(\frac{k_B T}{2\pi\hbar^2}\right)^{3/2} (m_e m_h)^{3/4} e^{-E_g/2k_B T}$$

so that $\exp(-E_F^e / k_B T) = (m_h m_e)^{3/4} \exp(-E_g / 2k_B T)$
and

$$E_F^e = \frac{E_g}{2} - \frac{3}{4} k_B T \ln(m_h m_e).$$

If $m_e = m_h$, we obtain the simple result that $E_F^e = E_g/2$ which says that the Fermi level lies in the middle of the energy gap. However, if the masses are not equal, E_F will lie closer to the band edge with higher curvature, thereby enhancing the Boltzmann factor term in the thermal excitation process, to compensate for the lower density of states for the higher curvature band.

If however $m_e \ll m_h$, the Fermi level approaches the conduction band edge and the full Fermi functions have to be considered. In this case

$$n_e = \frac{1}{2\pi^2}\left(\frac{2m_e}{\hbar^2}\right)^{3/2} \int_{E_c}^{\infty} \frac{(E-E_c)^{1/2}\, dE}{\exp[(E-E_F^e)/k_B T]+1} \equiv N_e F_{1/2}\left(\frac{E_F^e - E_e}{k_B T}\right)$$

where E_c is the bottom of the conduction band and the "effective electron density" is in accordance with equation given by

$$N_e = 2\left(\frac{m_e k_B T}{2\pi\hbar^2}\right)^{3/2}$$

and the Fermi integral is written as

$$F_j(\eta) = \frac{1}{j!}\int_0^{\infty} \frac{x^j\, dx}{\exp(x-\eta)+1}.$$

We can take $F_j(\eta)$ from the tables in Blakemore, "Semiconductor Physics" (Appendix B). For the semiconductor limit ($\eta < -4$), then $F_j(\eta) \rightarrow \exp(\eta)$. Clearly, when $F_j(\eta)$ is required to describe the carrier density, then $F_j(\eta)$ is also needed to describe the conductivity. These re¯nements are important for a detailed solution of the transport properties of semiconductors over the entire temperature range of interest.

Table: Occupation of impurity states in the ensemble.

j	N_j	spin	Ej
1	0	–	0
2	1	↑	$-E_d$
3	1	↓	$-E_d$
4	2	↑↓	$-E_d + E_{\text{Coulomb}}$

DONOR AND ACCEPTOR DOPING OF SEMICONDUCTORS

In general a semiconductor has electron and hole carriers due to the presence of impurities as well as from thermal excitation processes. For many applications, impurities are intentionally introduced to generate carriers: donor impurities to generate electrons in n-type material and acceptor impurities to generate holes in p-type material. Assuming for the moment that each donor contributes one electron to the conduction band, then the donors can contribute an excess carrier concentration up to N_d, where N_d is the donor impurity concentration. Similarly, if every acceptor contributes one hole to the valence band, then the excess hole concentration will be N_a, where N_a is the acceptor impurity concentration.

In general, the semiconductor is partly compensated, with both donor and acceptor impurities present. Furthermore, at finite temperatures, the donor and acceptor levels will be partially occupied so that somewhat less than the maximum charge will be released as mobile charge into the conduction and valence bands. The density of electrons bound to a donor site n_d is found from the ensemble as

$$\frac{n_d}{N_d} = \frac{\Sigma_j N_j e^{-(E_j - \mu N_j)/k_B T}}{\Sigma_j e^{-(E_j - \mu N_j)/k_B T}}$$

where E_j and N_j are respectively the energy and number of electrons that can be placed in state j, and μ is the chemical potential (Fermi energy). The system can be found in one of three states: one where no electrons are present (hence no contribution is made to the energy), and two states (one with spin +, the other with spin –) corresponding to the donor energy E_d, where E_d is a positive energy.

Placing two electrons in the same energy state would result in a very high energy because of the Coulomb repulsion between the two electrons; therefore this possibility is neglected in practical calculations. Writing either $N_j = 0, 1$ for the 3 states of importance, we obtain for the relative ion concentration of occupied donor sites

$$\frac{n_d}{N_d} = \frac{2e^{-(\varepsilon_d - \mu)/k_B T}}{1 + 2e^{-(\varepsilon_d - \mu)/k_B T}} = \frac{1}{1 + \frac{1}{2}e^{(\varepsilon_d - \mu)/k_B T}} = \frac{1}{1 + \frac{1}{2}e^{-(E_d - E_F^e)/k_B T}}$$

in which E_d and E_F^e are positive numbers and the zero of energy is taken at the bottom of the conduction band. The energy ε_d denotes the energy for the donor level.

Consequently, the concentration of electrons thermally ionized into the conduction band will be

$$N_d - n_d = \frac{N_d}{1 + 2e(E_d - E_F^e)/k_B T} = n_e - n_h$$

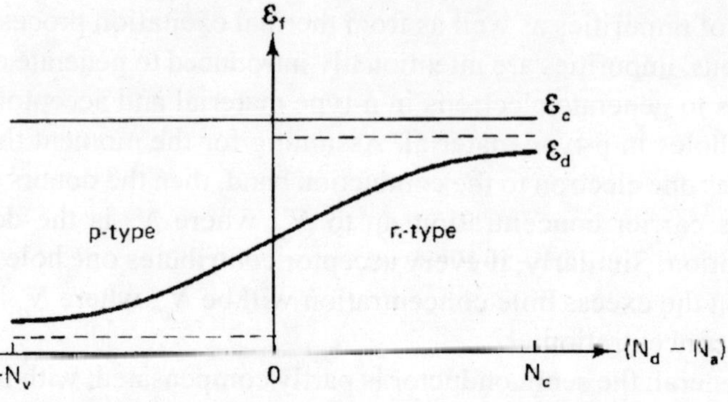

Fig. Variation of the Fermi energy with donor and acceptor concentrations. For a heavily doped n-type semiconductor E_F is close to the donor level E_d. The bottom of the conduction band E_c and the donor energy E_d. This plot is made assuming almost all the donor and acceptor states are ionized.

where n_e and n_h are the mobile electron and hole concentrations. At low temperatures, where $E_d \sim k_B T$, almost all of the carriers in the conduction band will be generated by the ionized donors, so that $n_h \ll n_e$ and $(N_d - n_d); n_e$. The Fermi level will then adjust itself so that from equations the following equation determines E_F^e:

$$n_e = 2\left(\frac{m_e k_B T}{2\pi\hbar^2}\right)^{3/2} e^{-E_F^e/k_B T} \simeq \frac{N_d}{1+2e(E_d - E_F^e)/k_B T}.$$

Solution of equation shows that the presence of the ionized donor carriers moves the Fermi level up above the middle of the band gap and close to the bottom of the conduction band. For the donor impurity problem, the Fermi level will be close to the position of the donor level E_d. The position of the Fermi level also varies with temperature. Figure assumes that almost all the donor electrons (or acceptor holes) are ionized and are in the conduction band, which is typical of temperatures where the n-type (or p-type) semiconductor would be used.

Here T_1 denotes the temperature at which the thermal excitation of intrinsic electrons and holes become important, and T_1 is normally a high temperature. In contrast, T_2 is normally a very low temperature and denotes the temperature below which donor-generated electrons begin to freeze out in impurity level bound states and no longer contribute to conduction. In the temperature range $T_2 < T < T_1$, the Fermi level falls as T increases according to

$$E_F = E_c - k_B T \ln(N_c/N_d)$$

where $N_c = 2m_e k_B T = (2\pi\hbar^2)$. The temperature dependence of the carrier concentration in the intrinsic range ($T > T_1$), the saturation range ($T_2 < T < T_1$), and finally the low temperature range ($T < T_2$) where carriers freeze out into bound states in the impurity band at E_d. The plot is as a function of $(1/T)$ and the corresponding temperature values are shown on the upper scale of the figure.

For the case of acceptor impurities, an ionized acceptor level releases a hole into the valence band, or alternatively, an electron from the valence band gets excited into an acceptor level, leaving a hole behind. At very low temperature, the acceptor levels are filled with holes. Because of hole-hole Coulomb repulsion, we can place no more than one hole in each acceptor level. A singly occupied hole can have either spin up or spin down. Thus for the acceptor levels, a formula analogous to equation for donors is obtained for the occupation of an acceptor level

$$\frac{n_d}{N_d} = \frac{1}{1+\frac{1}{2}e^{-(E\alpha-E_F^h)k_BT}}$$

so that the essential symmetry between holes and electrons is maintained.

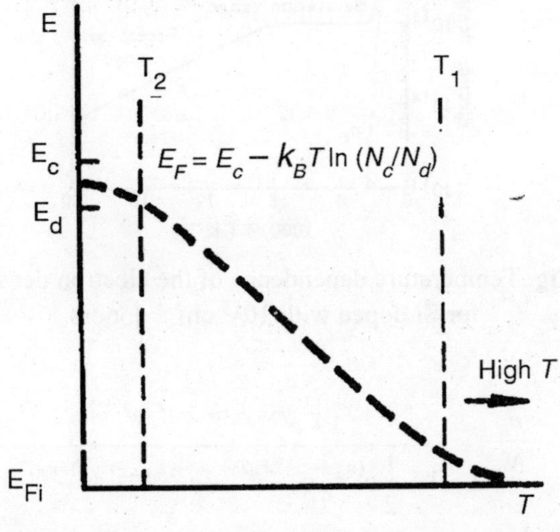

Fig. Temperature dependence of the Fermi energy for an n-type doped semiconductor. See the text for definitions of T_1 and T_2. Here E_{Fi} is the position of the Fermi level in the high temperature limit where the thermal excitation of carriers far exceeds the electron density contributed by the donor impurities.

A situation which commonly arises for the acceptor levels relates to the degeneracy of the valence bands for group IV and III-V compound semiconductors. We will illustrate the degenerate valence band in the case where spin-orbit interaction is considered and the two degenerate levels are only weakly coupled, so that we can approximate the impurity acceptor levels by hydrogenic acceptor levels for the heavy hole $\varepsilon_{a,h}$ and light hole $\varepsilon_{a,l}$ bands.

In this case the split-off band does not contribute significantly. The density of holes bound to both types of acceptor sites is given by

$$\frac{n_d}{N_d} = \frac{\Sigma_j N_j e^{-(E_j - \mu N_j)/k_B T}}{\Sigma_j e^{-(E_j - \mu N_j)/k_B T}}$$

Using the same arguments as above, we obtain:

$$\frac{n_d}{N_d} = \frac{2e^{-(\varepsilon_{a,l} - \mu)/k_B T} + 2e^{-(\varepsilon_{a,h} - \mu)/k_B T}}{1 + 2e^{-(\varepsilon_{a,l} - \mu)/k_B T} + 2e^{-(\varepsilon_{a,h} - \mu)/k_B T}}$$

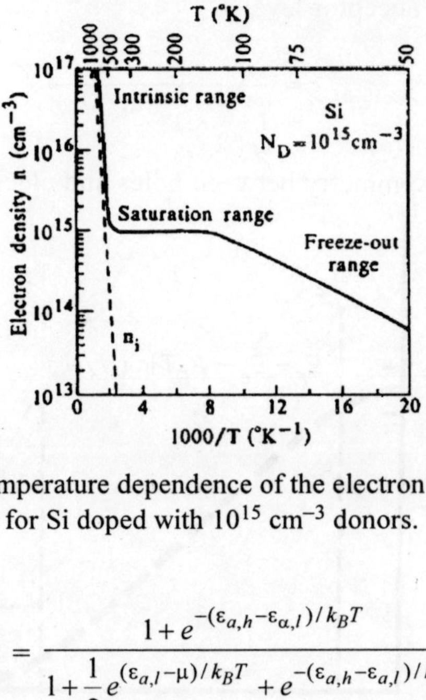

Fig. Temperature dependence of the electron density for Si doped with 10^{15} cm^{-3} donors.

so that

$$\frac{n_d}{N_d} = \frac{1 + e^{-(\varepsilon_{a,h} - \varepsilon_{a,l})/k_B T}}{1 + \frac{1}{2} e^{(\varepsilon_{a,l} - \mu)/k_B T} + e^{-(\varepsilon_{a,h} - \varepsilon_{a,l})/k_B T}}$$

If the thermal energy is large in comparison to the difference between the acceptor levels for the heavy and light hole bands, then

$$[(\varepsilon_{a,h} - \varepsilon_{a,l})/k_B T] \ll 1$$

and
$$\exp[-(\varepsilon_{a,h} - \varepsilon_{a,l})/k_B T] \simeq 1$$

so that
$$\frac{n_a}{N_a} \simeq \frac{1}{1 + \frac{1}{4} e^{-(\varepsilon_{a,l} - \mu)/k_B T}} \frac{1}{1 + \frac{1}{4} e^{-(E_a - E_F^h)/k_B T}}$$

where E_a and E_F^h are the positive values corresponding to $\varepsilon_{a,l}$ and μ, respectively. From Equation, the temperature dependence of E_F can be calculated for the case of doped semiconductors. The doping dependence of E_F for *p*-doped semiconductors as well as n-doped semiconductors.

CHARACTERIZATION OF SEMICONDUCTORS

In describing the electrical conductivity of semiconductors, it is customary to write the conductivity as

$$\sigma = n_e|e|\mu_e + n_h|e|\mu_h$$

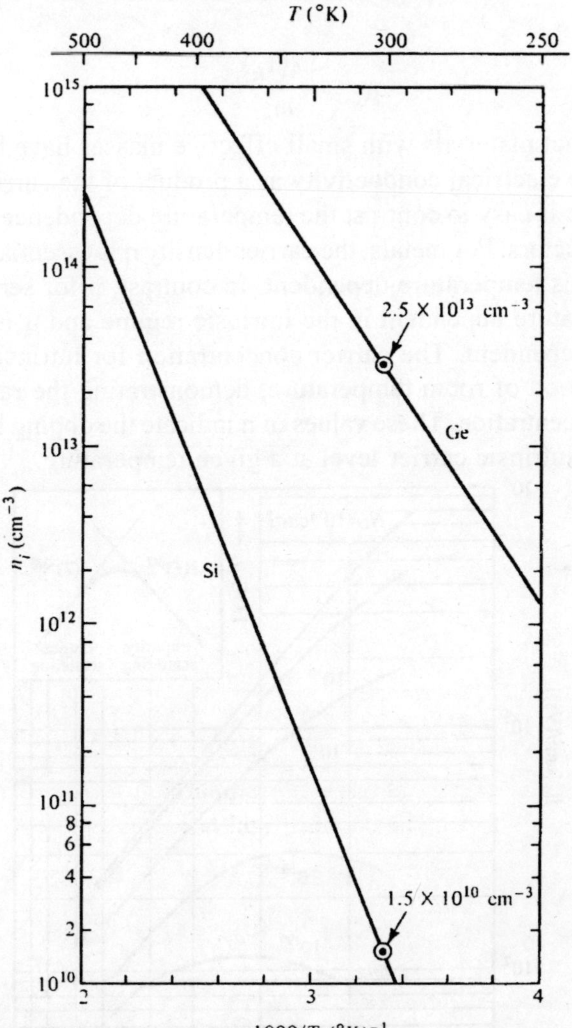

Fig. Temperature dependence of the electron concentration for intrinsic Si and Ge in the range 250 < *T* < 500 *K*.

in which n_e and n_h are the carrier densities for the carriers, and μ_e and μ_h are their mobilities. We have shown in equation that for cubic materials the static conductivity can under certain approximations be written as

$$\sigma = \frac{ne^2\tau}{m^*}$$

for each carrier type, so that the mobilities and effective masses are related by

$$\sigma_e = \frac{|e|\langle\tau_h\rangle}{m_h}$$

and

$$\mu_h = \frac{|e|\langle\tau_h\rangle}{m_h}$$

which show that materials with small effective masses have high mobilities. By writing the electrical conductivity as a product of the carrier density with the mobility, it is easy to contrast the temperature dependence of σ for metals and semiconductors. For metals, the carrier density n is essentially independent of T, while μ is temperature dependent. In contrast, n for semiconductors is highly temperature dependent in the intrinsic regime and μ is relatively less temperature dependent. The carrier concentration for intrinsic Si and Ge in the neighborhood of room temperature, demonstrating the rapid increase of the carrier concentration. These values of n indicate the doping levels necessary to exceed the intrinsic carrier level at a given temperature.

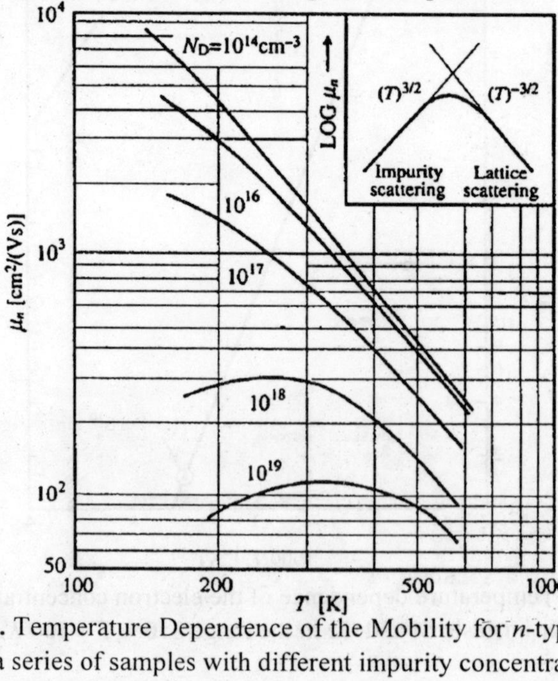

Fig. Temperature Dependence of the Mobility for *n*-type Si for a series of samples with different impurity concentrations

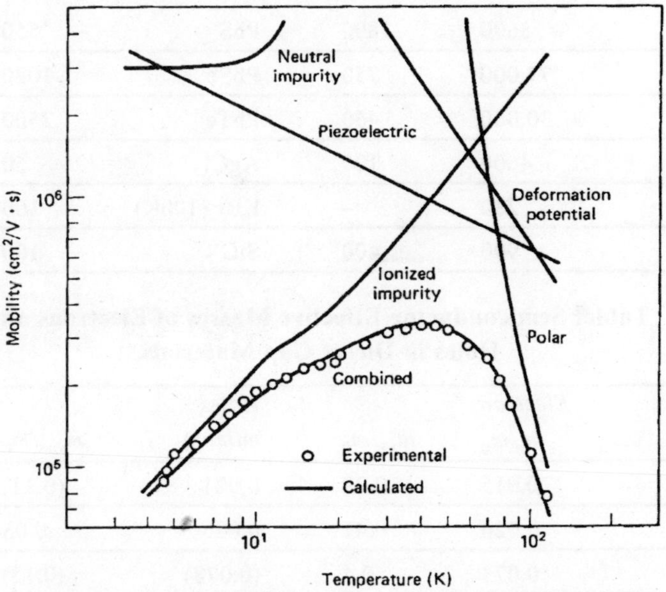

Fig. Temperature dependence of the mobility for n-type GeAs showing the separate and combined scattering processes.

The mobility for Ge samples with various impurity levels. The dependence expected for the intrinsic sample is also indicated. the observed temperature dependence can be explained by the different temperature dependences of the impurity scattering and phonon scattering mechanisms.

A table of typical mobilities for semiconductors is given in Table. By way of comparison, μ for copper at room temperature is 35cm^2/volt-sec. When using conductivity formulae in esu units, remember that the mobility is expressed in cm^2 /statvolt-sec and that all the numbers in Table have to be multiplied by 300 to match the units given in the notes.

In the characterization of a semiconductor for device applications, we are expected to provide information on the carrier density and mobility, preferably as a function of temperature. Such plots. When presenting characterization data in condensed form, the carrier density and mobility of semiconductors are traditionally given at 300 K and 77 K. Other information of values in semiconductor physics are values of the effective masses and of the energy gaps

Table: Mobilities for some Typical Semiconductors at Room Temperature in Units of cm^2/V-sec.

Crystal	*Electrons*	*Holes*	*Crystal*	*Electrons*	*Holes*
Diamond	1800	1200	GaAs	8000	300
Si	1350	480	GaSb	5000	1000

Ge	3600	1800	PbS	550	600
InSb	77,000	750	PbSe	1020	930
InAs	30,000	460	PbTe	2500	1000
InP	4600	100	AgCl	50	–
AlAs	280	–	KBr (100K)	100	–
AlSb	900	400	SiC	100	10–20

Table: Semiconductor Effective Masses of Electrons and Holes in Direct Gap Materials.

Crystal	*Electron*		*Holes*		*Spin-orbit*
	m_e/m_0	m_{hh}/m_0	mlh/m_0	m_{soh}/m_0	Δ(eV)
InSb	0.015	0.39	0.021	(0.11)	0.82
InAs	0.026	0.41	0.025	0.08	0.43
InP	0.073	0.4	(0.078)	(0.15)	0.11
GaSb	0.047	0.3	0.06	(0.14)	0.80
GaAs	0.066	0.5	0.082	0.17	0.34
Cu_2O	0.99	–	0.58	0.69	0.13

Table: Semiconductor Energy Gaps between the Valence and Conduction bands.

		Eg, eV				*Eg, eV*	
Crystal	*Gap*[a]	*0 K*	300 *K*	*Crystal*	*Gap*[a]	*0 K*	300 *K*
Diamond	*i*	5.4	–	HgTe[b]	*d*	{0.30	{
Si	*i*	1.17	1.11	PbS	*d*	0.286	0.34-0.37
Ge	*i*	0.744	0.66	PbSe	*i*	0.165	0.27
®Sn	*d*	0.00	0.00	PbTe	*i*	0.190	0.29
InSb	*d*	0.23	0.17	CdS	*d*	2.582	2.42
InAs	*d*	0.43	0.36	CdSe	*d*	1.840	1.74
InP	*d*	1.42	1.27	CdTe	*d*	1.607	1.44
GaP	*i*	2.32	2.25	ZnO	–	3.436	3.2
GaAs	*d*	1.52	1.43	ZnS	–	3.91	3.6
GaSb	*d*	0.81	0.68	SnTe	*d*	0.3	0.18
AlSb	*i*	1.65	1.6	AgCl	–	–	3.2
SiC(hex)	*i*	3.0	–AgI	G	–	2.8	

Te	*d*	0.33	$-Cu_2O$	d	2.172	–	
ZnSb	–	0.56	0.56	TiO_2	–	3.03	–

The indirect gap is labelled by i, and the direct gap is labelled by *d*. *HgTe* is a semimetal and the bands overlap, hence a negative gap.

THERMAL TRANSPORT

THERMAL TRANSPORT

The electrons in solids not only conduct electricity but also conduct heat, as they transfer energy from a hot junction to a cold junction. Just as the electrical conductivity characterizes the response of a material to an applied voltage, the thermal conductivity likewise characterizes the material with regard to heat flow. In fact the electrical conductivity and thermal conductivity are coupled, since thermal conduction also transports charge and electrical conduction transports energy. This coupling gives rise to thermo-electricity. The thermal conductivity for metals, semiconductors and insulators and then consider the coupling between electrical and thermal transport which gives rise to thermoelectric phenomena.

THERMAL CONDUCTIVITY

Thermal transport, like electrical transport follows from the Boltzmann equation. We will first derive a general expression for the electronic contribution to the thermal conductivity using Boltzmann's equation. We will then apply this general expression to find the thermal conductivity for metals and then for semiconductors.

The total thermal conductivity $\vec{k}$ of any material is, of course, the superposition of the electronic part $\vec{k}_e$ with the lattice part $\vec{k}_L$:

$$\vec{k} = \vec{k}_e + \vec{k}_L$$

We now consider the calculation of the electronic contribution to the thermal conductivity. The application of a temperature gradient to a solid gives rise to a flow of heat.

We define $\vec{U}$ as the thermal current which is driven by the heat energy $E - E_F$, which is the excess energy of an electron above the equilibrium energy E_F. Neglecting time dependent effects, we define $\vec{U}$ as

$$\vec{U} = \frac{1}{4\pi^3} \int \vec{v}(E - E_F) f(\vec{r},\vec{k}) d^3k$$

where the distribution function $f(\vec{r},\vec{k})$ is related to the Fermi function f_0 by $f = f_0 + f_1$. Under equilibrium conditions there is no thermal current density

$$\int \vec{v}(E - E_F) d^3k = 0$$

so that the thermal current is driven by the thermal gradient which causes a departure from the equilibrium distribution:

$$\vec{U} = \frac{1}{4\pi^3} \int \vec{v}(E - E_F) f_1 \, d^3k$$

where the electronic contribution to the thermal conductivity tensor $\vec{k}_e$ is defined by the relation

$$\vec{U} = -\vec{k}_e \cdot \frac{\partial T}{\partial \vec{r}}$$

Assuming no explicit time dependence for the distribution function, the function f_1 representing the departure of the distribution from equilibrium is found from solution of Boltzmann's equation

$$\vec{v} \cdot \frac{\partial f}{\partial \vec{r}} + \dot{\vec{k}} \cdot \frac{\partial f}{\partial \vec{k}} = -\frac{f_1}{\tau}.$$

In the absence of an electric field, $\dot{\vec{k}} = 0$ and the drift velocity $\vec{v}$ is found from the equation

$$\vec{v} \cdot \frac{\partial f}{\partial \vec{r}} = -\frac{f_1}{\tau}.$$

Using the linear approximation for the term $\partial f / \partial \vec{r}$ in the Boltzmann equation, we obtain

$$\frac{\partial f}{\partial \vec{r}} \cong \frac{\partial f_0}{\partial \vec{r}} = \frac{\partial}{\partial \vec{r}} \left[\frac{1}{1 + e^{(E-E_F)/k_BT}} \right]$$

$$= \left\{ -\frac{e^{(E-E_F)/k_BT}}{[1 + e^{(E-E_F)/k_BT}]^2} \right\} \left\{ -\frac{1}{k_BT} \frac{\partial E_F}{\partial \vec{r}} - \frac{(E - E_F)}{k_BT^2} \frac{\partial T}{\partial \vec{r}} \right\}$$

$$= \left\{ k_BT \frac{\partial f_0}{\partial E} \right\} \left\{ -\frac{1}{k_BT} \frac{\partial T}{\partial \vec{r}} \right\} \left\{ \frac{\partial E_F}{\partial T} + \frac{(E - E_F)}{T} \right\} = -\frac{\partial f_0}{\partial E} \left\{ \frac{\partial E_F}{\partial T} + \frac{(E - E_F)}{T} \right\} \frac{\partial T}{\partial \vec{r}}$$

We will now give some typical values for these two terms for semiconductors and metals.

For semiconductors, we evaluate the expression in equation by referring to equation

$$E_F = \frac{1}{2} E_g - \frac{3}{4} k_BT \ln(m_h / m_e)$$

from which

$$\frac{\partial E_F}{\partial T} \sim \frac{3}{4} k_B \ln(m_h / m_e)$$

showing that the temperature dependence of E_F arises from the inequality of the valence and conduction band effective masses. If $m_h = m_e$, which would be the case of strongly coupled "mirror" bands, then $\partial E_F/\partial T$ would vanish. For a significant mass difference such as $m_h/m_e = 2$, we obtain $\partial E_f/\partial T \sim 0.5k_B$ from equation. For a band gap of 0.5 eV and the Fermi level in the middle of the gap, we obtain for the other term in equation

$$[(E - E_F)/T] \approx [0.5/(1/40)]k_B = 20k_B$$

where $k_BT \approx 1/40$ eV at room temperature. Thus for a semiconductor, the term $(E - E_F)/T$ is much larger than the term $(\partial E_F/\partial T)$.

For a metal with a spherical Fermi surface, the following relation

$$E_F = E_F^0 - \frac{\pi^2}{12}\frac{(k_BT)^2}{E_F^0}$$

is derived in standard textbooks on statistical mechanics, so that at room temperature and assuming that for a typical metal $E_F^0 = 5$eV, we obtain from equation

$$\left|\frac{\partial E_F}{\partial T}\right| = \frac{\pi^2}{6}\frac{(k_BT)}{E_F^0} k_B \approx \frac{10}{6}\frac{\left(\frac{1}{40}\right)}{5} k_B \approx 8\times10^{-3} k_B$$

Thus, for both semiconductors and metals, the term $(E - E_F)/T$ tends to dominate over $(\partial E_F/\partial T)$, though there can be situations where the term $(\partial E_F/\partial T)$ cannot be neglected. In this presentation, we will temporarily neglect the term ∇E_F in equation in calculation of the electronic contribution to the thermal conductivity, but we will include this term formally in our derivation of thermoelectric effects in §5.3.

Typically, the electron energies of importance in any transport problem are those within k_BT of the Fermi energy so that for many applications for metals, we can make the rough approximation,

$$\frac{E - E_F}{T} \approx k_B$$

though the results given in this section are derived without the above approximation for $(E - E_F)/T$. Rather, all integrations are carried out in terms of the variable $(E - E_F)/T$.

We return now to the solution of the Boltzmann equation in the relaxation time approximation $\frac{\partial f}{\partial \vec{r}} = \left(-\frac{\partial f_0}{\partial f}\right)\left(\frac{E - E_F}{T}\right)\left(\frac{\partial T}{\partial \vec{r}}\right)$

Solution of the Boltzmann equation yields

$$f_1 = -\tau\vec{v}\cdot\left(\frac{\partial f}{\partial \vec{r}}\right) = \tau\vec{v}\left(\frac{\partial f_0}{\partial E}\right)\left(\frac{E-E_F}{T}\right)\frac{\partial T}{\partial \vec{r}}$$

Substitution of f_1 in the equation for the thermal current

$$\vec{U} = \frac{1}{4\pi^3}\int \vec{v}(E-F_F)\, f_1 d^3k$$

then results in

$$\vec{U} = \frac{1}{4\pi^3 T}\left(\frac{\partial T}{\partial \vec{r}}\right)\cdot \int \tau\vec{v}\vec{v}(E-E_F)^2\left(\frac{\partial f_0}{\partial E}\right)d^3k\,.$$

Using the Definition of the thermal conductivity tensor $\vec{k}_e$ given by equation we write the electronic contribution to the thermal conductivity $\vec{k}_e$ as

$$\vec{k}_e = \frac{-1}{4\pi^3 T}\int \tau\vec{v}\vec{v}(E-E_F)^2\left(\frac{\partial f_0}{\partial E}\right)d^3k$$

where $d^3k = d^2S\, dk_\perp = d^2S\, dE/|\partial E/\partial \vec{k}\,| = d^2S dE\,/(\hbar v)$ is used to exploit our knowledge of the dependence of the distribution function on the energy.

Thermal Conductivity for Metals

In the case of a metal, the integral for $\vec{k}_e$ given by equation can be evaluated easily by converting the integral over phase space $\int d^3k$ to an integral over $\int dE\, d^2S_F$ in order to exploit the d-function property of $-(\partial f_0/\partial E)$.

$$\int G(E)\left(-\frac{\partial f_0}{\partial E}\right)dE = G(E_F) + \frac{\pi^2}{6}(k_B T)^2\left[\frac{\partial^2 G}{\partial E^2}\right]_{E_F} + \cdots$$

It is necessary to consider the expansion given in equation in solving equation since $G(E_F)$ vanishes at $E = E_F$ for the integral defined in equation for $\vec{k}_e$. To solve the integral equation of equation we make the identi¯cation of

$$G(E) = g(E)(E-E_F)^2$$

where

$$g(E) = \frac{1}{4\pi^3}\int \tau\vec{v}\vec{v}\,d^2S\,/\,\hbar v$$

so that $G(E_F) = 0$ and $(\partial G/\partial E)|_E F = 0$ while

$$\left[\frac{\partial^2 G}{\partial E^2}\right]_{E_F} = G''(E_F) = 2g(E_F).$$

These relations will be used again in connection with the calculation of the thermopower in §5.3.1. For the case of the thermal conductivity for a metal we then obtain

$$\overleftrightarrow{k}_e = \frac{\pi^2}{3}(k_B T)^2 g(E_F) = \frac{(k_B T)^2}{12\pi\hbar}\int \tau \vec{v}\vec{v}\frac{d^2 S_F}{v}$$

where the integration is over the Fermi surface. We immediately recognize that the integral appearing in equation is the same as that for the electrical conductivity

$$\overleftrightarrow{\sigma} = \frac{e^2}{4\pi\hbar}\int \tau \vec{v}\vec{v}\frac{d^2 S_F}{v}$$

so that the electronic contribution to the thermal conductivity and the electrical conductivity tensors are proportional to each other

$$\overleftrightarrow{k}_e = \overleftrightarrow{\sigma}T\left(\frac{\pi^2 k_B^2}{3e^2}\right)$$

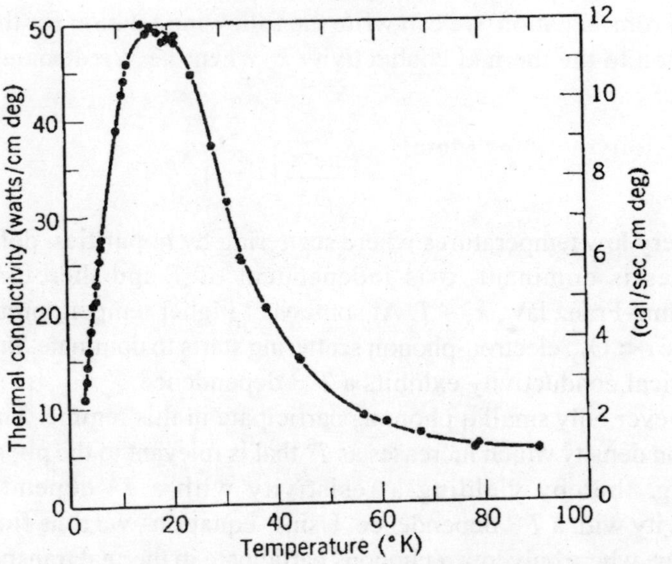

Fig. The temperature dependence of the thermal conductivity of copper. Note that both *k* and T are plotted on linear scales. At low temperatures where the phonon density is low, the thermal transport is by electrons predominantly, while at high temperatures, thermal transport by phonons becomes more important.

and equation is known as the Wiedemann–Franz Law. The physical basis for this relation is that in electrical conduction each electron carries a charge e and experiences an electrical force $e\vec{E}$ so that the electrical current per unit field is e^2. In thermal conduction, each electron carries a unit of thermal energy $k_B T$ and experiences a thermal force $k_B \partial T/\partial \vec{r}$ so that the heat current per unit thermal gradient is proportional to $k_B^2 T$. Therefore the ratio of $|k_e| = |\sigma|$ must be on the order of $(k_B^2 T/e^2)$. The Wiedemann–Franz law suggests that the ratio $k_e/(\sigma T)$ should be a constant (called the Lorenz constant)

$$\left|\frac{k_e}{\sigma T}\right| = \frac{\pi^2}{3}\left(\frac{k_B}{e}\right)^2 = 2.45\times10^{-8} \quad \text{watt ohm/deg}^2.$$

The ratio $k_e/(\sigma T)$ is approximately constant for all metals at high temperatures $T > \Theta_D$ and at very low temperatures $T \ll \Theta_D$, where Θ_D is the Debye temperature. The derivation of the Wiedemann–Franz Law depends on the relaxation time approximation, which is valid at high temperatures $T > \Theta_D$ where the electron scattering is dominated by the quasi-elastic phonon scattering process and is valid also at very low temperatures $T \ll \Theta_D$ where phonon scattering is unimportant and the dominant scattering mechanism is impurity and lattice defect scattering, both of which tend to be elastic scattering processes. The temperature dependence of the thermal conductivity of a metal is given. From equation we can write the following relation for the electronic contribution to the thermal conductivity k_e when the Wiedemann–Franz law is satisfied

$$k_e = \left(\frac{ne^2\tau}{m^*}\right) T \frac{\pi^2}{3}\left(\frac{k_B}{e}\right)^2$$

At very low temperatures where scattering by impurities, defects, crystal boundaries is dominant, σ is independent of T and therefore from the Wiedemann–Franz law, $k_e \sim T$. At somewhat higher temperatures, but still in the regime $t \ll \Theta_D$, electron-phonon scattering starts to dominate. In this regime, the electrical conductivity exhibits a T^{-5} dependence.

However only small q phonons participate in this regime. Thus it is only the phonon density which increases as T^3 that is relevant to the phonon-electron scattering, thereby yielding a resistivity with a T^3 dependence and a conductivity with a T^{-3} dependence. Using equation , we thus find that in the low T range where only low q phonons participate in thermal transport k_e should show a T^{-2} dependence. At high T where all the phonons contribute to thermal transport we have $\sigma \sim 1/T$ so that k_e becomes independent of T. Since $\Theta_D \sim$ 300 K for Cu, this temperature range far exceeds the upper limit.

Thermal Conductivity for Semiconductors

For the case of non-degenerate semiconductors, the integral for $\vec{k}_e$ in equation is evaluated by replacing $(E - E_F) \rightarrow E$, since in a semiconductor the electrons that can conduct heat must be in the conduction band, and the lowest energy an electron can have in the thermal conduction process is at the conduction band minimum. Then the thermal conductivity for a non-degenerate semiconductor can be written as

$$\vec{k}_e = \frac{1}{4\pi^3 T}\int \tau \vec{v}\vec{v}\, E^2\left(-\frac{\partial f_0}{\partial E}\right)d^3k$$

For intrinsic semiconductors, the distribution function can be approximated by the Maxwell-Boltzmann distribution so that $(\partial f_0 / \partial E) \rightarrow -(1/k_BT)e^{(-|E_F^e|/k_BT}\,e^{-(E/k_BT)}$

For a parabolic band we have $E = \hbar^2k^2/2m^*$, so that the volume element in reciprocal space can be written

$$\int d^3k = \int 4\pi k^2 dk = \int_0^\infty 2\pi(2m^*/\hbar^2)^{3/2} E^{1/2} dE\,,$$

and $\vec{v} = (1/\hbar)(\partial E/\partial \vec{k}) = \hbar\vec{k}/m$. Assuming a constant relaxation time, we then substitute all these terms into equation for $\vec{k}_e$ and integrate to obtain

$$k_{exx} = (1/4\pi^3 T)\int \tau v_x^2 E^2\left[(k_BT)^{-1}e^{-|E_F^e|(/k_BT)}e^{-E/(k_BT)}\right]2\pi(2m^*/\hbar^2)^{3/2}E^{1/2}dE$$

$$= \left[k_B(k_BT)\tau/(3\pi^2 m^*)\right](2m^* k_BT/\hbar^2)^{3/2}e^{-|E_F^e|/(k_BT)}\int_0^\infty x^{7/2}e^{-x}\,dx$$

where $\int_0^\infty x^{7/2}e^{-x}\,dx = 105\sqrt{\pi}/8$, from which it follows that k_{exx} has a temperature dependence of the form $T^{5/2}e^{-|E_F^e|/(k_BT)}$ in which the exponential term is dominant for temperatures of physical interest, where $k_BT \ll |E_F^e|$. We note from equation that for a semiconductor the temperature dependence
of the electrical conductivity is given by

$$\sigma_{xx} = \frac{2e^2\tau}{m^*}\left(\frac{m^* k_BT}{2\pi\hbar^2}\right)^{3/2} e^{-|E_F^e|/(k_BT)}$$

Assuming cubic symmetry, we can write the conductivity tensor as

$$\vec{\sigma} = \begin{pmatrix} \sigma_{xx} & 0 & 0 \\ 0 & \sigma_{xx} & 0 \\ 0 & 0 & \sigma_{xx} \end{pmatrix}$$

so that the electronic contribution to the thermal conductivity of a semiconductor can be written as

$$k_{exx} = \left(\frac{35}{2}\right)\left(\frac{k_B^2}{e^2}\right)\sigma_{xx}T$$

where

$$\sigma_{xx} = ne^2\tau/m_{xx} = ne\,\mu_{xx}$$

and we note that the coefficient (35/2) for this calculation for semiconductors corresponds to ($\pi^2/3$) for metals.

Except for numerical constants, the formal results relating the electronic contribution to the thermal conductivity k_{exx} and σ_{xx} are similar for metals and semiconductors, with the electronic thermal conductivity and electrical conductivity being proportional.

A major difference between semiconductors and metals is the magnitude of the electrical conductivity and hence of the electronic contribution to the thermal conductivity.

Since σ_{xx} is much smaller for semiconductors than for metals, k_e for semiconductors is relatively unimportant and the thermal conductivity is dominated by the lattice contribution k_L.

Thermal Conductivity for Insulators

In the case of insulators, heat is only carried by phonons (lattice vibrations). The thermal conductivity in insulators therefore depends on phonon scattering mechanisms. The lattice thermal conductivity is calculated from kinetic theory and is given

$$k_L = \frac{C_p \bar{v}_q \Lambda_{\text{ph}}}{3}$$

where C_p is the heat capacity, v_q is the average phonon velocity and Λ_{ph} is the phonon mean free path.

The total thermal conductivity of a solid is given as the sum of the lattice contribution k_L and the electronic contribution k_e. For metals the electronic contribution dominates, while for insulators and semiconductors the phonon contribution dominates. Let us now consider the temperature dependence of k_{exx}.

At very low T in the defect scattering range, the heat capacity has a dependence $C_p \propto T^3$ while vq and Λ_{ph} are almost independent of T.

As T increases and we enter the phonon-phonon scattering regime due to normal scattering processes and involving only low q phonons, C_p is still increasing with T but more slowly that T^3 while $\bar{v}_q$ remains independent of T and Λ_{ph}.

As T increases further, the thermal conductivity increases more and more gradually and eventually starts to decrease because of phonon-phonon events where the density of phonons available for scattering depends on $[\exp(\hbar\omega/k_BT) - 1]$.

This causes a peak in $k_L(T)$.

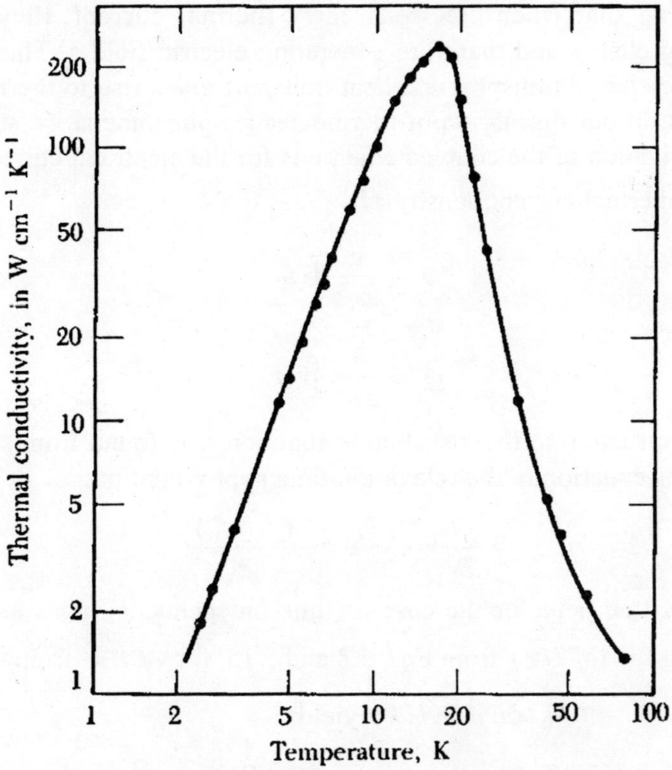

Fig. Temperature dependence of the thermal conductivity of a highly purified insulating crystal of NaF. Note that both k and T are plotted on a log scale, and that the peak in k occurs at low temperature (~ 17 K). The temperature dependence of k is further discussed in §6.4.

The decrease in $k_L(T)$ becomes more pronounced as C_p becomes independent of T and Λ_{ph} continues to be proportional to $[\exp(\hbar\bar{\omega}/k_BT) - 1]$ where $\bar{\omega}$ is a typical phonon frequency. As T increases further we eventually enter the $T \gg \Lambda_D$ regime where Λ_D is the Debye temperature. In this regime, the temperature dependence of Λ_{ph} simply becomes $\Lambda_{ph} \sim (1/T)$. For $k(T)$ for NaF we see that the peak in k occurs at about 18 K where the complete Bose factor $[\exp(\hbar\bar{\omega}/k_BT) - 1]$ must be used to describe the T dependence of k_L. For much of the temperature range, only low q phonons participate in the thermal conduction process. At higher temperatures where larger q phonons contribute to thermal conduction, umklapp processes become important in the phonon scattering process.

THERMOELECTRIC PHENOMENA

In many metals and semiconductors there exists a coupling between the electrical current and the thermal current. This coupling can be appreciated by observing that when electrons carry thermal current, they are also transporting charge and therefore generating electric field s. This coupling between the charge transport and heat transport gives rise to thermoelectric phenomena. In our discussion of thermoelectric phenomena we start with a general derivation of the coupled equations for the electrical current density $\vec{j}$ and the thermal current density $\vec{U}$:

$$\vec{j} = \frac{e}{4\pi^3}\int \vec{v} f_1 d^3k$$

$$\vec{U} = \frac{1}{4\pi^3}\int \vec{v}(E - E_F) f_1 d^3k$$

and the perturbation to the distribution function f_1 is found from solution of Boltzmann's equation in the relaxation time approximation:

$$\vec{v}\cdot\frac{\partial f}{\partial \vec{r}} + \dot{\vec{k}}\cdot\frac{\partial f}{\partial \vec{k}} = -\frac{(f - f_0)}{\tau},$$

which is written here for the case of time independent forces and field s. Substituting for $(\partial f/\partial \vec{r})$ from Eqs. 5.8 and 5.15, for $\partial f/\partial \vec{k}$ from equation , and for $\partial f/\partial \vec{k} = (\partial f_0/\partial E)(\partial E/\partial \vec{k})$ yields

$$\vec{v}\cdot(\partial f_0/\partial E)\left[(e\vec{E} - \nabla E_F) - \frac{(E - E_F)}{T}(\vec{\nabla} T)\right] = -\frac{f_1}{\tau},$$

so that the solution to the Boltzmann equation in the presence of an electric field and a temperature gradient is

$$f_1 = \vec{v}\tau\cdot(\partial f_0/\partial E)\left\{[(E - E_F)/T]\vec{\nabla} T - e\vec{E} + \vec{\nabla} E_F\right\}.$$

The electrical and thermal currents in the presence of both an applied electric field and a temperature gradient can thus be obtained by substituting f_1 into $\vec{j}$ and $\vec{U}$ in Eqs. 5.39 and 5.40 to yield expressions of the form: $\vec{j} = e^2\vec{k}_0\cdot(\vec{E} - \frac{1}{e}\vec{\nabla} E_F) - (e/T)\vec{k}_1\cdot\vec{\nabla} T$

and $\vec{U} = e\vec{k}_1\cdot\left(\vec{E} - \frac{1}{e}\vec{\nabla} E_F\right) - (1/T)\vec{k}_2\cdot\vec{\nabla} T$

where $\vec{k}_0$ is related to the conductivity tensor $\vec{\sigma}$ by

$$\vec{k}_0 = \frac{1}{4\pi^3}\int \tau\vec{v}\vec{v}(-\partial f_0/\partial E)\,d^3k = \frac{\vec{\sigma}}{e^2}$$

and

$$\vec{k}_1 = \frac{1}{4\pi^3}\int \tau\vec{v}\vec{v}(E-E_F)\left(-\frac{\partial f_0}{\partial E}\right)d^3k,$$

and $\vec{k}_2$ is related to the thermal conductivity tensor $\vec{k}_e$ by

$$\vec{k}_2 = \frac{1}{4\pi^3}\int \tau\vec{v}\vec{v}(E-E_F)^2(-\partial f_0/\partial E)d^3k = T\vec{k}_e$$

Note that the integrands for $\vec{k}_1$ and $\vec{k}_2$ are both related to that for $\vec{k}_0$ by introducing factors of $(E - E_F)$ and $(E - E_F)^2$, respectively. Note also that the same integral $\vec{k}_1$ occurs in the expression for the electric current $\vec{j}$ induced by a thermal gradient $\vec{\nabla}T$ and in the expression for the thermal current $\vec{U}$ induced by an electric field $\vec{E}$. The motion of charged carriers across a temperature gradient results in a flow of electric current expressed by the term $-(e/T)(\vec{k}_1)\cdot\vec{\nabla}T$. This term is the origin of thermoelectric effects.

The discussion up to this point has been general. If specific boundary conditions are imposed, we obtain a variety of thermoelectric effects such as the Seebeck effect, the Peltier effect and the Thomson effect. We now define the conditions under which each of these thermoelectric effects occur.

We define the thermopower $\vec{S}$ (Seebeck coefficient) and the Thomson coefficient T_b under conditions of zero current flow. Then referring to equation, we obtain under open circuit conditions

$$\vec{j} = 0 = e^2\vec{k}_0\cdot(\vec{E} - \frac{1}{e}\vec{\nabla}E_F) - (e/T)\vec{k}_1\cdot\vec{\nabla}T$$

so that the Seebeck coefficient $\vec{S}$ is defined by

$$\vec{E} - \frac{1}{e}\vec{\nabla}E_F = (1/eT)\vec{k}_0^{-1}\cdot\vec{k}_1\cdot\vec{\nabla}T \equiv \vec{S}\cdot\vec{\nabla}T$$

the Definition for the Thomson coefficient T_b $\vec{E} = \left(\frac{1}{e}\frac{\partial E_F}{\partial T} + \vec{S}\right)\vec{\nabla}T \equiv \vec{T}_b\cdot\vec{\nabla}T$

where $T\frac{\partial}{\partial T}\vec{S}$

For many thermoelectric systems of interest, $\vec{S}$ has a linear temperature dependence, and in this case it follows from equation that $\vec{T}_b$ and $\vec{S}$ are almost equivalent for practical purposes. Therefore the Seebeck and Thomson coefficients are used almost interchangeably in the literature.

From equation we have

$$\vec{S} = (1/eT)\vec{k}_0^{-1}\cdot\vec{k}_1$$

which is simplified by assuming an isotropic medium, yielding the scalar quantities

$$S = (1/eT)\,(k_1/k_0).$$

However in an anisotropic medium the tensor components of $^{\$}S$ are found from

$$S_{ij} = (1/eT)\,(k_0^{-1})_{i\alpha}\,(k_1)_{\alpha j},$$

where the Einstein summation convention is assumed. The thermopower or Seebeck effect in an n-type semiconductor is at the hot junction the Fermi level is higher than at the cold junction. Electrons will move from the hot junction to the cold junction in an attempt to equalize the Fermi level, thereby creating an electric field which can be measured in terms of the open circuit voltage V.

Another important thermoelectric coefficient is the Peltier coefficient $\overleftrightarrow{\Pi}$, defined as the proportionality between $\vec{U}$ and $\vec{j}$

$$\vec{U} \equiv \overleftrightarrow{\Pi} \cdot \vec{j}$$

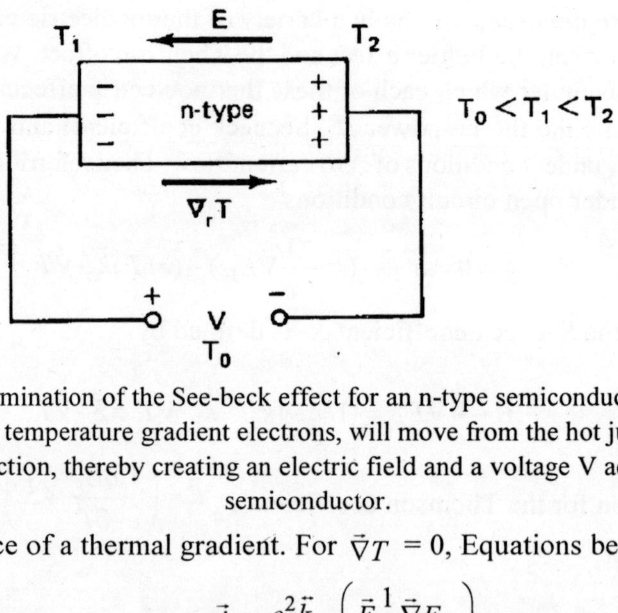

Fig. Determination of the See-beck effect for an n-type semiconductor. In the presence of a temperature gradient electrons, will move from the hot junction to the cold junction, thereby creating an electric field and a voltage V across the semiconductor.

in the absence of a thermal gradient. For $\vec{\nabla}T = 0$, Equations become

$$\vec{j} = e^2 \overleftrightarrow{k}_0 \cdot \left(\vec{E} - \frac{1}{e} \vec{\nabla} E_F \right)$$

$$\vec{U} = e \overleftrightarrow{k}_1 \cdot \left(\vec{E} - \frac{1}{e} \vec{\nabla} E_F \right)$$

so that

$$\vec{U} = (1/e)\overleftrightarrow{k}_1 \cdot (\overleftrightarrow{k}_0)^{-1} \cdot \vec{j}$$

$$= \overleftrightarrow{\Pi} \cdot \vec{j}$$

where

$$\overleftrightarrow{\Pi} = (1/e)\overleftrightarrow{k}_1 \cdot (\overleftrightarrow{k}_0)^{-1}$$

Comparing Eqs. 5.53 and 5.60 we see that $\overleftrightarrow{\Pi}$ and $\overleftrightarrow{S}$ are related by

$$\overleftrightarrow{\Pi} = T\overleftrightarrow{S},$$

where T is the temperature. For isotropic materials the Peltier coefficient thus becomes a scalar, and is proportional to the thermopower S :

$$\Pi = \frac{1}{e}(k_1 / k_0) = TS,$$

while for anisotropic materials the tensor components of $\overleftrightarrow{\Pi}$ can be found in analogy with equation. We note that both $\overleftrightarrow{S}$ and $\overleftrightarrow{\Pi}$ exhibit a linear dependence on e and therefore depend explicitly on the sign of the carrier, and measurements of $\overleftrightarrow{S}$ or $\overleftrightarrow{\Pi}$ can be used to determine whether transport is dominated by electrons or holes.

We have already considered the evaluation of k_0 in treating the electrical conductivity and $\overleftrightarrow{k}_2$ in treating the thermal conductivity. To treat thermoelectric phenomena we need now to evaluate $\overleftrightarrow{k}_1$

$$\overleftrightarrow{k}_1 = \frac{1}{4\pi^3}\int \tau \vec{v}\vec{v}(E - E_F)(-\partial f_0 / \partial E) d^3k \, .$$

In §5.3.1 we evaluate $\overleftrightarrow{k}_1$ for the case of a metal and in §5.3.2 we evaluate $\overleftrightarrow{k}_1$ for the case of the electrons in an intrinsic semiconductor.

Thermoelectric Phenomena in Metals

All thermoelectric effects in metals depend on the tensor $\overleftrightarrow{k}_1$ which we evaluate below for the case of a metal. We can then obtain the thermopower

$$\overleftrightarrow{S} = \frac{1}{eT}(\overleftrightarrow{k}_1 \cdot \overleftrightarrow{k}_0^{-1})$$

or the Peltier coefficient

$$\overleftrightarrow{\Pi} = \frac{1}{e}(\overleftrightarrow{k}_1 \cdot \overleftrightarrow{k}_0^{-1})$$

or the Thomson coefficient

$$\overleftrightarrow{T}_b = T\frac{\partial}{\partial T}\overleftrightarrow{S}.$$

To evaluate $\overleftrightarrow{k}_1$ for metals we wish to exploit the δ-function behaviour of $(-\partial f_0/\partial E)$. This is accomplished by converting the integration d^3k to an integration over dE and a constant energy surface, $d^3k = d^2S\, dE/\hbar v$. From Fermi statistics we have the general relation

$$\int G(E)\left(-\frac{\partial f_0}{\partial E}\right)dE = G(E_F) + \frac{\pi^2}{6}(k_B T)^2\left[\frac{\partial^2 G}{\partial E^2}\right]_{E_F} + \cdots$$

For the integral in equation which defines $\overleftrightarrow{k}_1$, we can write

$$G(E) = g(E)\,(E - E_F)$$

where

$$g(E) = \frac{1}{4\pi^3}\int \tau\vec{v}\vec{v}\, d^2S/v$$

and the integration in equation is carried out over a constant energy surface at energy E.

Differentiation of G(E) then yields

$$G'(E) = g'(E)(E - E_F) + g(E)$$
$$G''(E) = g''(E)\,(E - E_F) + 2g'(E).$$

Evaluation at $E = E_F$ yields

$$G(E_F) = 0$$
$$G''(E_F) = 2g'(E_F).$$

We therefore obtain

$$\overleftrightarrow{k}_1 = \frac{\pi^2}{3}(k_B T)^2\, g'(E_F).$$

We interpret $g'(E_F)$ in equation to mean that the same integral $\vec{k}_0$ which determines the conductivity tensor is evaluated on a constant energy surface E, and $g'(E_F)$ is the energy derivative of that integral evaluated at the Fermi energy E_F. The temperature dependence of $g'(E_F)$ is related to the temperature dependence of τ, since v is essentially temperature independent. For example, the acoustic phonon scattering in the high temperature limit $T \gg \Theta_D$ yields a temperature dependence $\tau \sim T^{-1}$ so that $\overleftrightarrow{k}_1$ in this important case for metals will be proportional to T.

For a spherical constant energy surface $E = \hbar^2k^2/2m^*$ and assuming a relaxation time τ that is independent of energy, we can readily evaluate equation to obtain

$$g(E) = \frac{\tau}{3\pi^2 m^*}\left(\frac{2m^*}{\hbar^2}\right)^{3/2} E^{3/2}$$

$$g'(E_F) = \frac{\tau}{2\pi^2 m^*}\left(\frac{2m^*}{\hbar^2}\right)^{3/2} E_F^{1/2}$$

and

$$k_1 = \frac{\tau}{6m^*}\left(\frac{2m^*}{\hbar^2}\right)^{3/2} E_F^{1/2}(k_B T)^2$$

Using the same approximations, we can write for k_0:

$$k_0 = \frac{\tau}{3\pi^2 m^*}\left(\frac{2m^*}{\hbar^2}\right)^{3/2} E_F^{1/2}$$

so that from equation we have for the Seebeck coefficient

$$S = \frac{\pi^2 k_B}{2e}\frac{k_B T}{E_F}.$$

From equation we see that S exhibits a linear dependence on T and a sensitivity to the sign of the carriers. We note from equation that a low carrier density implies a large S value.

Thus degenerate (heavily doped with $n \sim 10^{18} - 10^{19}$ /cm^3) semiconductors tend to have higher thermopowers than metals.

Thermopower for Semiconductors

We evaluate $\overleftrightarrow{k}_1$ for electrons in an intrinsic or lightly doped semiconductor for illustrative purposes. Intrinsic semiconductors are not important for practical thermoelectric devices since the contributions of electrons and holes to $\overleftrightarrow{k}_1$ are of opposite signs and tend to cancel. Thus it is only heavily doped semiconductors with a single carrier type that are important for thermoelectric applications.

The evaluation of the general expression for the integral $\overleftrightarrow{k}_1$

$$\overleftrightarrow{k}_1 = \frac{1}{4\pi^3}\int \tau\vec{v}\vec{v}(E - E_F)\left(-\frac{\partial f_0}{\partial E}\right)d^3k$$

is different for semiconductors and metals. An intrinsic semiconductor we need to make the substitution $(E - E_F) \rightarrow E$ in equation , since only conduction electrons can carry heat. The equilibrium distribution function for an intrinsic semiconductor can be written as

$$f_0 = e^{-E/(k_BT)}e^{-|E_F^e|/(k_BT)}$$

so that

$$\frac{\partial f_0}{\partial E} = -\frac{1}{k_BT}e^{-E/(k_BT)}e^{-|E_F^e|/(k_BT)}$$

To evaluate d^3k we need to assume a model for $E(\vec{k})$. For simplicity, assume a simple parabolic band

$$E = \hbar^2k^2/2m^*.$$

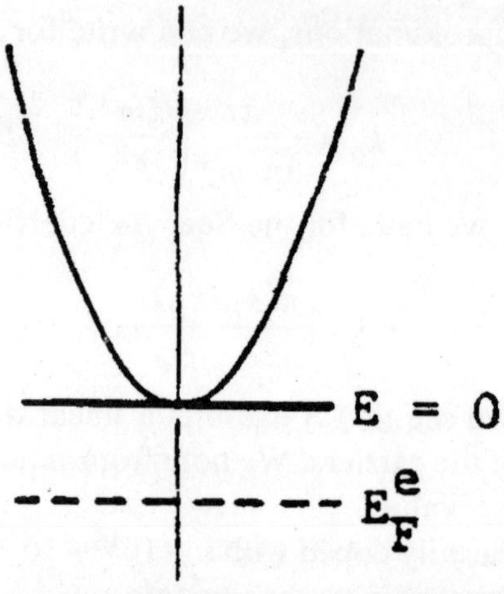

Fig. Schematic E vs k diagram, showing that $E = 0$ is the lowest electronic energy for heat conduction.

$$d^3k = 4\pi k^2 dk$$

so that

$$d^3k = 2\pi\left(\frac{2m^*}{\hbar^2}\right)^{3/2} E^{1/2}dE$$

and also

$$\vec{v} = \frac{1}{\hbar}(\partial E/\partial\vec{k}) = k\vec{\hbar}/m^*$$

Substitution into the equation for $\vec{k}_1$ for a semiconductor with the simple $E = \hbar^2k^2/2m^*$ dispersion relation then yields upon integration

$$k_{1xx} = \frac{5\tau k_BT}{m^*}(m^*k_BT/2\pi\hbar^2)^{3/2}e^{-|E_F^e|/(k_BT)}.$$

This expression is valid for a semiconductor for which the Fermi level is far from the band edge $(E - E_F) \gg k_BT$. The thermopower is then found by substitution

$$S = \frac{1}{eT}(k_{1xx} / k_{0xx})$$

where the expression for $k_{0xx} = \sigma_{xx/e2}$ is given by equation. We thus obtain the result

$$S = \frac{5}{2}\frac{k_B}{e}$$

which is a constant independent of temperature, independent of the band structure, but sensitive to the sign of the carriers. In an actual intrinsic semiconductor, the contribution of both electrons and holes to k_1 must be found. Likewise the calculation for k_{0xx} would also include contributions from both electrons and holes. Since the contribution to $(1/e)k_1xx$ for holes and electrons are of opposite sign, we can from equation expect that S for holes will cancel S for electrons for an intrinsic semiconductor.

Materials with a high thermopower or Seebeck coefficient are heavily doped degenerate semiconductors for which the Fermi level is close to the band edge and the complete Fermi function must be used.

Since S depends on the sign of the charge carriers, thermoelectric materials are doped either heavily doped n-type or heavily doped p-type semiconductors to prevent cancellation of the contribution from electrons and holes, as occurs in intrinsic semiconductors which because of thermal excitations of carriers have equal concentrations of electrons and holes.

Effect of Thermoelectricity on the Thermal Conductivity

From the coupled equations it is seen that the proportionality between the thermal current $\vec{U}$ and the temperature gradient $\vec{\nabla}T$ in the absence of electrical current ($\vec{j} = 0$) contains terms related to $\overleftrightarrow{k}_1$. We now solve Eqs. 5.44 and 5.45 to find the contribution of the thermoelectric terms to the electronic thermal conductivity. When $\vec{j} = 0$, equation becomes

$$\left(\vec{E} - \frac{1}{e}\vec{\nabla}E_F\right) = \frac{1}{eT}\overleftrightarrow{k}_0^{-1} \cdot \overleftrightarrow{k}_1 \cdot \vec{\nabla}T$$

so that

$$\vec{U} = -(1/T)\left[\overleftrightarrow{k}_2 \cdot \overleftrightarrow{k}_1 \cdot \overleftrightarrow{k}_0^{-1} \cdot \overleftrightarrow{k}_1\right] \cdot \vec{\nabla}T$$

where $\overleftrightarrow{k}_0$, $\overleftrightarrow{k}_1$, and $\overleftrightarrow{k}_2$ are given by Eqs. 5.46, 5.47, and 5.48, respectively, or

$$\overleftrightarrow{k}_0 = \frac{1}{4\pi^3\hbar}\int \tau\vec{v}\vec{v}\frac{d^2S_F}{v},$$

$$\overleftrightarrow{k}_1 = \frac{\pi^2}{3}(k_BT)^2\left(\frac{\partial\overleftrightarrow{k}_0}{\partial E}\right)_{E_F}$$

and

$$\vec{\vec{k}}_2 = \frac{(k_B T)^2}{12\pi\hbar} \int \tau \vec{v}\vec{v} \frac{d^2 S_F}{v}$$

We now evaluate the contribution to the thermal conductivity from the thermoelectric coupling effects for the case of a metal having a simple dispersion relation

$$E = \hbar^2 k^2 / 2m^*$$

In this case where τ is considered to be independent of E, Eqs. 5.91 and 5.90, respectively, provide expressions for k_1 and k_0 from which

$$\frac{1}{T} k_1 k_0^{-1} k_1 = \frac{\pi^4 n\tau}{4m^*} k_B^2 T \left(\frac{k_B T}{E_F}\right)^2$$

so that from Eqs. 5.28 and 5.94 the total electronic thermal conductivity for the metal becomes

$$k_e = \frac{\pi^2 n\tau}{3m^*} k_B^2 T \left[1 + \frac{3\pi^2}{4}\left(\frac{k_B T}{E_F}\right)^2\right]$$

For typical metals $(T/T_F) \sim (1/30)$ at room temperature so that the thermoelectric correction term is less than 1%.

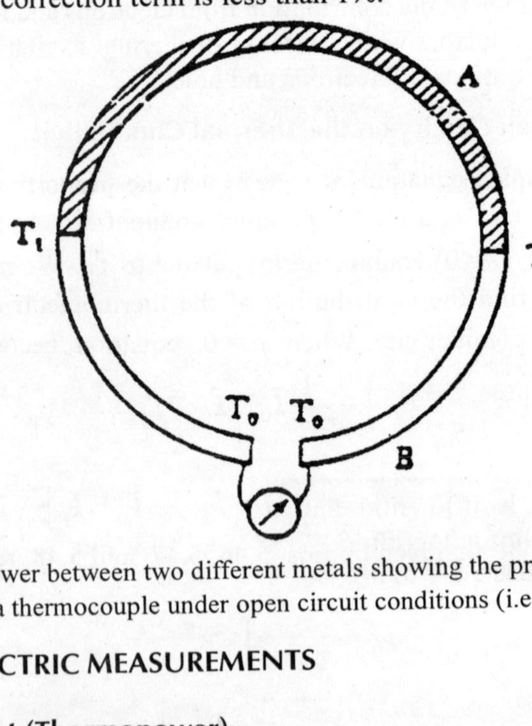

Fig. Thermopower between two different metals showing the principle of operation of a thermocouple under open circuit conditions (i.e., $j = 0$).

THERMOELECTRIC MEASUREMENTS

Seebeck Effect (Thermopower)

The thermopower S as defined in equation and is the characteristic coefficient in the Seebeck effect, where a metal subjected to a thermal gradient $\vec{\nabla}T$ exhibits an electric field $\vec{E} = S\vec{\nabla}T$, or an open-circuit voltage V under conditions of no current flow.

In the application of the Seebeck effect to thermocouple operation, we usually measure the difference in thermopower $S_A - S_B$ between two different metals A and B by measuring the open circuit voltage V_{AB}. This voltage can be calculated from

$$V_{AB} = \oint \vec{E} \cdot d\vec{r} = -I \oint S \frac{\partial T}{\partial \vec{r}} d\vec{r}$$

$$= \int_{T_0}^{T_1} S_B dT + \int_{T_1}^{T_2} S_A dT + \int_{T_2}^{T_0} S_B dT$$

$$= \int_{T_0}^{T_1} (S_A - S_B) dT .$$

w ith $T_1 \neq T_2$, an open-circuit potential difference V_{AB} can be measured and equation shows that V_{AB} is independent of the temperature T_0. Thus if T_1 is known and V_{AB} is measured, then temperature T_2 can be found from the calibration table of the thermocouple. From the simple expression of equation

$$S = \frac{\pi^2 k_B}{2e} \frac{k_B T}{E_F}$$

a linear dependence of S on T is predicted for simple metals. For actual thermocouples used for temperature measurements, the S (T) dependence is approximately linear, but is given by an accurate calibration table to account for small deviations from this linear relation. Thermocouples are calibrated at several fixed temperatures and the calibration table comes from a fit of these thermal data to a polynomial function that is approximately linear in T.

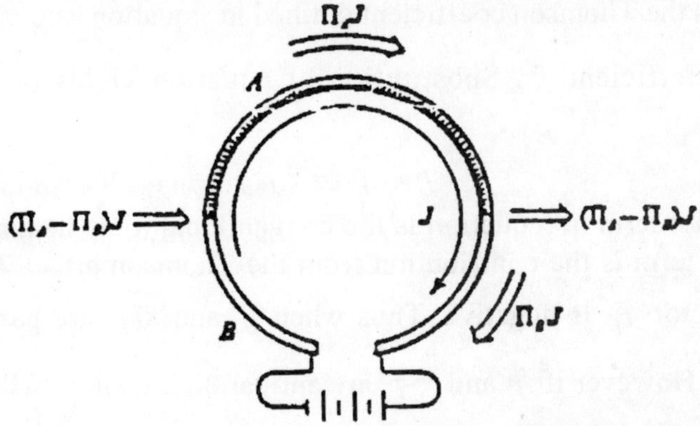

Fig. A heat engine based on the Peltier Effect with heat introduced at one junction and extracted at another under the conditions of no temperature gradient ($\nabla T = 0$).

Peltier Effect

The Peltier effect is the observation of a thermal current $\vec{U} = \ddot{\Pi} \cdot \vec{j}$ in the presence of an electric current $\vec{j}$ with no thermal gradient ($\vec{\nabla} T = 0$) so that

$$\ddot{\Pi} = T\ddot{S}.$$

The Peltier effect measures the heat generated (or absorbed) at the junction of two dissimilar metals held at constant temperature, when an electric current passes through the junction. Sending electric current around a circuit of two dissimilar metals cools one junction and heats another and is the basis of the operation of thermoelectric coolers. This thermoelectric effect is represented schematically. Because of the similarities between the Peltier coefficient and the Seebeck coefficient, materials exhibiting a large Seebeck effect also show a large Peltier effect. Since both $\ddot{S}$ and $\ddot{\Pi}$ are proportional to (1/*e*), the sign of $\ddot{S}$ and $\ddot{\Pi}$ is negative for electrons and positive for holes in the case of degenerate semiconductors. Reversing the direction of $\vec{j}$, will interchange the junctions where heat is generated (absorbed).

Thomson Effect

Assume that we have an electric circuit consisting of a single metal conductor. The power generated in a sample, such as an n-type semiconductor, is

$$P = \vec{j} \cdot \vec{E}$$

where the electric field can be obtained from equation as

$$\vec{E} = (\ddot{\sigma}^{-1}) \cdot \vec{j} - \ddot{T}_b \cdot \vec{\nabla} T$$

where $\ddot{T}_b$ is the Thomson coefficient defined in equation and is related to the Seebeck coefficient $\ddot{S}$. Substitution of equation yields the total power dissipation

$$P = j \cdot (\ddot{\sigma}^{-1}) \cdot \vec{j} - \vec{j} \cdot \ddot{T}_b \cdot \vec{\nabla} T$$

The first term in equation is the conventional joule heating term while the second term is the contribution from the Thomson effect. For an *n*-type semiconductor $\ddot{T}_b$ is negative. Thus when $\vec{j}$ and $\vec{\nabla} T$ are parallel, heating will result. However if $\vec{j}$ and $\vec{\nabla} T$ are antiparallel, cooling will occur. Thus reversal of the direction of $\vec{j}$ without changing the direction of $\vec{\nabla} T$ will reverse the sign of the Thomson contribution.

Likewise, a reversal in the direction of $\vec{\nabla}T$ keeping the direction of $\vec{j}$ unchanged will also reverse the sign of the Thomson contribution.

Thus, if either (but not both) the direction of the electric current or the direction of the thermal gradient is reversed, an absorption of heat from the surroundings will take place. The Thomson effect is utilized in thermoelectric refrigerators which are useful as practical low temperature laboratory coolers. We see a schematic diagram explaining the operation of a thermoelectric cooler. We see that for a degenerate n-type

semiconductor where $\vec{T}_b$, $\vec{S}$, and $\vec{\Pi}$ are all negative, when $\vec{j}$ and $\vec{\nabla}T$ are antiparallel then cooling occurs and heat is extracted from the cold junction and transferred to the heat sink at temperature T_H. For the p-type leg, all the thermoelectric coefficients are positive, so equation shows that cooling occurs when $\vec{j}$ and $\vec{\nabla}T$ are parallel. Thus both the *n*-type and *p*-type legs in a thermoelectric element contribute to cooling in a thermoelectric cooler.

The Kelvin Relations

The three thermoelectric effects are related and these relations were first derived by Lord Kelvin after he became a Lord and changed his name from Thomson to Kelvin. The Kelvin relations are based on arguments of irreversible thermodynamics and relate Π, S, and T_b.

If we define the thermopower $S_{AB} = S_B - S_A$ and the Peltier coefficient similarly $\Pi_{AB} = \Pi_A - \Pi_B$ for material A joined to material B, then we obtain the first Kelvin relation:

$$S_{AB} = \frac{\Pi_{AB}}{T}.$$

The Thomson coefficient T_b is defined by

$$T_b = T\frac{\partial S}{\partial T}$$

which allows determination of the Seebeck coefficient at temperature T_0 by integration of the above equation

$$S(T_0) = \int_0^{T_0}\left[\frac{T_b(T)}{T}\right]dT.$$

Furthermore, from the above Definitions, we deduce the second Kelvin relation

$$T_{b,A} - T_{b,B} = T\frac{\partial S_{AB}}{\partial T} = T\frac{\partial S_A}{\partial T} - T\frac{\partial S_B}{\partial T}$$

from which we obtain an expression relating all three thermoelectric coefficients

$$T_{b'A} = T\frac{\partial S_A}{\partial T} = T\frac{\partial(\Pi_A/T)}{\partial T} = \frac{\partial \Pi_A}{\partial T} - \frac{\Pi_A}{T} = \frac{\partial \Pi_A}{\partial T} - S_A$$

Thermoelectric Figure of Merit

A good thermoelectric material for cooling applications must have a high thermoelectric figure of merit, Z, which is defined by

$$Z = \frac{S^2\sigma}{k}$$

where S is the thermoelectric power (Seebeck coefficient), σ is the electrical conductivity, and k is the thermal conductivity. In order to achieve a high Z, one requires a high thermoelectric power S, a high electrical conductivity σ to maintain high carrier mobility, and a low thermal conductivity k to retain the applied thermal gradient.

In general, it is difficult in practical systems to increase Z for the following reasons: increasing S for simple materials also leads to a simultaneous decrease in σ, and an increase in σ leads to a comparable increase in the electronic contribution to k because of the Wiedemann-Franz law. So with known conventional solids, a limit is rapidly obtained where a modification to any one of the three parameters S, σ, or k adversely affects the other transport coefficients, so that the resulting Z does not vary significantly. Currently, the materials with the highest Z are Bi_2Te_3 alloys such as $Bi_0.5\ Sb_1.5\ Te_3$ with $ZT \sim 1$ at 300K.

Only small increases in Z have been achieved in the last two to three decades. Research on thermoelectric materials has therefore been at a low level since about 1960. Since 1994, new interest has been revived in thermoelectricity with the discovery of new materials: skutterudites –$CeFe_{4-x}Co_xSb_{12}$ or $LaFe_{4-x}Co_xSb_{12}$ for $0 < x < 4$, which offer promise for higher Z values in bulk materials, and low dimensional systems (quantum wells, quantum wires) which offer promise for enhanced Z relative to bulk Z values in the same material. Thus thermoelectricity has again become an active research field.

PHONON DRAG EFFECT

For a simple metal such as an alkali metal one would expect the thermopower S to be given by the simple expression in equation , and to be negative since the carriers are electrons. This is true at room temperature for all of the alkali metals except Li. Furthermore, S is positive for the noble metals Au, Ag and Cu. This can be understood by recalling the complex Fermi surfaces for these metals, where we note that copper in fact exhibits hole orbits in the extended zone. In general, with multiple carrier types as occur in semiconductors, the interpretation of thermopower data can become complicated.

Another complication which must also be considered, especially at low temperatures, is the phonon drag effect. In the presence of a thermal gradient, the phonons will diffuse and "drag" the electrons along with them because of the electron-phonon interaction. For a simple explanation of phonon drag, consider a gas of phonons with an average energy density E_{ph}=V. Using kinetic theory, we find that the phonon gas exerts a pressure on the electron gas.

$$P = \frac{1}{3}\left(\frac{E_{ph}}{V}\right)$$

In the presence of a thermal gradient, the electrons are subject to a force density

$$F_x/V = -dP/dx = -\frac{1}{3V}\left(\frac{dE_{ph}}{dT}\right)\frac{dT}{dx}.$$

To prevent the flow of current, this force must be balanced by the electric force. Thus, for an electron density n, we obtain

$$-neE_x + F_x/V = 0$$

giving a phonon-drag contribution to the thermopower. Using the Definition of the Seebeck coefficient for an open circuit system, we can write

$$S_{ph} = \frac{E_x}{(dT/dx)} \approx -\left(\frac{1}{3enV}\right)\frac{dE_{ph}}{dT} = \frac{C_{ph}}{3en}$$

where C_{ph} is the phonon heat capacity per unit volume. Although this is only a rough approximate derivation, it predicts the correct temperature dependence, in that the phonondrag contribution is important at temperatures where the phonon specific heat is large. The total thermopower is a sum of the diffusion contribution (considered in §5.4.1) and the phonon drag term S_{ph}.

The phonon drag effect depends on the electron-phonon coupling; at higher temperatures where the phonon-phonon coupling (Umklapp processes) becomes more important than the electron-phonon coupling, phonon drag effects become less important.

Chapter 3

Electron and Phonon Scattering

ELECTRON SCATTERING

The transport properties of solids depend on the availability of carriers and on their scattering rates. Electron scattering brings an electronic system which has been subjected to external perturbations back to equilibrium. Collisions also alter the momentum of the carriers as the electrons are brought back into equilibrium. Electron collisions can occur through a variety of mechanisms such as electron-phonon, electron-impurity, electron-defect, and electron-electron scattering processes.

In principle, the collision rates can be calculated from scattering theory. To do this, we introduce a transition probability $S(\vec{k},\vec{k}')$ for scattering from a state $\vec{k}$ to a state $\vec{k}'$. Since electrons obey the Pauli principle, scattering will occur from an occupied to an unoccupied state. The process of scattering from $\vec{k}$ to $\vec{k}'$ decreases the distribution function $f(\vec{r},\vec{k},t)$ and depends on the probability that $\vec{k}$ is occupied and that $\vec{k}'$ is unoccupied. The process of scattering from $\vec{k}'$ to ~k increases $f(\vec{r},\vec{k},t)$ and depends on the probability that state $\vec{k}'$ is occupied and state $\vec{k}$ is unoccupied. We will use the following notation for describing the scattering process:

- f_k is the probability that an electron occupies a state ~k ² [1 ¡ f_k] is the probability that state $\vec{k}$ is unoccupied
- $S(\vec{k},\vec{k}')$ is the probability per unit time that an electron in state $\vec{k}$ will be scattered to state $\vec{k}'$
- $S(\vec{k}',\vec{k})$ is the probability per unit time that an electron in state $\vec{k}'$ will be scattered back into state $\vec{k}$.

Using these Definitions, the rate of change in the distribution function due to collisions can be written as:

$$\left.\frac{\partial f(\vec{r},\vec{k},t)}{\partial t}\right|_{\text{collisions}} = \int d^3k'[f_{k'}(1-f_k)S(\vec{k}',\vec{k}) - f_k(1-f_{k'})S(\vec{k},\vec{k}')]$$

where d^3k' is a volume element in $\vec{k}'$ space. The integration in equation is over k space and the spherical coordinate system, together with the arbitrary force $\vec{F}$ responsible for the scattering event that introduces a perturbation

$$f_{k'} = f_{0\vec{k}} + \frac{\partial f_{0\vec{k}}}{\partial E}\frac{\hbar}{m^*}\vec{k}\cdot\vec{F} + \cdots$$

Using Fermi's Golden Rule for the transition probability per unit time between states $\vec{k}$ and $\vec{k}'$ we can write

$$S(\vec{k},\vec{k}') \simeq \frac{2\pi}{\hbar}\left|H_{\vec{k}\vec{k}'}\right|^2 \{\delta[E(\vec{k})] - \delta[E(\vec{k}')]\}$$

where the matrix element of the Hamiltonian coupling states $\vec{k}$ and $\vec{k}'$ is

$$H_{\vec{k}\vec{k}'} = \frac{1}{N}\int_V \psi^*_{\vec{k}}(\vec{r})\nabla V \psi_{\vec{k}'}(\vec{r})\, d^3r$$

in which N is the number of unit cells in the sample and ∇V is the perturbation Hamiltonian responsible for the scattering event associated with the force $\vec{F}$.

At equilibrium $f_k = f_0(E)$ and the principle of detailed balance applies

$$S(\vec{k}',\vec{k})f_0(E')[1-f_0(E)] = S(\vec{k},\vec{k}')\, f_0(E)[1-f_0(E')]$$

so that the distribution function does not experience a net change via collisions when in the equilibrium state

$$\left.(\partial f(\vec{r},\vec{k},t)/\partial t)\right|_{\text{collisions}} = 0.$$

We define collisions as elastic when $E(\vec{k}') = E(\vec{k})$ and in this case $f_0(E') = f_0(E)$ so that $S(\vec{k}',\vec{k}) = S(\vec{k},\vec{k}')$ Collisions for which $E(\vec{k}') \neq E(\vec{k})$ are termed inelastic. The term quasi-elastic is used to characterize collisions where the percentage change in energy is small. For our purposes here, we shall consider $S(\vec{k},\vec{k}')$ as a known function which can be calculated quantum mechanically by a detailed consideration of the scattering mechanisms which are important for a given practical case; this statement is true in principle, but in practice $S(\vec{k},\vec{k}')$ is usually specified in an approximate way.

The return to equilibrium depends on the frequency of collisions and the effectiveness of a scattering event in randomizing the motion of the electrons. Thus, small angle scattering is not as effective in restoring a system to equilibrium as large angle scattering. For this reason we distinguish between τ_D, the time for the system to be restored to equilibrium, and τ_c, the time between collisions. These times are related by

$$\tau_D = \frac{\tau_c}{1-\cos\theta}$$

where θ is the mean change of angle of the electron on collision. The time τD is the quantity which enters into Boltzmann's equation while 1/τc determines the actual scattering rate.

The mean free time between collisions, τ_c, is related to several other quantities of interest: the mean free path l_f, the scattering cross section σ_d, and the concentration of scattering centers N_c by

$$\tau_c = \frac{1}{N_c \sigma_d v}$$

where v is the drift velocity given by

$$v = \frac{l_f}{\tau_c} = \frac{1}{N_c \sigma_d \tau_c}$$

and is in the direction of the electron transport. The drift velocity is of course very much smaller in magnitude than the instantaneous velocity of the electron at the Fermi level, which is typically of magnitude $v_F \sim 10^8$ cm/sec. Electron scattering centers include phonons, impurities, dislocations, the crystal surface, etc.

The most important electron scattering mechanism for both metals and semiconductors is electron-phonon scattering (scattering of electrons by the thermal motion of the lattice), though the scattering process for metals differs in detail from that in semiconductors.

In the case of metals, much of the Brillouin zone is occupied by electrons, while in the case of semiconductors, most of the Brillouin zone is unoccupied and represents states into which electrons can be scattered. In the case of metals, electrons are scattered from one point on the Fermi surface to another point, and a large change in momentum occurs, corresponding to a large change in $\vec{k}$. In the case of semiconductors, changes in wave vector from $\vec{k}$ to $-\vec{k}$ normally correspond to a very small change in wave vector, and thus changes from $\vec{k}$ to $-\vec{k}$ can be accomplished much more easily in the case of semiconductors. By the same token, small angle scattering (which is not so efficient for returning the system to equilibrium) is especially important for semiconductors where the change in wavevector is small. Since the scattering processes in semiconductors and metals are quite different, they will be discussed separately below. Scattering probabilities for more than one scattering process are additive and therefore so are the reciprocal scattering time or scattering rates:

$$(\tau^{-1})_{\text{total}} = \sum_i \tau_i^{-1}$$

since $1/\tau$ is proportional to the scattering probability. Metals have large Fermi wavevectors k_F, and therefore large momentum transfers Δk can occur as a result of electronic collisions. In contrast, for semiconductors, k_F is small and so also is Δk on collision.

SCATTERING PROCESSES IN SEMICONDUCTORS

Electron-Phonon Scattering

Electron-phonon scattering is the dominant scattering mechanism in crystalline semiconductors except at very low temperatures. Conservation of energy in the scattering process, which creates or absorbs a phonon of energy $\hbar\omega(\vec{q})$, is written as:

$$E_i - E_f = \pm\hbar\omega(\vec{q}) = \frac{\hbar^2}{2m^*}(k_i^2 - k_f^2),$$

where E_i is the initial energy, E_f is the final energy, k_i the initial wavevector, and k_f the final wavevector. Here, the "+" sign corresponds to the creation of phonons (the phonon emission process), while the "–" sign corresponds to the absorption of phonons. Conservation of momentum in the scattering by a phonon of wavevector $\vec{q}$ yields

$$\vec{k}_i - \vec{k}_f = \pm\vec{q}\,.$$

For semiconductors, the electrons involved in the scattering event generally remain in the vicinity of a single band extremum and involve only a small change in $\vec{k}$ and hence only low phonon $\vec{q}$ vectors participate. The probability that an electron makes a transition from an initial state i to a final state f is proportional to:

(a) The availability of final states for electrons,
(b) The probability of absorbing or emitting a phonon,
(c) The strength of the electron-phonon coupling or interaction.

The first factor, the availability of final states, is proportional to the density of final electron states $\rho(E_f)$ times the probability that the final state is unoccupied. (This occupation probability for a semiconductor is assumed to be unity since the conduction band is essentially empty.) For a parabolic band $\rho(E_f)$ is :

$$\rho(E_f) = \frac{(2m^*)^{3/2} E_f^{1/2}}{2\pi^2\hbar^3} = (2m^*)^{3/2} \frac{[E_i \pm \hbar\omega(\vec{q})]^{1/2}}{2\pi^2\hbar^3},$$

where equation has been employed and the "+" sign corresponds to absorption of a phonon and the "–" sign corresponds to phonon emission.

The probability of absorbing or emitting a phonon is proportional to the electron-phonon coupling $G(\vec{q})$ and to the phonon density $n(\vec{q})$ for absorption, and $[1 + n(\vec{q})]$ for emission, where $n(\vec{q})$ is given by the Bose-Einstein factor

$$n(\vec{q}) = \frac{1}{e^{\frac{\hbar\omega(\vec{q})}{k_B T}} - 1}.$$

Combining the terms in Eqs. 6.13 and 6.14 gives a scattering probability (or $1/\tau_c$) proportional to a sum over final states

$$\frac{1}{\tau_c} \sim \frac{(2m^*)^{3/2}}{2\pi^2\hbar^3} \sum_{\vec{q}} G(\vec{q}) \left[\frac{[E_i + \hbar\omega(\vec{q})]^{1/2}}{e^{\frac{\hbar\omega(\vec{q})}{k_B T}} - 1} + \frac{[E_i - \hbar\omega(\vec{q})]^{1/2}}{1 - e^{\frac{-\hbar\omega(\vec{q})}{k_B T}}} \right]$$

where the first term in the big bracket of equation corresponds to phonon absorption and the second term to phonon emission. If $E_i < \hbar\omega(\vec{q})$, only the phonon absorption process is energetically allowed.

The electron-phonon coupling coefficient $G(\vec{q})$ in equation depends on the electronphonon coupling mechanism. There are three important coupling mechanisms in semiconductors which we briefly describe below: electromagnetic coupling, piezoelectric coupling, and deformation-potential coupling.

Electromagnetic coupling is important only for semiconductors where the charge distribution has different signs on neighboring ion sites. In this case, the oscillatory electric field can give rise to oscillating dipole moments associated with the motion of neighboring ion sites in the LO mode. The electromagnetic coupling mechanism is important in coupling electrons to longitudinal optical phonon modes in III-V and II-VI compound semiconductors, but does not contribute in the case of silicon. To describe the LO mode we can use the Einstein approximation, since $\omega(\vec{q})$ is only weakly dependent on $\vec{q}$ for the optical modes. In this case $\hbar\omega_0 \gg k_B T$ and $\hbar\omega_0 \gg E$, so that from equation the collision rate is proportional to

$$\frac{1}{\tau_c} \sim \frac{m^{*3/2} (\hbar\omega_0)^{1/2}}{e^{\hbar\omega_0 / k_B T} - 1}.$$

Thus, the collision rate depends on the temperature T, the LO phonon frequency ω_0 and the electron effective mass m^*. The corresponding mobility for the optical phonon scattering is

$$\mu = \frac{e\langle\tau\rangle}{m^*} \sim \frac{e(e^{\hbar\omega_0/k_BT} - 1)}{m^{*5/2}(\hbar\omega_0)^{1/2}}$$

Thus for optical phonon scattering, the mobility μ is independent of the electron energy E and decreases with increasing temperature.

As in the case of electromagnetic coupling, piezoelectric coupling is important in semiconductors which are ionic or partly ionic. If these crystals lack inversion symmetry, then acoustic mode vibrations generate regions of compression and rarefaction in a crystal which lead to electric field s. The piezoelectric scattering mechanism is thus associated with the coupling between electrons and phonons arising from these electromagnetic

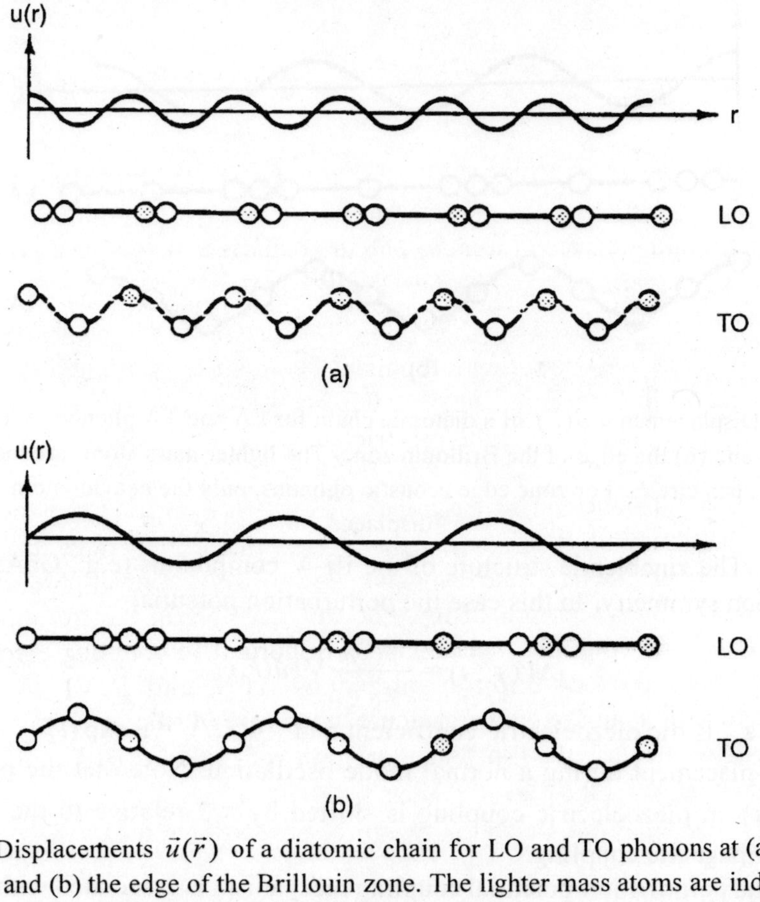

Fig. Displacements $\vec{u}(\vec{r})$ of a diatomic chain for LO and TO phonons at (a) the centre and (b) the edge of the Brillouin zone. The lighter mass atoms are indicated by open circles. For zone edge optical phonons, only the lighter atoms are displaced.

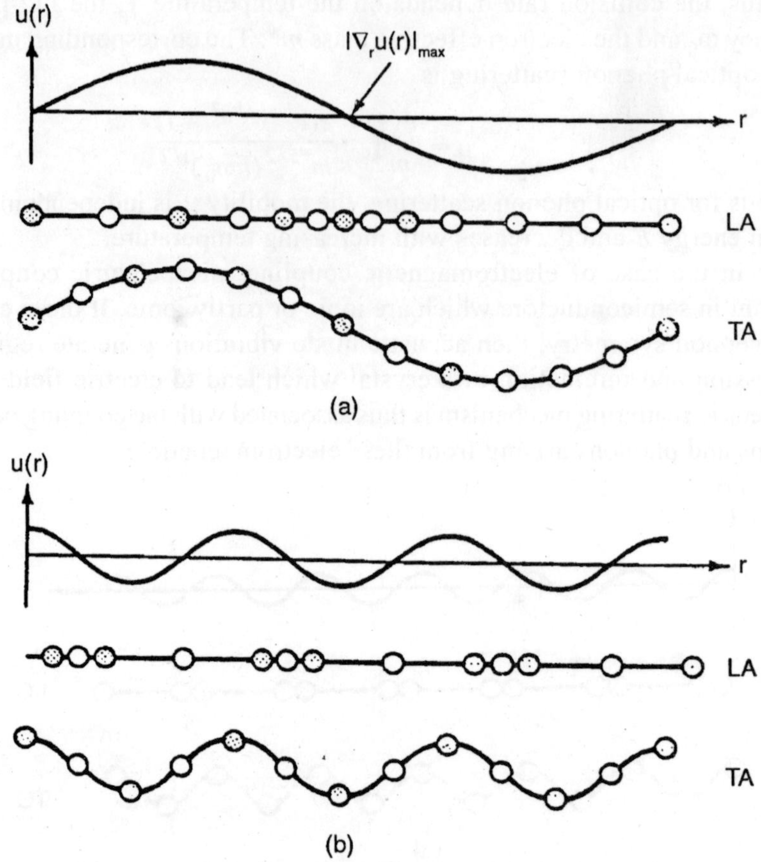

Fig. Displacements $\vec{u}(\vec{r})$ of a diatomic chain for LA and TA phonons at (a) the centre and (b) the edge of the Brillouin zone. The lighter mass atoms are indicated by open circles. For zone edge acoustic phonons, only the heavier atoms are displaced.

fields. The zincblende structure of the III–V compounds (e.g., GaAs) lacks inversion symmetry. In this case the perturbation potential

$$\Delta V(\vec{r}\,,t) = \frac{-ie\varepsilon_{pz}}{\epsilon_0\, q}\vec{\nabla}\cdot\vec{u}(\vec{r},t)$$

where ε_{pz} is the piezoelectric coefficient and $\sim\vec{u}(\vec{r},t) = u\exp(i\vec{q}\cdot\vec{r}-\omega t)$ is the displacement during a normal mode oscillation. Note that the phase of $\Delta V(\vec{r},t)$ in piezoelectric coupling is shifted by $\pi/2$ relative to the case of electromagnetic coupling.

The deformation-potential coupling mechanism is associated with energy shifts of the energy band extrema caused by the compression and rarefaction of crystals during acoustic mode vibrations. The deformation potential scattering mechanism is important in crystals like silicon which have inversion symmetry (and hence no piezoelectric scattering coupling) and have the same species on each site (and hence no electromagnetic coupling). The longitudinal acoustic modes are important for phonon coupling in *n*-type Si and Ge where the conduction band minima occur away from $\vec{k} = 0$.

For deformation potential coupling, it is the LA acoustical phonons that are most important though contributions by LO optical phonons still make a contribution. For the acoustic phonons, we have the condition $\hbar\omega \ll k_B T$ and $\hbar\omega \ll E$, while for the optical phonons it is usually the case that $\hbar\omega \ll k_B T$ at room temperature. For the range of acoustic phonon modes of interest, $G(\vec{q}) \sim q$, where q is the phonon wave vector and ! $\sim q$ for acoustic phonons. Furthermore for the LA phonon branch the phonon absorption process will depend on $n(q)$ in accordance with the Bose factor

$$\frac{1}{e^{\hbar\omega / k_B T} - 1} \simeq \frac{1}{\left[1 + \dfrac{\hbar\omega}{k_B T} + \cdots\right] - 1} \sim \frac{k_B T}{\hbar\omega} \sim \frac{k_B T}{q},$$

while for phonon emission

$$\frac{1}{1 - e^{\hbar\omega / k_B T}} \simeq \frac{1}{\left[1 - \dfrac{\hbar\omega}{k_B T} + \cdots\right] - 1} \sim \frac{k_B T}{\hbar\omega} \sim \frac{k_B T}{q} .$$

Therefore, in considering both phonon absorption and phonon emission, the factors

$$G(\vec{q})[e^{\hbar\omega / k_B T} - 1]^{-1}$$

and

$$G(\vec{q})[1 - e^{-\hbar\omega / k_B T}]^{-1}$$

are independent of *q* for the *LA* branch. Consequently for the acoustic phonon scattering process, the carrier mobility μ decreases with increasing *T* according to

$$\mu = \frac{e\langle\tau\rangle}{m^*} \sim m^{*-5/2}\, E^{-1/2} (k_B T)^{-1} .$$

For the optical LO contribution, we have $G(\vec{q})$ independent of $\vec{q}$ but an $E^{1/2}$ factor is introduced by equation for both phonon absorption and emission, leading to the same basic dependence as given by equation . Thus, we find that the temperature and energy dependence of the mobility μ is different for the various electron-phonon coupling mechanisms. These differences in the E and T dependences can thus be used to identify which scattering mechanism is dominant in specific semiconducting samples. Furthermore, when explicit account is taken of the energy dependence of τ, departures from the strict Drude model $\sigma = ne^2\tau/m^*$ can be expected.

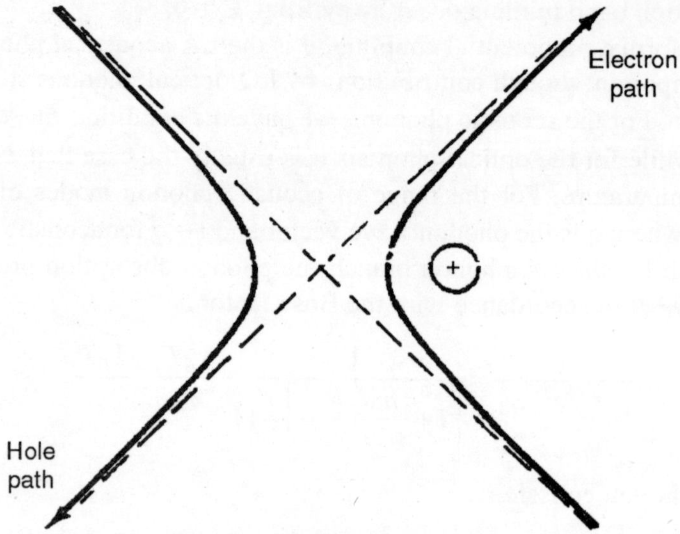

Fig. Trajectories of electrons and holes in ionized impurity scattering. The scattering centre is at the origin.

Ionized Impurity Scattering

As the temperature is reduced, phonon scattering becomes less important so that in this regime, ionized impurity scattering and other defect scattering mechanisms can become dominant. Ionized impurity scattering can also be important in heavily doped semiconductors over a wider temperature range because of the larger defect density. This scattering mechanism involves the deflection of an electron with velocity v by the Coulomb field of an ion with charge Ze, as modified by the dielectric constant ε of the medium and by the screening of the impurity ion by free electrons. Most electrons are scattered through small angles as they are scattered by ionized impurities. The perturbation potential is given by

$$\Delta V(\vec{r}) = \frac{\pm Ze^2}{4\pi\varepsilon_0 r}$$

and the $\pm$ signs denote the different scattering trajectories for electrons and holes. In equation the screening of the electron by the semiconductor environment is handled by the static dielectric constant of the semiconductor

$^2{}_0$. Because of the long-range nature of the Coulomb interaction, screening by other free carriers and by other ionized impurities could be important. Such screening effects are further discussed in §6.2.4.

The scattering rate $1/\tau_I$ due to ionized impurity scattering is given to a good approximation by the Conwell–Weisskopf formula

$$\frac{1}{\tau_I} \sim \frac{Z^2 N_I}{m^{*1/2} E^{3/2}} \ell n \left\{ 1 + \left[\frac{4\pi\varepsilon E}{Z e^2 N_I^{1/3}} \right]^2 \right\}$$

in which N_I is the ionized charged impurity density. The Conwell–Weisskopf formula works quite well for heavily doped semiconductors. We note here that $\tau_I \sim E^{3/2}$, so that it is the low energy electrons that are most effected by ionized impurity scattering.

Neutral impurities also introduce a scattering potential, but it is much weaker than that for the ionized impurity. Free carriers can polarize a neutral impurity and interact with the resulting dipole moment, or can undergo an exchange interaction. In the case of neutral impurity scattering, the perturbation potential is given by

$$\Delta V(\vec{r}) \simeq \frac{\hbar^2}{m^*} \left(\frac{r_B}{r^5} \right)^{1/2}$$

where r_B is the ground state Bohr radius of the electron in a doped semiconductor and r is the distance of the electron to the impurity scattering centre.

Other Scattering Mechanisms

Other scattering mechanisms in semiconductors include:

(a) Neutral impurity centers—these make contributions at very low temperatures, and are mentioned in §6.2.2.

(b) Dislocations — these defects give rise to anisotropic scattering at low temperatures.

(c) Boundary scattering by crystal surfaces — this scattering becomes increasingly important the smaller the crystal size.

(d) Intervalley scattering from one equivalent conduction band minimum to another. This scattering process requires a phonon with large q and consequently results in a relatively large energy transfer.

(e) Electron-electron scattering – similar to charged impurity scattering. This mechanism can be important in distributing energy and momentum among the electrons in the solid and thus can act in conjunction with other scattering mechanisms in establishing equilibrium.

(f) Electron-hole scattering — depends on having both electrons and holes present. Because the electron and hole motions induced by an applied electric field are in opposite directions, electron-hole scattering tends to reverse the direction of the incident electrons and holes.

Screening Effects in Semiconductors

In the vicinity of a charged impurity or an acoustic phonon, charge carriers are accumulated or depleted by the scattering potential, giving rise to a charge density

$$\rho(\vec{r}) = e[n(\vec{r}) - p(\vec{r}) + N_a^-(\vec{r}) - N_d^+(\vec{r})] = en^*(\vec{r})$$

where $n(\vec{r}), p(\vec{r}), N_a^-(\vec{r}), N_d^+(\vec{r})$, and $n^*(\vec{r})$ are, respectively, the electron, hole, ionized acceptor, ionized donor, and effective total carrier concentrations as a function of distance r to the scatterer. We can then write expressions for these quantities in terms of their excess charge above the uniform potential in the absence of impurities:

$$n(\vec{r}) = n + \delta n(\vec{r})$$

$$N_d^+(\vec{r}) = N_d^+ + \delta N_d^+(\vec{r}),$$

and similarly for the holes and acceptors. The space charge $\rho(\vec{r})$ is related to the perturbing potential by Poisson's equation

$$\nabla^2\phi(\vec{r}) = -\frac{\rho(\vec{r})}{\varepsilon_0}.$$

Approximate relations for the excess concentrations are

$$\delta n(\vec{r})/n \simeq -e\phi(\vec{r})/(k_B T)$$

$$\delta N_d^+(\vec{r})/N_d^+ \simeq -e\phi(\vec{r})/(k_B T)$$

and similar relations for the holes. Substitution of equation into

$$\nabla^2\phi(\vec{r}) = \frac{n^* e^2}{\epsilon_0\, k_B T}\phi(\vec{r}).$$

We define an effective Debye screening length λ such that

$$\lambda^2 = \frac{\epsilon_0\, k_B T}{n^* e^2}$$

For a spherically symmetric potential equation becomes

$$\frac{d^2}{dr^2}(r\phi(r)) = \frac{r\phi(r)}{\lambda^2}$$

which yields a solution

$$\phi(r) = \frac{Ze^2}{4\pi \in_0 r} e^{-e/\lambda} .$$

Thus, the screening effect produces an exponential decay of the scattering potential $\phi(r)$ with a characteristic length λ that depends through equation on the effective electron concentration. When the concentration gets large, λ decreases and screening becomes more effective.

When applying screening effects to the ionized impurity scattering problem, we Fourier expand the scattering potential to take advantage of the overall periodicity of the lattice

$$\Delta V(\vec{r}) = \sum_G A_G \exp(i\vec{G}\cdot\vec{r})$$

where the Fourier coefficients are given by

$$A_G = \frac{1}{V}\int_V \tilde{N}V(\vec{r})\exp(-i\vec{G}\cdot\vec{r})d^3r$$

and the matrix element of the perturbation Hamiltonian in equation becomes

$$H_{\vec{k},\vec{k}'} = \frac{1}{N}\sum_G \int_V e^{-i\vec{k}\cdot\vec{r}} u_k^*(r) A_G e^{-i\vec{G}\cdot\vec{r}} u_{k'}(r) d^3r .$$

We note that the integral in equation vanishes unless $\vec{k} - \vec{k}' = \vec{G}$ so that

$$H_{\vec{k},\vec{k}'} = \frac{A_G}{N}\int_V u_k^*(r) u_{k'}(r) d^3r$$

within the first Brillouin zone so that for parabolic bands $u_k(\vec{r}) = u_{k'}(\vec{r})$ and

$$H_{\vec{k},\vec{k}'} = A_{\vec{k}-\vec{k}'} .$$

Now substituting for the screening potential in equation we obtain

$$A_G = \frac{Ze^2}{4\pi \in_0 V}\int_V \exp(-i\vec{G}\cdot\vec{r})d^3r$$

where $d^3r = r^2 \sin\theta d\theta d\phi dr$ so that the angular integration gives 4π and the spatial integration gives

$$A_G = \frac{Ze^2}{\in_0 V|\vec{G}|^2}$$

and

$$H_{\vec{k},\vec{k}'} = \frac{Ze^2}{\in_0 V|\vec{k}-\vec{k}'|^2} .$$

When screening is included in considering the ionized impurity scattering mechanism the integration becomes

$$A_G = \frac{Ze^2}{4\pi \in_0 V} \int_V e^{-r/\lambda} e^{-i\vec{G}\cdot\vec{r}} \, d^3r = \frac{Ze^2}{\in_0 V\left[|\vec{G}|^2 + |1/\lambda|^2\right]}$$

and

$$H_{\vec{k},\vec{k}'} = \frac{Ze^2}{\in_0 V\left[|\vec{k}-\vec{k}'|^2 + |1/\lambda|^2\right]}$$

so that screening clearly reduces the scattering due to ionized impurity scattering. The discussion given here also extends to the case of scattering in metals, which is treated below.

Combining the various scattering mechanisms discussed above for semiconductors, the following picture emerges, where the effect of these processes on the carrier mobility is considered. Here it is seen that screening effects are important for carrier mobilities at low temperature.

ELECTRON SCATTERING IN METALS

Basically the same class of scattering mechanisms are present in metals as in semiconductors, but because of the large number of occupied states in the conduction bands of metals, the temperature dependences of the various scattering mechanisms are quite different.

Electron-Phonon Scattering

In metals as in semiconductors, the dominant scattering mechanism is usually electronphonon scattering. In the case of metals, electron scattering is mainly associated with an electromagnetic interaction of ions with nearby electrons, the longer range interactions being screened by the numerous mobile electrons.

For metals, we must therefore consider explicitly the probability that a state $\vec{k}$ is occupied $f_0(\vec{k})$ or unoccupied $[1 - f_0(\vec{k})]$.

The scattering rate is found by explicit consideration of the scattering rate into a state $\vec{k}$ and the scattering out of that state.

$$\left.\frac{\partial f}{\partial t}\right|_{\text{collisions}} \sim \frac{1}{\tau} \simeq$$

$$\sum_q G(\vec{q}) \left\{ \overbrace{[1 - f_0(\vec{k})]\left[\underbrace{f_0(\vec{k}-\vec{q})n(\vec{q})}_{\text{Phonon absorption}} + \underbrace{f_0(\vec{k}+\vec{q})[1+n(\vec{q})]}_{\text{Phonon emission}} \right]}^{\text{Scattering into } \vec{k}} \right.$$

$$-[f_0(\vec{k})]\Bigg[\overbrace{\underbrace{1-f_0(\vec{k}+\vec{q})]n(\vec{q})}_{\text{Phonon absorption}}+\underbrace{1-f_0(\vec{k}-\vec{q})][1+n(\vec{q})]}_{\text{Phonon emission}}}^{\text{Scattering into } \vec{k}}\Bigg]\Bigg\}$$

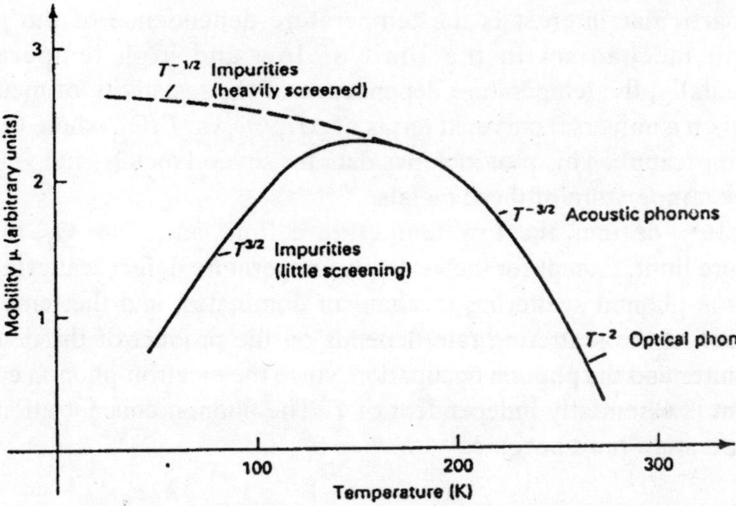

Fig. Typical Temperature Dependence of the Carrier Mobility in Semiconductors.

Here the first term in equation is associated with scattering electrons into an element of phase space at $\vec{k}$ with a probability given by $[1-f_0(\vec{k})]$ that state $\vec{k}$ is unoccupied and has contributions from both phonon absorption processes and phonon emission processes. The second term arises from electrons scattered out of state $\vec{k}$ and here, too, there are contributions from both phonon absorption processes and phonon emission processes. The equilibrium distribution function $f_0(\vec{k})$ for the electron is the Fermi distribution function while the function $n(\vec{q})$ for the phonons is the Bose distribution function . Phonon absorption depends on the phonon density $n(\vec{q})$, while phonon emission depends on the factor $\{1+n(\vec{q})\}$. These factors arise from the properties of the creation and annihilation operators for phonons (to be reviewed in recitation). The density of final states for metals is the density of states at the Fermi level which is consequently approximately independent

Symbol	Metal	$\Theta_D(K)$
⊕	Au	175
Δ	Na	202
⊕	Cu	333
	Al	395
•	Ni	472

of energy and temperature. The condition that, in metals, electron scattering takes place to states near the Fermi level implies that the largest phonon wave vector in an electron collision is $2k_F$ where k_F is the electron wave vector at the Fermi surface.

Of particular interest is the temperature dependence of the phonon scattering mechanism in the limit of low and high temperatures. Experimentally, the temperature dependence of the resistivity of metals can be plotted on a universal curve in terms of $\rho T/\rho\Theta_D$ vs. T/Θ_D where Θ_D is the Debye temperature. This plot includes data for several metals, and values for the Debye temperature of these metals.

$T \ll \Theta_D$ defines the low temperature limit and $T \gg \Theta_D$ the high temperature limit. Except for the very low temperature defect scattering limit, the electron-phonon scattering mechanism dominates, and the temperature dependence of the scattering rate depends on the product of the density of phonon states and the phonon occupation, since the electron-phonon coupling coefficient is essentially independent of T. The phonon concentration in the high temperature limit becomes

$$n(\vec{q}) = \frac{1}{\exp(\hbar\omega / k_B T) - 1} \approx \frac{k_B T}{\hbar\omega}$$

since $(\hbar\omega/k_B T) \ll 1$, so that from equation we have $1/\tau \sim T$ and $\sigma = ne\mu \sim T^{-1}$. In this high temperature limit, the scattering is quasi-elastic and involves large-angle scattering, since phonon wave vectors up to the Debye wave vector q_D are involved in the electron scattering where q_D is related to the Debye frequency ω_D and to the Debye temperature Θ_D according to

$$\hbar\omega_D = k_B\,\Theta_D = \hbar q_D v_q$$

where v_q is the velocity of sound.

We can interpret q_D as the radius of a Debye sphere in $\vec{k}$-space which defines the range of accessible $\vec{q}$ vectors for scattering, i.e., $0 < q < q_D$. The magnitude of q_D is comparable to the Brillouin zone dimensions but the energy change of an electron (ΔE) on scattering by a phonon will be less than $k_B\,\Theta_D$; 1/40eV so that the restriction of $(\Delta E)_{max}$; $k_B\,\Theta_D$ implies that the maximum electronic energy change on scattering will be small compared with the Fermi energy E_F.

We thus obtain that for $T > \Theta_D$ (the high temperature regime), $\Delta E < k_B T$ and the scattering will be quasi-elastic. In the opposite limit, $T \gg \Theta_D$, we have $\hbar\omega_q$; $k_B T$ (because only low frequency acoustic phonons are available for scattering) and in the low temperature limit there is the possibility that $\Delta E > k_B T$, which implies inelastic scattering.

In the low temperature limit, $T \ll \Theta_D$, the scattering is also small-angle scattering, since only low energy (low wave vector) phonons are available for scattering. At low temperature, the phonon density contributes a factor of T^3 to the scattering rate when the sum over phonon states is converted to an integral and $q^2\,dq$ is written in terms of the dimensionless variable $\hbar\omega_q/k_BT$ with $\omega = v_q q$.

Since small momentum transfer gives rise to small angle scattering, the small energy transfer we can write,

$$\left|\vec{k}_i - \vec{k}_f\right| \sim k_f(1-\cos\theta) \approx \frac{1}{2}k_f\theta^2 \approx \frac{1}{2}k_f(q/k_f)^2$$

so that another factor of q^2 appears in the integration over ~q when calculating $(1/\tau_D)$. Thus, the electron scattering rate at low temperature is predicted to be proportional to T^5 so that $\sigma \sim T^{-5}$ (Bloch–Grüneisen formula). Thus, when phonon scattering is the dominant scattering mechanism in metals, the following results are obtained:

$$\sigma \sim \Theta_D/T \qquad T \gg \Theta_D$$
$$\sigma \sim (\Theta_D/T)^5 \qquad T \ll \Theta_D$$

In practice, the resistivity of metals at very low temperatures is dominated by other scattering mechanisms such as impurities, etc., and electron-phonon scattering is relatively unimportant.

The possibility of umklapp processes further increases the range of phonon modes that can contribute to electron scattering in electron-phonon scattering processes. In an umklapp process, a non-vanishing reciprocal lattice vector can be involved in the momentum conservation relation.

The relation between the wave vectors for the phonon and for the incident and scattered electrons $\vec{G} = \vec{k} + \vec{q} + \vec{k}'$ is shown when crystal momentum is conserved for a non-vanishing reciprocal lattice vector $\vec{G}$. Thus, phonons involved in an umklapp process have large wave vectors with magnitudes of about 1/3 of the Brillouin zone dimensions.

Therefore, substantial energies can be transferred on collision through an umklapp process.

At low temperatures, normal scattering processes (i.e., normal as distinguished from umklapp processes) play an important part in completing the return to equilibrium of an excited electron in a metal, while at high temperatures, umklapp processes become more important.

This discussion is applicable to the creation or absorption of a single phonon in a particular scattering event. Since the restoring forces for lattice vibrations in solids are not strictly harmonic, anharmonic corrections to the restoring forces give rise to multiphonon processes where more than one phonon can be created or annihilated in a single scattering event. Experimental evidence for multiphonon processes is provided in both optical and transport studies. In some cases, more than one phonon at the same frequency are created (harmonics), while in other cases, multiple phonons at different frequencies (overtones) are involved.

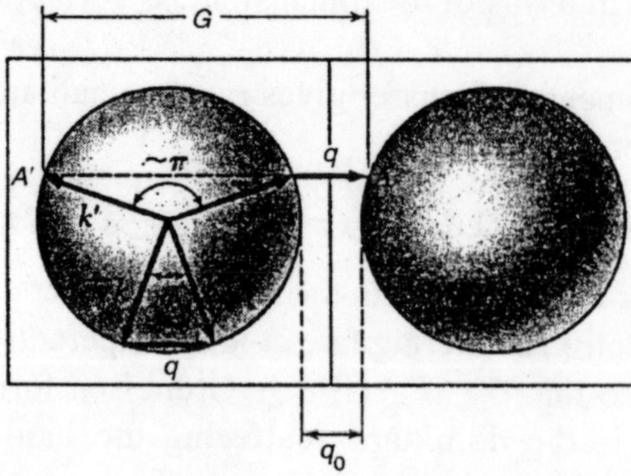

Fig. Schematic diagram showing the relation between the phonon wave vector $\vec{q}$ and the electron wave vectors $\vec{k}$ and $\vec{k}'$ in two Brillouin zones separated by the reciprocal lattice vector $\vec{G}$ (umklapp process).

Other Scattering Mechanisms in Metals

At very low temperatures where phonon scattering is of less importance, other scattering mechanisms become important, and

$$\frac{1}{\tau} = \sum_i \frac{1}{\tau}$$

where the sum is over all the scattering processes.

(a) *Charged imfpurity scattering:* The effect of charged impurity scattering (*Z* being the difference in the charge on the impurity site as compared with the charge on a regular lattice site) is of less importance in metals than in semiconductors because of screening effects by the free electrons.

(b) *Neutral impurities:* This process pertains to scattering centers having the same charge as the host. Such scattering has less effect on the transport properties than scattering by charged impurity sites, because of the much weaker scattering potential.

(c) Vacancies, interstitials, dislocations, size-dependent effects the effects for these defects on the transport properties are similar to those for semiconductors.

For most metals, phonon scattering is relatively unimportant at liquid helium temperatures, so that resistivity measurements at 4K provide a method for the detection of impurities and crystal defects. In fact, in characterizing the quality of a high purity metal sample, it is customary to specify the resistivity ratio ρ(300K)/ρ(4K). This quantity is usually called the residual resistivity ratio (RRR), or the residual resistance ratio. In contrast, a typical semiconductor is characterized by its conductivity and Hall coefficient at room temperature and at 77 K.

PHONON SCATTERING

Whereas electron scattering is important in electronic transport properties, phonon scattering is important in thermal transport, particularly for the case of insulators where heat is carried mainly by phonons. The major scattering mechanisms for phonons are phononphonon scattering, phonon-boundary scattering and defect-phonon scattering.

Phonon-phonon Scattering

Phonons are scattered by other phonons because of anharmonic terms in the restoring potential. This scattering process permits:

- Two phonons to combine to form a third phonon or
- One phonon to break up into two phonons.

In these anharmonic processes, energy and wavevector conservation apply:

$$\vec{q}_1 + \vec{q}_2 = \vec{q}_3 \qquad \text{normal processes}$$

or

$$\vec{q}_1 + \vec{q}_2 = \vec{q}_3 + \vec{Q} \qquad \text{umklapp processes}$$

where $\vec{Q}$ corresponds to a phonon wave vector of magnitude comparable to that of reciprocal lattice vectors. When umklapp processes are present, the scattered phonon wavevector $\vec{q}_3$ can be in a direction opposite to the energy flow, thereby giving rise to thermal resistance. Because of the high momentum transfer and the large phonon energies that are involved, umklapp processes dominate the thermal conductivity at high T.

The phonon density is proportional to the Bose factor so that the scattering rate is proportional to

$$\frac{1}{\tau_{ph}} \sim \frac{1}{(e^{\hbar\omega/(k_B T)} - 1)}$$

At high temperatures $T \gg \Theta_D$, the scattering time thus varies as T^{-1} since

$$\tau_{ph} \sim (e^{\hbar\omega/k_B T} - 1) \sim \hbar\omega/(k_B T)$$

while at low temperatures $T \sim \Theta_D$, an exponential temperature dependence for τ_{ph} is found

$$\tau_{ph} \sim e^{\hbar\omega/k_B T} - 1.$$

These temperature dependences are important in considering the lattice contribution to the thermal conductivity.

Phonon-Boundary Scattering

Phonon-boundary scattering is important at low temperatures where the phonon density is low. In this regime, the scattering time is independent of T. The thermal conductivity in this range is proportional to the phonon density which is in turn proportional to T^3. This effect combined with phonon-phonon scattering results in a thermal conductivity · for insulators with the general shape. The lattice thermal conductivity follows the relation

$$\kappa_L = C_p v_q \Theta_{ph}/3$$

where the phonon mean free path Λ_{ph} is related to the phonon scattering probability $(1/\tau_{ph})$ by

$$\tau_{ph} = \Lambda_{ph}/v_q$$

in which v_q is the velocity of sound and C_p is the heat capacity at constant pressure. Phonon-boundary scattering becomes more important as the crystallite size decreases.

Defect-Phonon Scattering

Defect-phonon scattering includes a variety of crystal defects, charged and uncharged impurity sites and different isotopes of the host constituents. The thermal conductivity, the scattering effects due to different isotopes of Li.

The low mass of Li makes it possible to see such effects clearly. Isotope effects are also important in graphite and diamond which have the highest thermal conductivity of any solid.

Electron-Phonon Scattering

If electrons scatter from phonons, the reverse process also occurs. When phonons impart momentum to electrons, the electron distribution is affected. Thus, the electrons will also carry energy as they are dragged also by the stream of phonons. This phenomenon is called phonon drag. In the case of phonon drag we must simultaneously solve the Boltzmann equations for the electron and phonon distributions which are coupled by the phonon drag term.

TEMPERATURE DEPENDENCE OF THE ELECTRICAL AND THERMAL CONDUCTIVITY

For the electrical conductivity, at very low temperatures, impurity, defect, and boundary scattering dominate. In this regime σ is independent of temperature. At somewhat higher temperatures but still far below Θ_D the electrical conductivity for metals exhibits a strong temperature dependence

$$\sigma \propto (\Theta_D/T)^5 \qquad T \ll \Theta_D.$$

At higher temperatures where $T \gg \Theta_D$, scattering by phonons with any q vector is possible and the formula

$$\sigma \sim (\Theta_D/T) \qquad T \gg \Theta_D$$

applies. We now summarize the corresponding temperature ranges for the thermal conductivity.

Although the thermal conductivity was formally discussed in, a meaningful discussion of the temperature dependence of κ depends on scattering processes. The total thermal conductivity κ in general depends on the lattice and electronic contributions, κ_L and κ_e, respectively. The temperature dependence of the lattice contribution is discussed in with regard to the various phonon scattering processes and their temperature dependence. For the electronic contribution, we must consider the temperature dependence of the electron scattering processes discussed. At very low temperatures, in the impurity scattering range, σ is independent of T and the same scattering processes apply for both the electronic thermal conductivity and the electrical conductivity so that $\kappa_e \propto T$ in the impurity scattering regime where σ ~ constant and the Wiedemann–Franz law is applicable. Copper, defect and boundary scattering are dominant below ~ 20 K, while phonon scattering becomes important at higher *T*.

At low temperatures $T \ll \Theta_D$, but with T in a regime where phonon scattering has already become the dominant scattering mechanism, the thermal transport depends on the electron-phonon collision rate which in turn is proportional to the phonon density. At low temperatures the phonon density is proportional to T^3. This follows from the proportionality of the phonon density of states arising from the integration of $\int q^2 dq$, and from the dispersion relation for the acoustic phonons $\omega = qv_q$

$$q = \omega/v_q = xkT/\hbar v_q$$

where $x = \hbar\omega/k_BT$. Thus in the low temperature range of phonon scattering where $T \ll \Theta_D$ and the Wiedemann–Franz law is no longer satisfied, the temperature dependence of τ is found from the product $T\ (T^{-3})$ so that $\kappa_e \propto T^{-2}$. One reason why the Wiedemann–Franz law is not satisfied in this temperature regime is that κ_e depends on the collision rate τ_c while σ depends on the time to reach thermal equilibrium, τ_D. At low temperatures where only low q phonons participate in scattering events the times τ_c and τ_D are not the same. At high *T* where $T \gg \Theta_D$ and the Wiedemann–Franz law applies, κ_e approaches a constant value corresponding to the regime where σ is proportional to $1/T$. This occurs at temperatures much higher than those shown. The decrease in κ above the peak value at ~17 K follows a $1/T^2$ dependence quite well. In addition to the electronic thermal conductivity, there is heat flow due to lattice vibrations. The phonon thermal conductivity mechanism is in fact the principal mechanism operative in semiconductors and insulators, since the electronic contribution in this case is negligibly small. Since κ_L contributes also to metals the total measured thermal conductivity for metals should exceed the electronic contribution $(\pi^2 k_B^2 T\sigma)/(3e^2)$. In good metallic conductors of high purity, the electronic thermal conductivity dominates and the phonon contribution tends to be small. On the other hand, in conductors where the thermal conductivity due to phonons makes a significant contribution to the total thermal conductivity, it is necessary to separate the electronic and lattice contributions before applying the Wiedemann–Franz law to the total κ. With regard to the lattice contribution, κ_L at very low temperatures is dominated by defect and boundary scattering processes. From the relation

$$\kappa_{\mathrm{L}} = \frac{1}{3} C_p v_q \Lambda_{\mathrm{ph}}$$

we can determine the temperature dependence of κ_L, since $C_p \sim T^3$ at low T while the sound velocity v_q and phonon mean free path Λ_{ph} at very low T are independent of T. In this regime the number of scatterers is independent of T.

In the regime where only low q phonons contribute to transport and to scattering, only normal scattering processes contribute. In this regime C_p is still increasing as T^3, v_q is independent of T, but $1/\Lambda_{\mathrm{ph}}$ increases in proportion to the phonon density of states. With increasing T, the temperature dependence of C_p becomes less pronounced and that for Θ_{ph} becomes more pronounced as more scatters participate, leading eventually to a decrease in κ_L.

We note that it is only the inelastic collisions that contribute to the decrease in Λ_{ph}, since elastic phonon-phonon scattering has a similar effect as impurity scattering for phonons. The inelastic collisions are of course due to anharmonic forces.

Eventually phonons with wavevectors large enough to support umklapp processes are thermally activated. Umklapp processes give rise to thermal resistance and in this regime κ_L decreases as $\exp(-\Theta_D/T)$. In the high temperature limit $T \gg \Theta_D$, the heat capacity and phonon velocity are both independent of T. The $\kappa_L \sim 1/T$ dependence arises from the $1/T$ dependence of the mean free path, since in this limit the scattering rate becomes proportional to $k_B T$.

Chapter 4

Magneto-transport Phenomena

Since the electrical conductivity is sensitive to the product of the carrier density and the carrier mobility rather than each of these quantities independently, it is necessary to look for different transport techniques to provide information on the carrier density n and the carrier mobility μ separately.

Magneto-transport provides us with such techniques, at least for simple cases, since the magnetoresistance is sensitive to the carrier mobility and the is sensitive to the carrier density. We return to the discussion of magneto-transport for lower dimensional systems later in the course particularly with regard to the quantum and giant magnetoresistance.

MAGNETO-TRANSPORT IN THE CLASSICAL REGIME ($\omega_c\tau < 1$)

The magnetoresistance and measurements, which are used to characterize semiconductors, are made in the weak magnetic field limit $\omega_c\tau \ll 1$ where the cyclotron frequency is given by

$$\omega_c = eB/(m^*c).$$

The cyclotron frequency ω_c is the angular frequency of rotation of a charged particle as it makes an orbit in a plane perpendicular to the magnetic field.

In the low field limit (defined by $\omega_c\tau \ll 1$) the carriers are scattered long before com pleting a single cyclotron orbit in real space, so that quantum effects are unimportant. In higher field s where $\omega_c\tau > 1$, quantum effects become important. In this limit (discussed in Part III of this course), the electrons complete cyclotron orbits and the resonance achieved by tuning the microwave frequency of a resonant cavity to coincide with ω_c allows us to measure the effective mass of electrons in semiconductors.

A simplified version of the magnetoresistance phenomenon can be obtained in terms of the $\vec{F} = m\vec{a}$ approach and is presented in §7.1.1. The virtue of the simplified approach is to introduce the concept of the Hall field and the general form of the magneto-conductivity tensor. A more general version of these results will then be given using the Boltzmann equation formulation. The advantage of the more general derivation is to put the derivation on a firmer foundation and to distinguish between the various effective masses which enter the transport equations: the cyclotron effective mass of equation the longitudinal effective mass along the magnetic field direction, and the dynamical effective mass which describes transport in an electric field.

CLASSICAL MAGNETO-TRANSPORT EQUATIONS

For the simplified $\vec{F} = m\vec{a}$ treatment, let the magnetic field $\vec{B}$ be directed along the *z* direction. Then writing $\vec{F} = m\vec{a}$ for the electronic motion in the plane perpendicular to $\vec{B}$ we obtain

$$\vec{F} = e(\vec{E} + \vec{v} \times \vec{B}/c) = m^* \dot{\vec{v}} + m^* \vec{v}/\tau$$

where $m^* \vec{v}/\tau$ is introduced to account for damping or electron scattering. For static electric and magnetic field s, there is no time variation in the problem so that $\dot{\vec{v}} = 0$ and thus the equation of motion reduces to

$$m^* v_x/\tau = e(E_x + v_y B/c)$$
$$m^* v_y/\tau = e(E_y - v_x B/c)$$

which can be written as

$$\frac{m^*}{\tau}(v_x + iv_y) = e(E_x + iE_y) - i\frac{eB}{c}(v_x + iv_y)$$

where *i* is the unit imaginary, so that $j_x + ij_y = ne(v_x + iv_y)$ becomes

$$(j_x + ij_y) = \left(\frac{ne^2\tau}{m^*}\right)\frac{(E_x + iE_y)}{1 + i\omega_c\tau}$$

where the cyclotron frequency is defined by equation The unit imaginary *i* is introduced into and because of the circular motion of the electron orbit in a magnetic field, suggesting circular polarization for field s and velocities.

Equating real and imaginary parts of yields,

$$j_x = \left(\frac{ne^2\tau}{m^*}\right)\left[\frac{E_x}{1 + (\omega_c\tau)} + \frac{\omega_c\tau E_y}{1 + (\omega_c\tau)^2}\right]$$

$$j_y =$$

Since $\vec{v} = v_z\hat{z}$ is parallel to $\vec{B}$ or $(\vec{v} \times \vec{B}) = 0$, the motion of an electron along the magnetic field experiences no force due to the magnetic field, so that

$$j_z = \frac{ne^2\tau}{m^*}E_z$$

Equations yield the magnetoconductivity tensor defined by $\vec{j} = \ddot{\sigma}_B \cdot \vec{E}$ in the presence of a magnetic field in the low field limit where $\omega_c \tau \ll 1$ and the classical approach given here is applicable. In this limit an electron in a magnetic field is accelerated by an electric field and follows Ohm's law (as in the case of zero magnetic field):

$$\vec{j} = \ddot{\sigma}_B \cdot \vec{E}$$

except that the magnetoconductivity tensor $\ddot{\sigma}_B$ depends explicitly on magnetic field and in accordance with Euation assumes the form

$$\ddot{\sigma}_B = \frac{ne^2 \tau / m^*}{1 + (\omega_c \tau)^2} \begin{pmatrix} 1 & \omega_c \tau & 0 \\ -\omega_c \tau & 1 & 0 \\ 0 & 0 & 1 + (\omega_c \tau)^2 \end{pmatrix}.$$

The magnetoresistivity tensor (which is more closely related to laboratory measurements) is defined as the inverse of the magnetoconductivity tensor

$$\ddot{\rho}_B = [\ddot{\sigma}_B]^{-1} = \frac{m^*}{ne^2 \tau} \begin{pmatrix} 1 & -\omega_c \tau & 0 \\ \omega_c \tau & 1 & 0 \\ 0 & 0 & 1 \end{pmatrix}$$

MAGNETORESISTANCE

The magnetoresistance is defined in terms of the diagonal components of the magnetoresistivity tensor given by equation

$$\Delta\rho/\rho \equiv (\rho(B) - \rho(0))/\rho(0)$$

and, in general, depends on $(\omega_c \tau)^2$ or on B^2. Since $\omega_c \tau = (e\tau/m^*c)B = \mu B/c$, the magnetoresistance provides information on the carrier mobility.

The longitudinal magnetoresistivity $\Delta\rho_{zz}/\rho_{zz}$ is measured with the electric field parallel to the magnetic field. On the basis of a spherical Fermi surface one carrier model, we have $E_z = j_z/\sigma_0$ from equation , so that there is no longitudinal magnetoresistivity in this case; that is, the resistivity is the same whether or not a magnetic field is present, since $\sigma_0 = ne^2\tau / m^*$. On the other hand, many semiconductors do exhibit longitudinal magnetoresistivity experimentally, and this effect arises from the non-spherical shape of their constant energy surfaces.

The transverse magnetoresistivity $\Delta\rho_{xx}/\rho_{xx}$ is measured with the current flowing in some direction (x) perpendicular to the magnetic field. With the direction of current flow along the x direction, then $j_y = 0$ and we can write from Eqs. 7.8 and 7.9 as

$$E_y = (\omega_c \tau) E_x$$

so that

$$j_x = \sigma_0 \left[\frac{E_x}{1+(\omega_c\tau)^2} + \frac{(\omega_c\tau)^2 E_x}{1+(\omega_c\tau)^2} \right] = \sigma_0 E_x$$

and again there is no transverse magnetoresistance for a material with a single carrier type having a spherical Fermi surface. Introduction of either a more complicated Fermi surface or more than one type of carrier results in a transverse magnetoresistance.

When the velocity distribution of carriers at a finite temperature is taken into account, a finite transverse magnetoresistance is also obtained. In a similar way, multi-valley semiconductors (having several electron or hole constant energy surfaces some of which are equivalent by symmetry) can also display a transverse magnetoresistance.

In all of these cases the magnetoresistance exhibits a B^2 dependence. The effect of two carrier types on the transverse magnetoresistance is discussed in some detail.

We note that the σ_{xy} and σ_{yx} terms arise from the presence of a magnetic field. The significance of these terms is further addressed in our discussion of the. We note that for non-spherical constant energy surfaces, Equation must be rewritten to reflect the fact that m^* is a tensor so that $\vec{v}$ and $\vec{E}$ need not be parallel, even in the absence of a magnetic field. This point is clarified to some degree in the derivation of magneto-transport effects given in using the Boltzmann Equation.

THE

If an electric current is flowing in a semiconductor transverse to an applied magnetic field, an electric field is generated perpendicular to the plane containing $\vec{j}$ and $\vec{B}$. This is known as the . Because the magnetic field acts to deflect the charge carriers transverse to their current flow, the Hall field is required to ensure that the transverse current vanishes.

Let x be the direction of current flow and z the direction of the magnetic field. Then the boundary condition for the is $j_y = 0$. From the magnetoconductivity tensor equation we have

$$j_y = \frac{ne^2\tau}{m^*}\left(\frac{1}{1+(\omega_c\tau)^2}\right)(E_y - \omega_c\tau E_x)$$

so that a non-vanishing Hall field

$$E_y = \omega_c\tau E_x$$

must be present to ensure the vanishing of j_y. It is convenient to define the Hall coefficient

$$R_{\text{Hall}} \equiv \frac{E_y}{j_x B_z} = \frac{\tau E_x (eB_z / m^* c)}{j_x B_z}.$$

Substitution of the Hall field into the expression for j_x then yields

$$j_x = \frac{ne^2\tau}{m^*[1+(\omega_c\tau)^2]}[E_x + \omega_c \tau E_y] = \frac{ne^2\tau[1+(\omega_c\tau)^2]E_x}{m^*[1+(\omega_c\tau)^2]}$$

or

$$j_x = \frac{ne^2\tau}{m^*}E_x = \sigma_{dc} E_x.$$

Substitution of this expression into the Hall coefficient yields

$$R_{\text{Hall}} = \frac{e\tau}{m^* c(ne^2\tau / m^*)} = \frac{1}{nec}$$

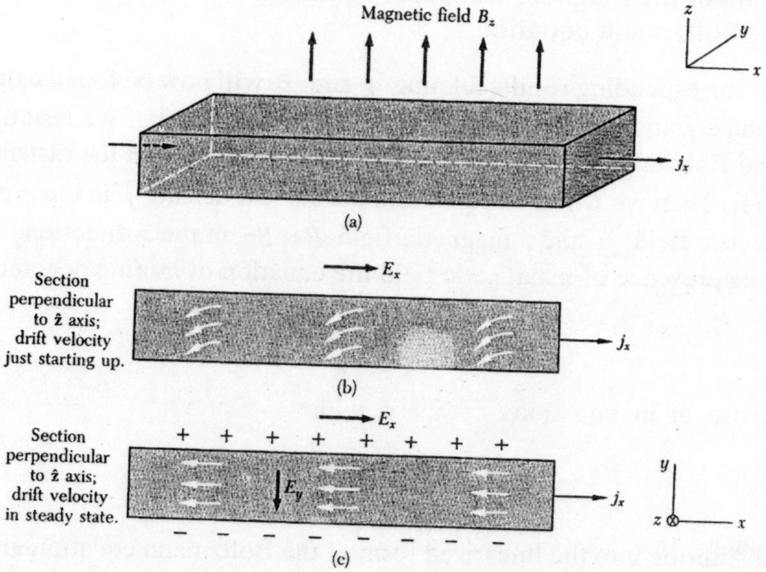

Fig. The standard geometry for the : a specimen of rectangular crosssection is placed in a magnetic field B_z. An electric field E_x applied across the end electrodes causes an electric current density j_x to flow down the bar. The drift velocity of the electrons immediately after the electric field is applied. The deflection in the y direction is caused by the magnetic field. Electrons accumulate on one face of the bar and a positive ion excess is established on the opposite face until, the transverse electric field (Hall field) just cancels the force due to the magnetic field.

The Hall coefficient is important because:

1. R_{Hall} depends only on the carrier density n, aside from universal constants
2. The sign of R_{Hall} determines whether conduction is by electrons ($R_{Hall} < 0$) or by holes ($R_{Hall} > 0$).

If the carriers are of one type we can relate the Hall mobility μ_{Hall} to R_{Hall}:

$$\mu = \frac{e\tau}{m^*} = \left(\frac{ne^2\tau}{m^*}\right)c\left(\frac{1}{nec}\right) = \sigma c R_{Hall}$$

We define

$$\mu_{Hall} \equiv \sigma c R_{Hall}$$

and μ_{Hall} carries the same sign as R_{Hall}. The resistivity component $\rho_{xy}/ ne\mu_{Hall}$ is called the Hall resistivity. A variety of new effects can occur in R_{Hall} when there is more than one type of carrier, as is commonly the case in semiconductors, and this is discussed..

Derivation of the Magneto-transport Equations from the Boltzmann Equation

The corresponding results relating $\vec{j}$ and $\vec{E}$ will now be found using the Boltzmann equation in the absence of temperature gradients. We first use the linearized Boltzmann equation given by equation to obtain the distribution function f_1. Then we will use f_1 to obtain the current density $\vec{j}$ in the presence of an electric field $\vec{E}$ and a magnetic field $\vec{B} = B\hat{z}$ in the *z*-direction.

In the presence of a magnetic field the equation of motion becomes

$$\hbar\dot{\vec{k}} = e\left(\vec{E} + \frac{1}{c}\vec{v}\times\vec{B}\right).$$

We use, as in equation , $f = f_0 + f_1$ with

$$\frac{\partial f_0}{\partial \vec{k}} = \frac{\partial f_0}{\partial E}\frac{\partial E(\vec{k})}{\partial \vec{k}} = \hbar\vec{v}\frac{\partial f_0}{\partial E}.$$

Substituting into the linearized form of the Boltzmann equation gives an equation for f_1:

$$e\left(\vec{E} + \frac{1}{c}\vec{v}\times\vec{B}\right)\cdot\left(\hbar\vec{v}\frac{\partial f_0}{\partial E} + \frac{\partial f_1}{\partial \vec{k}}\right) = -\frac{f_1}{\tau}.$$

In analogy with the case of zero field, we assume a solution for f_1 of the form

$$f_1 = -e\tau\vec{v}\cdot\vec{V}\frac{\partial f_0}{\partial E}$$

with where $\vec{V}$ is a vector to be determined in analogy with the solution for $B = 0$. The form of is motivated by the form suggested by the magnetoconductivity tensor.

For a simple parabolic band, $\vec{v} = \hbar\vec{k}/m*$, and substitution of equation gives

$$\frac{e}{\hbar c}(\vec{v} \times \vec{B}) \cdot \frac{\partial f_1}{\partial \vec{k}} = \frac{e^2\tau}{m^* c}(\vec{v} \times \vec{B}) \cdot \vec{V}\frac{\partial f_0}{\partial E}.$$

The following equation for $\vec{V}$ is then obtained from equation

$$\vec{v} \cdot \vec{E} - \frac{e\tau}{m^* c}(\vec{v} \times \vec{B}) \cdot \vec{V} = \vec{v} \cdot \vec{V}$$

where we have neglected a term $\vec{E} \cdot \vec{V}$ which is small (of order $|\vec{E}|^2$ if $|\vec{E}|$ is small). Equation 7.27 is equivalent to

$$v_x E_x + \omega_c \tau v_x V_y = v_x V_x$$

$$v_y E_y - \omega_c \tau v_y V_x = v_y V_y$$

which can be rewritten more compactly as:

$$\vec{V}_\perp = \left(\vec{E}_\perp - (e\tau/m^* c)[\vec{B} \times \vec{E}_\perp]\right)\left(1 + (e\tau B/m^* c)^2\right)^{-1}$$

$$V_z = E_z$$

where the notation "⊥" in equation denotes the component in the $x - y$ plane, perpendicular to $\vec{B}$. This solves the problem of finding f_1.

Now we can carry out the calculation of $\vec{j}$ in equation , using the new expression for f_1 given by Eqs., and 7.29. With the more detailed calculation using the Boltzmann equation, it is clear that the cyclotron mass governs the cyclotron frequency while the dynamic effective mass controls the coefficients ($ne^2\tau/m^*$) in Eqs. 7.6 and 7.7. We discuss in §7.5 how to calculate the cyclotron effective mass.

TWO CARRIER MODEL

We calculate both the and the transverse magnetoresistance for a two-carrier model. The geometry under which transport measurements are made $(\hat{j} \parallel \hat{x})$ imposes the condition $j_y = 0$. From the magnetoconductivity tensor of

equation $j_y = -\dfrac{\sigma_{01}\beta_1 E_x}{1+\beta_1^2} + \dfrac{\sigma_{01}E_y}{1+\beta_1^2} - \dfrac{\sigma_{02}\beta_2 E_x}{1+\beta_2^2} + \dfrac{\sigma_{02}E_y}{1+\beta_2^2} = 0$

where

$$\beta = \omega_c \tau$$

and the subscripts on σ_{0i} and β_i refer to the carrier index, $i = 1, 2$, so that $\sigma_{0i} = n_i e^2 \tau_i/m^*$ and $\beta_i = \omega_0 c\tau_i$. Solving equation yields a relation between E_y and E_x which defines the Hall field

$$E_y = E_x \left[\frac{\dfrac{\sigma_{01}/\beta_1}{1+\beta_1^2} + \dfrac{\sigma_{02}/\beta_2}{1+\beta_2^2}}{\dfrac{\sigma_{01}}{1+\beta_1^2} + \dfrac{\sigma_{02}}{1+\beta_2^2}} \right]$$

for a two carrier system. This basic equation is applicable to two kinds of electrons, two kinds of holes, or a combination of electrons and holes. The generalization of equation to more than two types of carriers is immediate. The magnetoconductivity tensor is found by substitution of equation into:

$$j_x = \frac{\sigma_{01}/E_x}{1+\beta_1^2} + \frac{\sigma_{01}/\beta_1 E_y}{1+\beta_1^2} + \frac{\sigma_{02}/E_x}{1+\beta_2^2} + \frac{\sigma_{02}/\beta_2 E_y}{1+\beta_2^2}$$

In general equation is a complicated relation, but simplifications can be made in the low field limit $\beta \ll 1$, where we can neglect terms in β^2 relative to terms in β. Retaining the lowest power in terms in β then yields

$$E_y = \left[\frac{\sigma_{01}\beta_1 + \sigma_{02}\beta_2}{\sigma_{01} + \sigma_{02}} \right]$$

$$j_x = (\sigma_{01} + \sigma_{02})E_x.$$

We thus obtain the following important relation for the Hall coefficient which is independent of magnetic field in this limit

$$\mathrm{R}_{\mathrm{Hall}} \equiv \frac{E_y}{j_x B_z} = \frac{\beta_1\sigma_{01} + \beta_2\sigma_{02}}{(\sigma_{01}+\sigma_{02})^2 B} = \frac{\mu_1\sigma_{01} + \mu_2\sigma_{02}}{c(\sigma_{01}+\sigma_{02})^2}$$

where we have made use of the relation between β and the mobility μ

$$\beta = \mu B/c = e\tau B/(m_c^*\, c) = \omega_c\tau.$$

This allows us to write $\mathrm{R}_{\mathrm{Hall}}$ in terms of the Hall coefficients R_i for each of the two types of carriers

$$\mathrm{R}_{\mathrm{Hall}} = \frac{R_1\sigma_{01}^2 + R_2\sigma_{02}^2}{(\sigma_{01}+\sigma_{02})^2}$$

since

$$\frac{\beta}{B} = R_{\mathrm{Hall}}\sigma\,,$$

where for each carrier type we have

$$\sigma_{0i} = \frac{n_i e^2 \tau_i}{m_i^*}$$

and

$$R_i = \frac{1}{n_i e_i c}$$

where $i = 1, 2$. We note in equation that $e_i = \pm|e|$ where $|e|$ is the magnitude of the charge on the electron. Thus electrons and holes contribute with opposite sign to R_{Hall} in equation . When more than one carrier type is present, it is not always the case that the sign of the Hall coefficient is the same as the sign of the majority carrier type. A minority carrier type may have a higher mobility, and the carriers with high mobility make a larger contribution per carrier to R_{Hall} than do the low mobility carriers.

The magnetoconductivity for two carrier types is obtained from equation upon substitution of equation and retaining terms in β^2. For the transverse magnetoconductance we obtain

$$\frac{\sigma_B(B) - \sigma_B(0)}{\sigma_B(0)} = \frac{2\sigma_{01}^2\beta_1^2 + 2\sigma_{02}^2\beta_2^2 + 2\sigma_{01}\sigma_{02}(\beta_1 + \beta_2)^2}{(\sigma_{01}\sigma_{02})^2}$$

which is an average of β_1 and β_2 weighted by conductivity components σ_{01} and σ_{02}. But since $\Delta\rho/\rho = -\Delta\sigma/\sigma$ we obtain the following result for the transverse magnetoresistance

$$\frac{\Delta\rho}{\rho} = \frac{2\sigma_{01}^2\beta_1^2 + 2\sigma_{02}^2\beta_2^2 + \sigma_{01}\sigma_{02}(\beta_1 + \beta_2)^2}{(\sigma_{01}\sigma_{02})^2}$$

We note that the magnetoconductivity tensor yields no longitudinal magnetoresistance for a spherical two-carrier model.

CYCLOTRON EFFECTIVE MASS

To calculate the magnetoresistance and explicitly for non-spherical Fermi surfaces, we need to derive a formula for the cyclotron frequency $\omega_c = eB/(m_c^* c)$ which is generally applicable for non-spherical Fermi surfaces. The cyclotron effective mass can be determined in either of two ways. The first method is the tube integral method for a schematic of the constant energy surfaces at energy E and $E + \Delta E$) which defines the cyclotron effective mass as

$$m_c^* = \frac{1}{2\pi} \oint \frac{\hbar d\kappa}{|v|} = \frac{\hbar^2}{2\pi} \oint \frac{d\kappa}{|\partial E / \partial k|}$$

where $d\kappa$ is an infinitesimal element of length along the contour and we can obtain m_c^* by direct integration. For the second method, we convert the line integral over an enclosed area, making use of $\Delta E = \Delta k(\Delta E/\partial k)$ so that

$$m_c^* = \frac{\hbar^2}{2\pi}\frac{1}{\Delta E}\oint(\Delta k)\,d\kappa = \frac{\hbar^2}{2\pi}\frac{\Delta A}{\Delta E}$$

where ΔA is the area of the strip indicated by the separation ΔE. Therefore we obtain the relation

$$m_c^* = \frac{\hbar^2}{2\pi}\frac{\partial A}{\partial E}$$

which gives the second method for finding the cyclotron effective mass.

For a spherical constant energy surface, we have $A = \pi k^2$ and $E(\vec{k}) = (\hbar^2 k^2)/(2m^*)$ so that $m^* = m^*$. For an electron orbit described by an ellipse in reciprocal space (which is appropriate for the general orbit in the presence of a magnetic field on an ellipsoidal constant energy surface at wave vector k_B along the magnetic field) we write

$$E(\vec{k}_\perp) = \frac{\hbar^2 k_1^2}{2m_1} + \frac{\hbar^2 k_2^2}{2m_2} = \frac{\hbar^2 k_0^2}{2m_0}$$

which defines the area A enclosed by the constant energy surface as

$$A = \pi k_1 k_2 = \pi k_0^2 \sqrt{m_1 m_2 / m_0}$$

where $(k_i^2 / m_i) = (k_0^2 / m_0)$. Then substitution in equation gives

$$m_c^* = \sqrt{m_1 m_2}\,.$$

This expression for m_c^* gives a clear physical picture of the relation between m_c^* and the electron orbit on a constant ellipsoidal energy surface in the presence of a magnetic field. Since finding the electron orbit requires geometrical calculation for a general magnetic field orientation, it is more convenient for computer computation to use the relation

$$m_c^* = \left(\frac{\det \vec{m}^*}{\hat{b}\cdot\vec{m}^*\cdot\hat{b}}\right)^{1/2}$$

for calculating m_c^* for ellipsoidal constant energy surfaces where $\hat{b}\cdot\vec{m}^*\cdot\hat{b}$ is the effective mass component along the magnetic field, det $\vec{m}^*$ denotes the determinant of the effective mass tensor $\vec{m}^*$, and $\hat{b}$ is avector along the magnetic field. One can show that for this case the Hall mobility is given by

$$\mu_{\text{Hall}} = \frac{e\tau}{m_c^*}$$

and

$$\mu_{\text{Hall}} \frac{B}{c} \equiv \omega_c \tau = \beta$$

so that μ_{Hall} involves the cyclotron effective mass.

EFFECTIVE MASSES FOR ELLIPSOIDAL FERMI SURFACES

The effective mass of carriers in a magnetic field is complicated by the fact that several effective mass quantities are of importance. These include the cyclotron effective mass m_c^* for electron motion transverse to the magnetic field and the longitudinal effective mass m_B^* for electron motion along the magnetic field

$$m_B^* = \hat{b} \cdot \overleftrightarrow{m}^* \cdot \hat{b}$$

obtained by projecting the effective mass tensor along the magnetic field. These motions are considered in finding f_1, the change in the electron distribution function due to forces and fields.

Returning to Equation the initial exposition for the current density calculated by the Boltzmann equation, we obtained the Drude formula

$$\overleftrightarrow{\sigma} = ne^2\tau\left(\frac{\overleftrightarrow{1}}{m^*}\right).$$

Thereby defining the drift mass tensor in an electric field. Referring to the magnetoresis tance and magneto-conductance tensors, we see the drift term $(ne^2\tau / m^*)$ which utilizes the drift mass tensor and terms in $(\omega_c\tau)$ which utilizes the effective cyclotron mass m_c^*. Where the Fermi surface for a semiconductor consists of ellipsoidal carrier pockets, then the drift effective mass components are found in accordance with the procedure outlined in for ellipsoidal carrier pockets. We conveniently use equation to determine the cyclotron effective mass for ellipsoidal carrier pockets.

DYNAMICS OF ELECTRONS IN A MAGNETIC FIELD

We relate the electron motion on a constant energy surface in a magnetic field to real space orbits. Consider first the case of $B = 0$. At a given k value, $E(\vec{k})$ and $v(\vec{k})$ are specified and each of the quantities is a constant of the motion, where $v(\vec{k}) = (1/\hbar)[\partial E(\vec{k})/\partial \vec{k}]$. If there are no forces on the system, $E(\vec{k})$ and $v(\vec{k})$ are unchanged with time. Thus, at any instant of time there is an equal probability that an electron will be found anywhere on a constant energy surface. The role of an external electric field $\vec{E}$ is to change the $\vec{k}$ vector on this constant energy surface according to the equation of motion

$$\hbar\dot{\vec{k}} = e\vec{E}$$

so that under a force $e\vec{E}$ the energy of the system is changed.

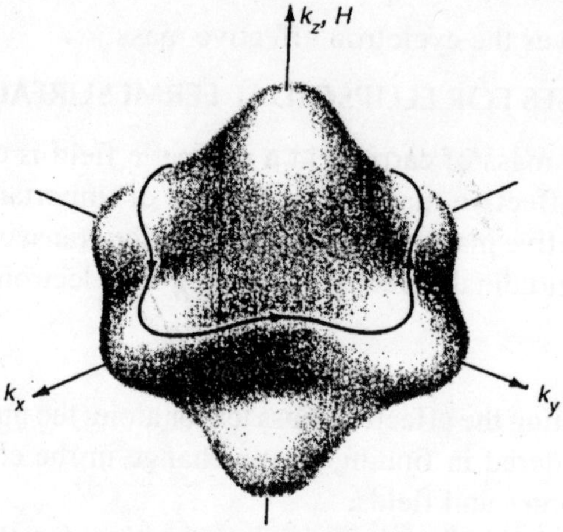

Fig. Schematic diagram of the motion of an electron along a constant energy (and constant k_z) trajectory in the presence of a magnetic field in the z-direction.

In a constant magnetic field (no electric field), the electron will move on a constant energy surface in $\vec{k}$ space in an orbit perpendicular to the magnetic field and following the equation of motion

$$\hbar\dot{\vec{k}} = \frac{e}{c}(\vec{v}\times\vec{B})$$

where we note that $|v_\perp|$ remains unchanged along the electron orbit. The electrons will execute the indicated orbit at a cyclotron frequency ω_c given by $\omega_c = eB/m_c^* c$ where m_c^* is the cyclotron effective mass.

For high magnetic field s, when $\omega_c\tau \gg 1$, an electron circulates many times around its semiclassical orbit before undergoing a collision. In this limit, there is interest in describing the orbits of carriers in $\vec{r}$ space. The solution to the semiclassical equations

$$\dot{\vec{r}} = \vec{v} = (1/\hbar)\partial E/\partial\vec{k}$$

$$\hbar\dot{\vec{k}} = e[\vec{E} + (1/c)\vec{v}\times\vec{B}]$$

is

$$\vec{r}_\perp = \vec{r} - \frac{\vec{B}}{B^2}(\vec{B}\cdot\vec{r})$$

in which $\vec{r} = \vec{r}_\parallel + \vec{r}_\perp$. We also note that

$$\vec{B}\times\hbar\dot{\vec{k}} = e\left[\vec{B}\times\vec{E} + (1/c)\vec{B}\times(\vec{v}\times\vec{B})\right]$$

$$= e\left[\vec{B}\times\vec{E}+(1/c)B^2\vec{v}-(\vec{B}\cdot\vec{v})\vec{B}\right]$$

Making use of Equation, we may write

$$\dot{\vec{r}}_\perp = \frac{c\hbar}{eB^2}\vec{B}\times\dot{\vec{k}}+\frac{c}{B^2}(\vec{E}\times\vec{B})$$

which upon integration yields

$$\vec{r}_\perp(t)-\vec{r}_\perp(0) = \frac{\hbar}{\omega_c B m_c^*}\vec{B}\times[\vec{k}(t)-\vec{k}(0)]+\vec{\omega}t$$

where

$$\vec{\omega} = \frac{c}{B^2}(\vec{E}\times\vec{B}).$$

From equation we see that the orbit in real space is $\pi/2$ out of phase with the orbit in reciprocal space.

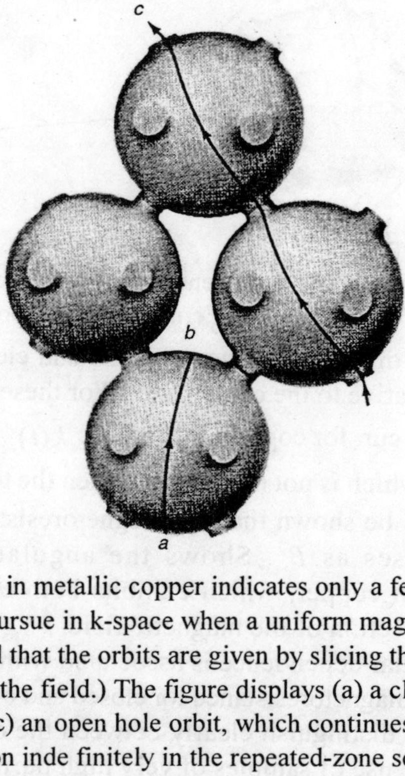

Fig. The electron orbits in metallic copper indicates only a few of the many types of orbits an electron can pursue in k-space when a uniform magnetic field is applied to a noble metal. (Recall that the orbits are given by slicing the Fermi surface with planes perpendicular to the field.) The figure displays (a) a closed electron orbit; (b) a closed hole orbit; (c) an open hole orbit, which continues in the same general direction inde finitely in the repeated-zone scheme.

For the case of closed orbits, after a long time $t\ \pi \approx \tau$. Then the second term $\vec{\omega}\ t$ of equation will dominate, giving a transverse or Hall current

$$\vec{j}_\perp \rightarrow \frac{ne}{B^2}(\vec{E}\times\vec{B})$$

where n is the electron density. Similarly the longitudinal current $\vec{j} \parallel \vec{E}$ will approach a constant value or saturate since $\vec{E} \perp \vec{j}$, and $\vec{E} \perp \vec{B}$.

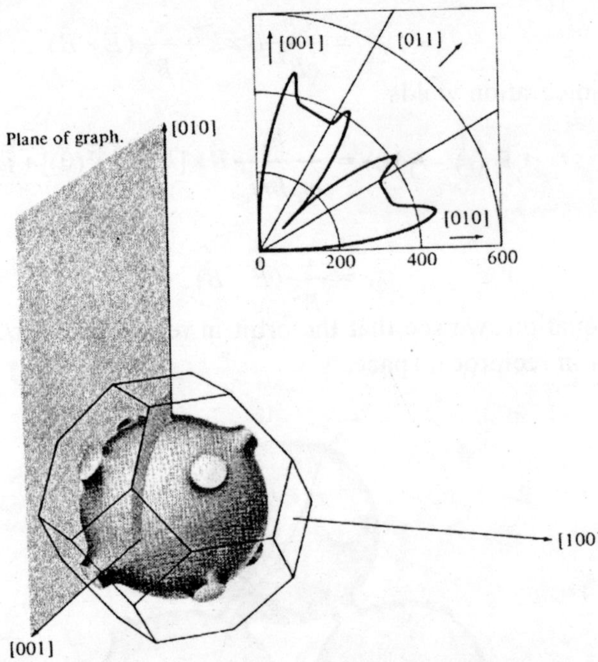

Fig. The spectacular directional dependence of the high- field magnetoresistance in copper is that characteristic of a Fermi surface supporting open orbits.

The situation is very different for the magnetic and electric field s applied in special directions relative to the crystal axes. For these special direction open electron orbits can occur, for copper. In this case $\vec{k}(t)-\vec{k}(0)$ has a component proportional to ~Et which is not negligible. When the term $[\vec{k}(t)-\vec{k}(0)]$ must be considered, it can be shown that the magnetoresistance does not saturate but instead increases as B^2. Shows the angular dependence of the magnetoresistance in copper, which exhibits both closed and open orbits depending on the directi:n of the magnetic field.

The strong angular dependence is associated with the large difference in the magnitude of the magnetoresistance for closed and open orbits. Large values of $\omega_c\tau$ are needed to distinguish clearly between the open and closed orbits, thereby requiring the use of samples of very high purity and low temperature (e.g., 4.2 K) operation. Its magnitude is fixed at 18 kilogauss, and its direction is varied continuously from [001] to [010]. The graph is a polar plot of the transverse magnetoresistance $[\rho(H) - \rho(0)]/\rho(0)$ vs. orientation of the field.

Chapter 5

Transport in Low Dimensional Systems

Transport phenomena in low dimensional systems such as in quantum wells (2D), quantum wires (1D), and quantum dots (0D) are dominated by quantum effects not included in the classical treatments based on the Boltzmann equation.

With the availability of experimental techniques to synthesize materials of high chemical purity and of nanometer dimensions, transport in low dimensional systems has become an active current research area.

OBSERVATION OF QUANTUM EFFECTS IN REDUCED DIMENSIONS

Quantum effects dominate the transport in quantum wells and other low dimensional systems such as quantum wires and quantum dots when the de Broglie wavelength of the electron

$$\lambda_{dB} = \frac{\hbar}{(2m^* E)^{1/2}}$$

exceeds the dimensions of a quantum structure of characteristic length $L_z(\lambda_{dB} > L_z)$ or likewise for tunneling through a potential barrier of length L_z. To get some order of magnitude estimates of the electron kinetic energies E below which quantum effects become important where a log-log plot of λ_{dB} vs E in equation is shown for

GaAs and InAs. From the plot we see that an electron energy of E ~ 0:1 eV for GaAs corresponds to a de Broglie wavelength of 200 Å. Thus wave properties for electrons can be expected for structures smaller than λ_{dB}.

To observe quantum effects, the thermal energy must also be less than the energy level separation, $k_B T < \Delta E$, where we note that room temperature corresponds to 25 meV.

Since quantum effects depend on the phase coherence of electrons, scattering can also destroy quantum effects. The observation of quantum effects thus requires that the carrier mean free path be much larger than the dimensions of the quantum structures (wells, wires or dots).

The limit where quantum effects become important has been given the name of mesoscopic physics. Carrier transport in this limit exhibits both particle and wave characteristics.

In this ballistic transport limit, carriers can in some cases transmit charge or energy without scattering.

The small dimensions required for the observation of quantum effects can be achieved by the direct fabrication of semiconductor elements of small dimensions (quantum wells, quantum wires and quantum dots). Another approach is the use of gates on a field effect transistor to define an electron gas of reduced dimensionality.

In this context, negatively charged metal gates can be used to control the source to drain current of a 2D electron gas formed near the GaAs/AlGaAs interface. A thin conducting wire is formed out of the 2D electron gas. Controlling the gate voltage controls the amount of charge in the depletion region under the gates, as well as the charge in the quantum wire.

Thus lower dimensional channels can be made in a 2D electron gas by using metallic gates.

DENSITY OF STATES IN LOW DIMENSIONAL SYSTEMS

We showed in equation that the density of states for a 2D electron gas is a constant for each 2D subband

$$g_{2D} = \frac{m^*}{\pi\hbar^2}$$

The inset is appropriate to the quantum well formed near a modulation doped GaAs-AlGaAs interface. In the diagram only the lowest bound state is occupied.

Using the same argument, we now derive the density of states for a 1D electron gas

$$N_{1D} = \frac{2}{2\pi}(k) = \frac{1}{\pi}(k)$$

which for a parabolic band $E = E_n + \hbar^2 k^2/(2m^*)$ becomes

$$N_{1D} = \frac{2}{2\pi}(k) = \frac{1}{\pi}\left(\frac{2m^*(E-E_n)}{\hbar^2}\right)^{1/2}$$

yielding an expression for the density of states $g_{1D}(E) = \partial N_{1D}/\partial E$

$$g_{1D}(E) = \frac{1}{2\pi}\left(\frac{2m}{\hbar^2}\right)^{1/2}(E-E_n)^{-1/2}.$$

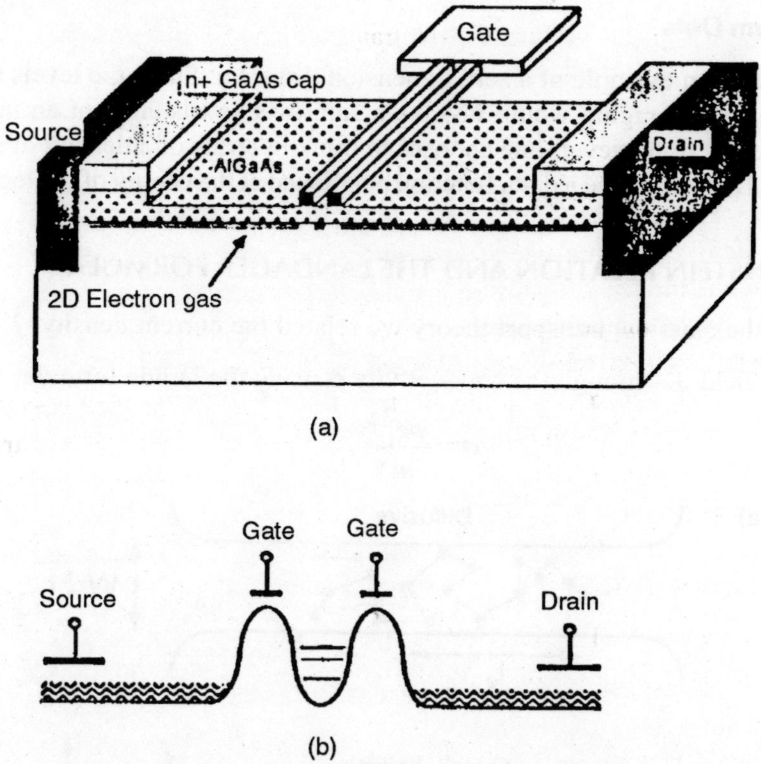

Fig. (a) Schematic diagram of a lateral resonant tunneling field -effect transistor which has two closely spaced fine finger metal gates;
(b) schematic of an energy band diagram for the device.

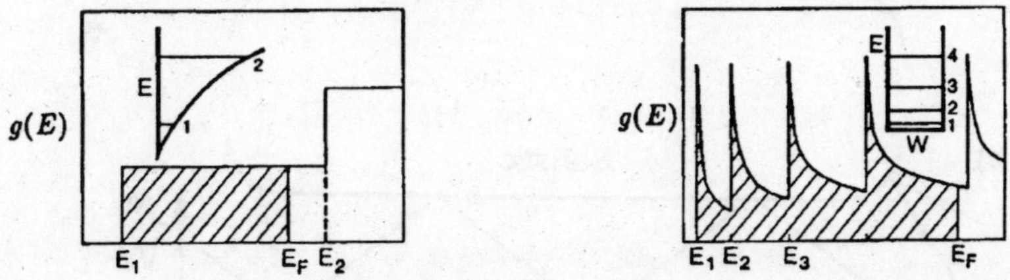

Fig. Density of states g(E) as a function of energy. (a) Quasi-2D density of states, with only the lowest subband occupied (hatched). Inset: Confinement potential perpendicular to the plane of the 2DEG. The discrete energy levels correspond to the bottoms of the first and second 2D subbands. (b) Quasi-1D density of states, with four 1D subbands occupied. Inset: Square-well lateral confinement potential with discrete energy levels indicating the 1D subband extrema.

The interpretation of this expression is that at each doubly confinedbound state level E_n there is a singularity in the density of states, where the first four levels are occupied.

Quantum Dots

This is an example of a zero dimensional system. Since the levels are all discrete any averaging would involve a sum over levels and not an integral over energy. If, however, one chooses to think in terms of a density of states, then the DOS would be a delta function positioned at the energy of the localized state.

THE EINSTEIN RELATION AND THE LANDAUER FORMULA

In the classical transport theory we related the current density $\vec{j}$ to the electric field $\vec{E}$ through the conductivity σ using the Drude formula

$$\sigma = \frac{ne^2\tau}{m^*}.$$

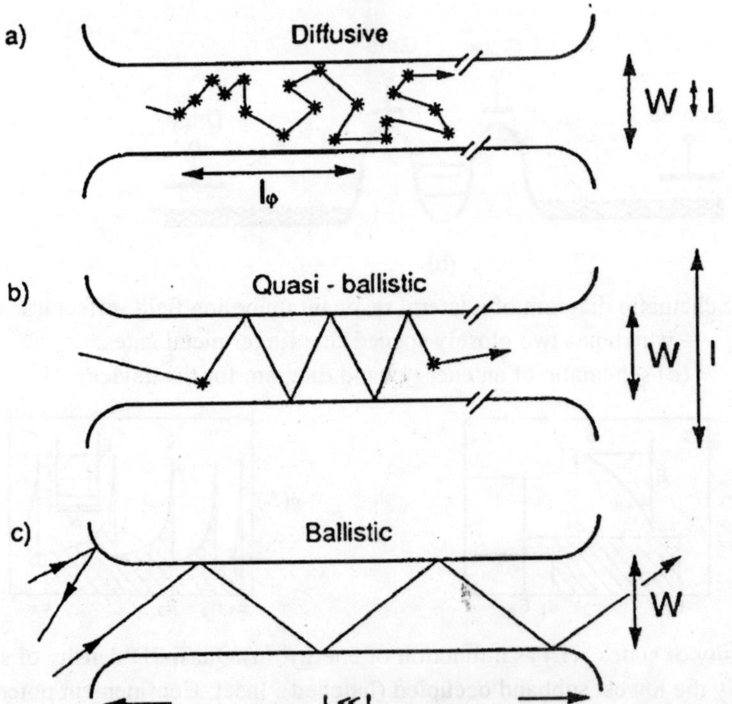

Fig. Electron trajectories characteristic of the diffusive ($\ell < W, L$), quasi-ballistic ($W < \ell < L$), and ballistic ($W, L < \ell$) transport regimes, for the case of specular boundary scattering. Boundary scattering and internal impurity scattering (asterisks) are of equal importance in the quasi-ballistic regime. A nonzero resistance in the ballistic regime results from backscattering at the connection between the narrow channel and the wide 2DEG regions. Taken from *H.* Van Houten et al. in "Physics and Technology of Submicron Structures".

The equation is valid when many scattering events occur within the path of an electron through a solid. As the dimensions of device structures become smaller and smaller, other regimes become important.

To relate transport properties to device dimensions it is often convenient to rewrite the Drude formula by explicitly substituting for the carrier density n and for the relaxation time τ. Writing $\tau = \ell/v_F$ where ℓ is the mean free path and v_F is the Fermi velocity, and writing $n/k_F^2/(2\pi)$ for the carrier density for a 2D electron gas (2DEG) we obtain

$$\sigma = \frac{k_F^2}{2\pi} e^2 \frac{\ell}{m^* v_F} = \frac{k_F^2 e^2 \ell}{2\pi\hbar k_F} = \frac{e^2}{h} k_F \ell$$

where e^2/h is a universal constant and is equal to $\sim (26\ \mathrm{k\Omega})^{-1}$.

Two general relations that are often used to describe transport in situations where collisions are not important within device dimensions are the Einstein relation and the Landauer formula. We discuss these relations below. The Einstein relation follows from the continuity equation

$$\vec{j} = eD\vec{\nabla}_n$$

where D is the diffusion coefficient and $\vec{\nabla}_n$ is the gradient of the carrier density involved in the charge transport. In equilibrium the gradient in the electrochemical potential $\vec{\nabla}_\mu$ is zero and is balanced by the electrical force and the change in Fermi energy

$$\vec{\nabla}_\mu = 0 = -e\vec{E} + \vec{\nabla}_n \frac{dE_F}{dn} = -e\vec{E} + \vec{\nabla}_n / g(E_F)$$

where $g(E_F)$ is the density of states at the Fermi energy. Substitution of equation yields

$$\vec{j} = eDg(E_F)e\vec{E} = \sigma\vec{E}$$

yielding the Einstein relation

$$\sigma = e^2 Dg(E_F)$$

which is a general relation valid for 3D systems as well as systems of lower dimensions.

The Landauer formula is an expression for the conductance G which is the proportionality between the current I and the voltage V,

$$I = GV.$$

For 2D systems the conductance and the conductivity have the same dimensions, and for a large 2D conductor we can write

$$G = (W/L)\sigma$$

where W and L are the width and length of the conducting channel in the urrent direction, respectively. If W and L are both large compared to the mean free path ℓ, then we are in the diffusive regime. However when we are in the opposite regime, the ballistic regime, where $\ell > W, L$, then the conductance is written in terms of the Landauer formula which is obtained from equations. Writing the number of quantum modes N, then $N\pi = k_F W$ or

$$k_F = \frac{N\pi}{W}$$

and noting that the quantum mechanical transition probability coupling one channel to another in the ballistic limit $|t_{\alpha;\beta}|^2$ is $\pi\ell/(2LN)$ per mode, we obtain the general Landauer formula

$$G = \frac{2e^2}{h}\sum_{\alpha,\beta}^{N}\left|e_{\alpha,\beta}\right|^2.$$

We will obtain the Landauer formula below for some explicit examples, which will make the derivation of the normalization factor more convincing.

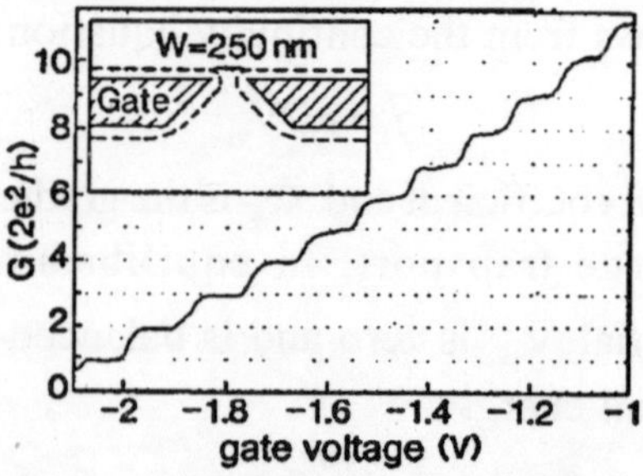

Fig. Point contact conductance as a function of gate voltage at 0.6 K, obtained from the raw data after subtraction of the background resistance. The conductance shows plateaus at multiples of $e^2/\pi\hbar$.

ONE DIMENSIONAL TRANSPORT AND QUANTIZATION OF THE BALLISTIC CONDUCTANCE

In the last few years one dimensional ballistic transport has been demonstrated in a two dimensional electron gas (2DEG) of a GaAs-GaAlAs heterojunction by constricting the electron gas to flow in a very narrow channel. Ballistic transport refers to carrier transport without scattering. As we show below, in the ballistic regime, the conductance of the 2DEG through the constriction shows quantized behaviour with the conductance changing in quantized steps of $(e^2/\pi\hbar)$ when the effective width of the constricting channel is varied by controlling the voltages of the gate above the 2DEG. We first give a derivation of the quantization of the conductance.

The current I_x flowing between source and drain due to the contribution of one particular 1D electron subband is given by

$$I_x = ne\delta v$$

where n is the carrier density (i.e., the number of carriers per unit length of the channel) and δv is the increase in electron velocity due to the application of a voltage V. The carrier density in 1D is

$$n = \frac{2}{2\pi} k_F = \frac{k_F}{\pi}$$

and the gain in velocity δv resulting from an applied voltage V is

$$eV = \frac{1}{2} m^* (v_F + \delta v)^2 - \frac{1}{2} m^* v_F^2 = m^* v_F \delta v + \frac{1}{2} m^* (\delta v)^2.$$

Retaining only the first order term in equation yields

$$\delta v = eV / m^* v_F$$

so that from equation we get for the source-drain current

$$I_x = \frac{k_F}{\pi} e \frac{eV}{m^* v_F} = \frac{e^2}{\pi\hbar} V$$

since $\hbar k_F = m^* v_F$. This yields a conductance per 1D electron subband G_i of

$$G_i = \frac{e^2}{\pi\hbar}$$

or summing over all occupied subbands i we obtain

$$G = \sum_i \frac{e^2}{\pi\hbar} = \frac{ie}{\pi\hbar}$$

Two experimental observations of these phenomena were simultaneously published. The experiments by Van Wees et al. were done using ballistic point contacts on a gate structure placed over a two-dimensional electron gas. The width W of the gate (in this case 2500Å) defines the effective width W^0 of the conducting electron channel, and the applied gate voltage is varied in order to control the effective width W^0. Superimposed on the raw data for the resistance vs gate voltage is a collection of periodic steps after subtracting off the background resistance of 400.

There are several conditions necessary to observe perfect $2e^2/h$ quantization of the 1D conductance. One requirement is that the electron mean free path l_e be much greater than the length of the channel L. This limits the values of channel lengths to $L < 5{,}000$ Å even though mean free path values are much larger, $l_e = 8.5$ µm. It is important to note, however, that $l_e = 8.5$µm is the mean free path for the 2D electron gas. When the channel is formed, the screening effect of the 2D electron gas is no longer present and the effective mean free path becomes much shorter.

A second condition is that there are adiabatic transitions at the inputs and outputs of the channel. This minimizes reflections at these two points, an important condition for the validity of the Landauer formula. A third condition requires the Fermi wavelength $\lambda_F = 2\pi/k_F$ (or $k_F L > 2\pi$) to satisfy the relation $\lambda_F < L$ by introducing a sufficient carrier density (3.6×10^{11}cm^{-2}) into the channel. Finally, it is necessary that the thermal energy $k_B T \ll Ej - Ej_{-1}$ where $Ej - E_{j-1}$ is the subband separation between the j and j ¡ 1 one dimensional energy levels. Therefore, the quantum conductance measurements are done at low temperatures ($T < 1$ *K*).

The point contacts were made on high-mobility molecular-beam-epitaxygrown GaAs-AlGaAs heterostructures using electron beam lithography. The electron density of the material is 3.6×10^{11}/cm^2 and the mobility is 8.5×10^5cm^2/V s (at 0.6 K). These values were obtained directly from measurements of the devices themselves. For the transport measurements, a standard Hall bar geometry was defined by wet etching. At a gate voltage of $V_g = -0.6$ V the electron gas underneath the gate is depleted, so that conduction takes place through the point contact only. At this voltage, the point contacts have their maximum effective width W ^{0} max, which is about equal to the opening W between the gates. By a further decrease (more negative) of the gate voltage, the width of the point contacts can be reduced, until they are fully pinched off at $V_g = -2.2$V.

The results agree well with the appearance of conductance steps that are integral multiples of $e^2/\pi h$, indicating that the conductance depends directly on the number of 1D subbands that are occupied with electrons. To check the validity of the proposed explanation for these steps in the conductance, the effective width W^0 for the gate was estimated from the voltage $V_g = -0.6$ V to be 3600 Å, which is close to the geometric value for W.

We see that the average conductance varies linearly with V_g which in turn indicates a linear relation between the effective point contact width W' and V_g. From the 16 observed steps and a maximum effective point contact width W'_{max} = 3600 Å, an estimate of 220 Å is obtained for the increase in width per step, corresponding to $\lambda_F/2$. Theoretical work done by Rolf Landauer nearly 20 years ago shows that transport through the channel can be described by summing up the conductances for all the possible transmission modes, each with a well defined transmission coefficient t_{nm}. The conductance of the 1D channel can then be described by the Landauer formula

$$G = \frac{e^2}{\pi h} \sum_{n,m=1}^{N_c} |t_{nm}|^2$$

where N_c is the number of occupied subbands. If the conditions for perfect quantization described earlier are satisfied, then the transmission coefficient reduces to $|t_{nm}|^2 / \delta_{nm}$. This corresponds to purely ballistic transport with no scattering or mode mixing in the channel (i.e., no back reflections).

A more explicit derivation of the Landauer formula for the special case of a 1D system can be done as follows. The current flowing in a 1D channel can be written as

$$I_j = \int_{E_i}^{E_f} eg_j(E)v_z(E)T_j(E)dE$$

where the electron velocity is given by

$$m^*v_z = \hbar k_z$$

and

$$E = E_j + \frac{\hbar^2 k_z^2}{2m^*}$$

while the 1D density of states is from equation given by

$$g_j(E) = \frac{(2m^*)^{1/2}}{h(E - E_g)^{1/2}}$$

and $T_j(E)$ is the probability that an electron injected into subband j with energy E will get across the 1D wire ballistically. Substitution of equation then yields

$$I_j = \frac{2e}{h}\int_{E_i}^{E_j} T_j(E)dE = \frac{2e^2}{h}T_j\Delta V$$

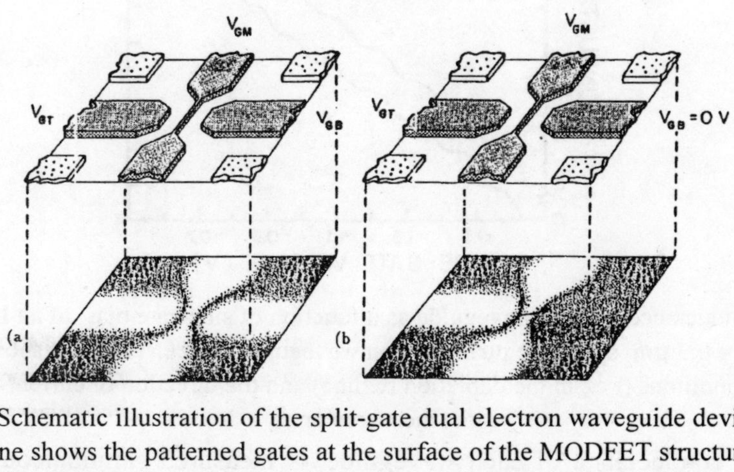

Fig. (a) Schematic illustration of the split-gate dual electron waveguide device. The top plane shows the patterned gates at the surface of the MODFET structure. The bottom plane shows the implementation of two closely spaced electron waveguides when the gates (indicated by V_{GT}, V_{GB} and V_{GM}) are properly biased. Shading represents the electron concentration. The four ohmic contacts which allow access to the inputs and outputs of each waveguide. (b) Schematic of the "leaky" electron waveguide implementation.

The bottom gate is grounded ($V_{GB} = 0$) so that only one waveguide is in an "on" state. V_{GM} is fixed such that only a small tunneling current crosses it. The current flowing through the waveguide as well as the tunneling current (depicted by arrows) are monitored simultaneously.

where we have noted that the potential energy difference is the difference between initial and final energies $e\Delta V = E_f - E_i$. Summing over all occupied states j we then obtain the Landauer formula

$$G = \frac{2e^2}{h} \sum_j T_j .$$

BALLISTIC TRANSPORT IN 1D ELECTRON WAVEGUIDES

Another interesting quantum-effect structure proposed and implemented the appropriate negative biases on patterned gates at the surface, two electron waveguides can be formed at the heterointerface of a MODFET structure. The two electron waveguides are closely spaced over a certain length and their separation is controlled by the middle gate bias (V_{GM}).

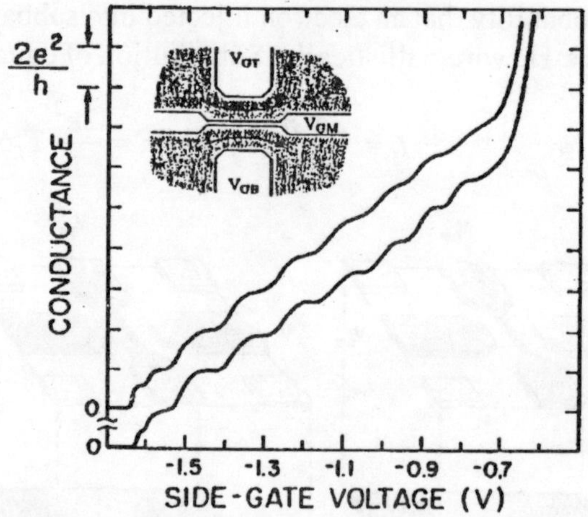

Fig. Conductance of each waveguide as a function of side gate bias of an L = 0:5 μm, W = 0:3 μm split-gate dual electron waveguide device. The inset shows the biasing conditions (i.e., in the depletion regime) and the direction of current flow for the measurements.

The conductance of each waveguide, is measured simultaneously and independently as a function of the side gate biases (V_{GT} and V_{GB}) and each show the quantized $2e^2/h$ conductance steps.

Such a device can be used to study 1D coupled electron waveguide interactions. An electron directional coupler based on such a structure has also been proposed. Since each gate can be independently accessed, various other regimes can be studied in addition to the coupled waveguide regime. One interesting regime is that of a "leaky" electron.

For such a scheme, one of the side gates is grounded so that the 2D electron gas underneath it is unaffected. The middle gate is biased such that only a small tunneling current can flow from one waveguide to the other in the 2D electron gas.

The other outer gate bias V_{GT} is used to sweep the subbands in the waveguide through the Fermi level.

In such a scheme, there is only one waveguide which has a thin side wall barrier established by the middle gate bias. The current flowing through the waveguide as well as the current leaking out of the thin middle barrier are independently monitored.

The $I - V_{GS}$ characteristics for the leaky electron waveguide implementation. Conductance steps of order $2e^2/h$ are observed for the current flowing.

However what is unique to the leaky electron experiment is that the tunneling current leaking from the thin side wall is monitored.

Very strong oscillations in the tunneling current are observed as the Fermi level is modulated in the waveguide.

We show below that the tunneling current is directly tracing out the 1D density of states of the waveguide.

An expression for the tunneling current I_S2 flowing through the sidewall of the waveguide can be obtained by the following integral,

$$I_{S2} = e\sum_j \int_{-\infty}^{\infty} v_{\perp j}(E - E_j) g_{1Dj}(E - E_j) T_j(E - E_b)$$
$$[f(E - E_F - e\Delta V, T) - f(E - E_F, T)]\, dE$$

where we have accounted for the contribution to the current from each occupied subband j. Here E_j is the energy at the bottom of the jth subband and E_b is the height of the tunneling energy barrier. The normal velocity against the tunneling barrier is $v_{\perp j}(E) = \hbar k_{\perp j} / m^*$.

The transmission coefficient, $T_j(E - E_b)$, to first order, is the same for the different 1D electron subbands since the barrier height relative to the Fermi level E_F is fixed.

The Fermi function, f, gives the distribution of electrons as a function of temperature and applied bias ΔV between the input of the waveguide and the 2DEG on the other side of the tunneling barrier.

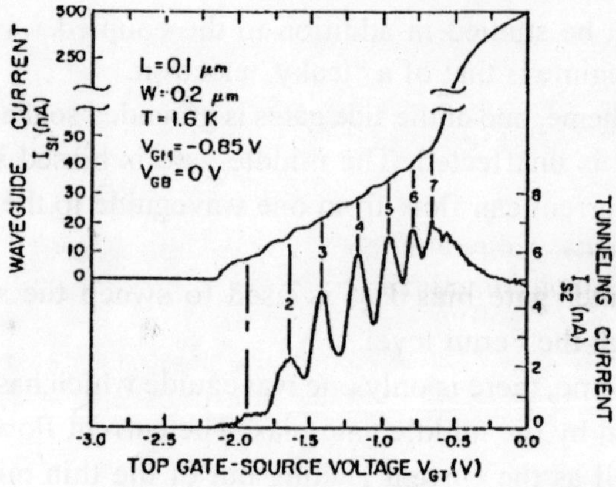

Fig. The current vs gate-source voltage characteristics of a leaky electron waveguide implemented with the proper biases for the device. For this case one of the sidegates is grounded so that only one leaky electron waveguide is on. I_{S1} is the current flowing through the waveguide and I_{S2} is the current tunneling through the thin middle side barrier. The bias voltage V_{DS} between the contacts is 100 μV.

For low enough temperatures and small ΔV,

$$I_{S2} \cong \frac{e^2\hbar}{m^*}\sum_j k_{\perp j}T_j(E_F - E_b)g_{1D,j}(E_F - E_j)\Delta V$$

where $k_{\perp j}$ is a constant for each subband and has a value determined by the confining potential. The tunneling current I_{S2} is proportional to $g_{1D,j}$ $(E_F - E_j)$, the 1D density of states.

Increasing V_{GT} to less negative values sweeps the subbands through the Fermi level and since $k_{\perp j}$ and T_j $(E_F - E_b)$ are constant for a given subband, the 1D density of states $g_{1D,j}$ summed over each subband, can be extracted from measurement of I_{S2}.

Shows a schematic of the cross-section of the waveguide for three different values of the gate source voltage V_{GT}.

The middle gate bias V_{GM} is fixed and is the same for all three cases. The parabolic potential is a result of the fringing field s and is characteristic of the quantum well for split-gate defined channels. By making V_{GT} more negative, the sidewall potential of the waveguide is raised with respect to the Fermi level.

Fewer energy levels below the Fermi level (i.e.,fewer occupied subbands) which corresponds to a very negative V_{GT}, only the first subband is occupied. At less negative values of V_{GT}, the second subband becomes occupied and then the third and so on.

There is no tunneling current until the first level has some carrier occupation.

The tunneling current increases as a new subband crosses below the Fermi-level.

The difference between E_F in the quantum well (on the left) and in the metallic contact (on the right) is controlled by the bias voltage between the input and output contacts of the waveguide V_{DS}.

In fact, a finite voltage V_{DS} and a finite temperature gives rise to lifetime effects and a broadening of the oscillations in I_{S2}.

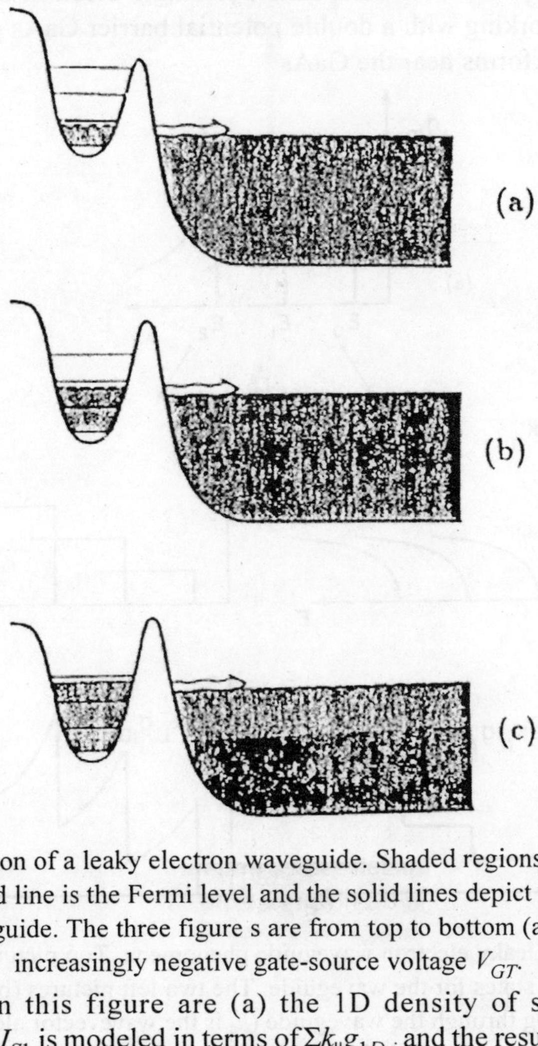

Fig. Cross-section of a leaky electron waveguide. Shaded regions represent electrons. The dashed line is the Fermi level and the solid lines depict the energy levels in the waveguide. The three figure s are from top to bottom (a), (b), (c) at increasingly negative gate-source voltage V_{GT}.

Included in this figure are (a) the 1D density of states g_{1D}. The measurement of I_{S1} is modeled in terms of $\Sigma k_{\|} g_{1D,j}$ and the results for I_{S1} (VGT) which relate to the steps in the conductance can be used to extract g_{1D}. The multiplication of $k_{\perp,j}$ and $g_{1D,j}$ to obtain $\Sigma k_{\perp j} g_{1D,j}$ summed over occupied levels which is measured by the tunneling current I_{S2}.

The 1D density of states g_1D can then be extracted from either I_{S1} or I_{S2}.

SINGLE ELECTRON CHARGING DEVICES

By making even narrower channels it has been possible to observe single electron charging in a nanometer field -effect transistor. In studies, a metal barrier is placed in the middle of the channel and the width of the metal barrier and the gap between the two constricted gates are of very small dimensions.

The first experimental observation of single electron charging was by Meirav et al., working with a double potential barrier GaAs device. The two dimensional gas forms near the GaAs

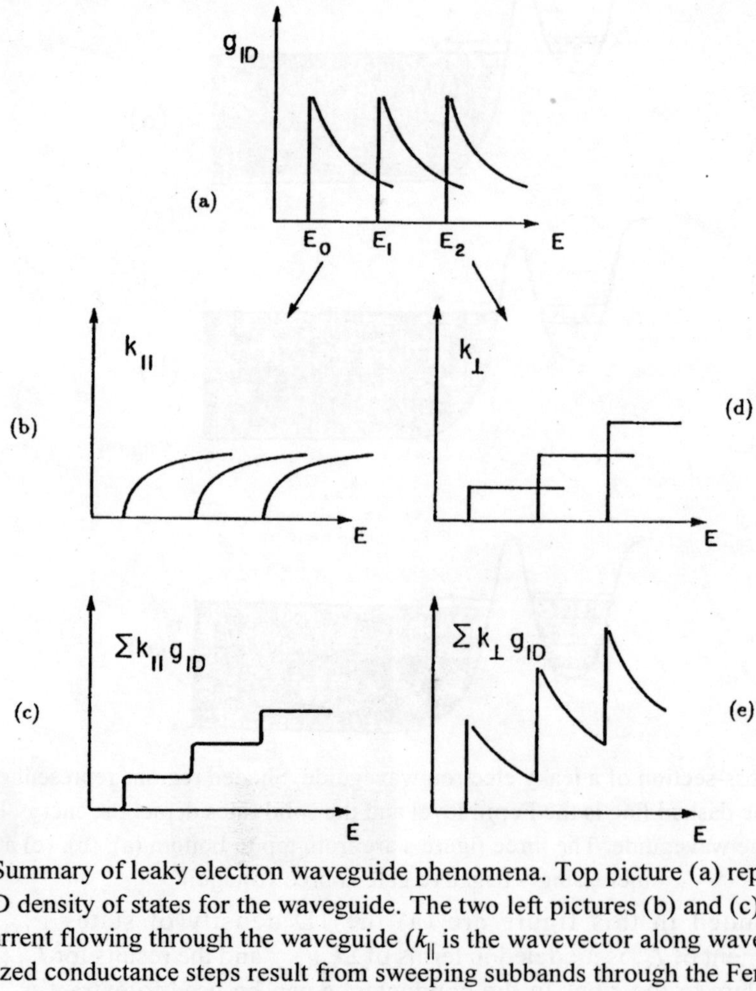

Fig. Summary of leaky electron waveguide phenomena. Top picture (a) represents the 1D density of states for the waveguide. The two left pictures (b) and (c) are for the current flowing through the waveguide ($k_{\parallel}$ is the wavevector along waveguide). Quantized conductance steps result from sweeping subbands through the Fermi level as I_{S1}. The two right hand pictures (d) and (e) are for the tunneling current ($k_{\perp}$ is transverse wavevector). Oscillations in the tunneling current I_{S2} arise from sweeping each subband through the Fermi level.

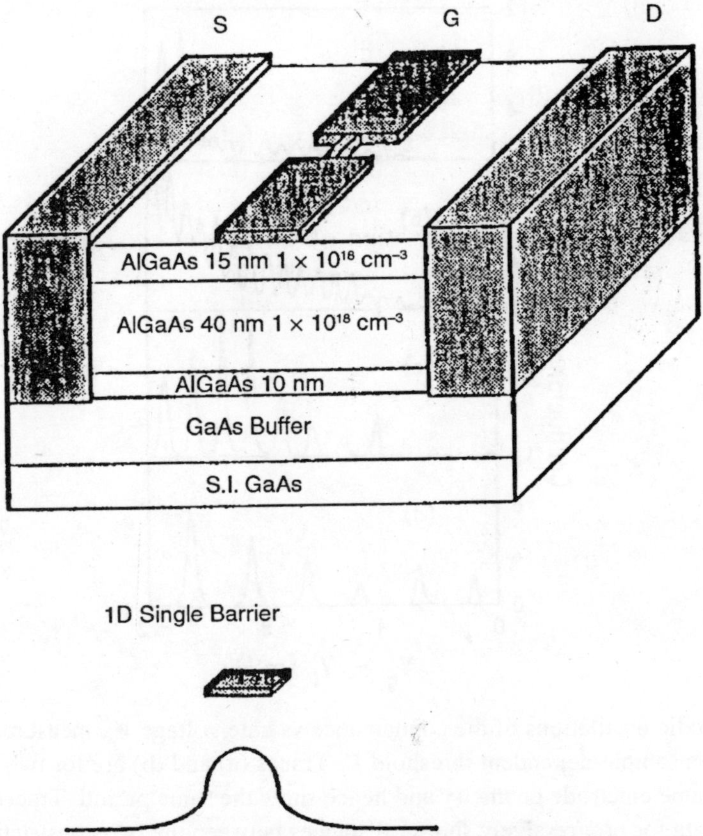

Fig. A split gate nanometer field -effect transistor. In the narrow channel a 1D electron gas forms when the gate is biased negatively.

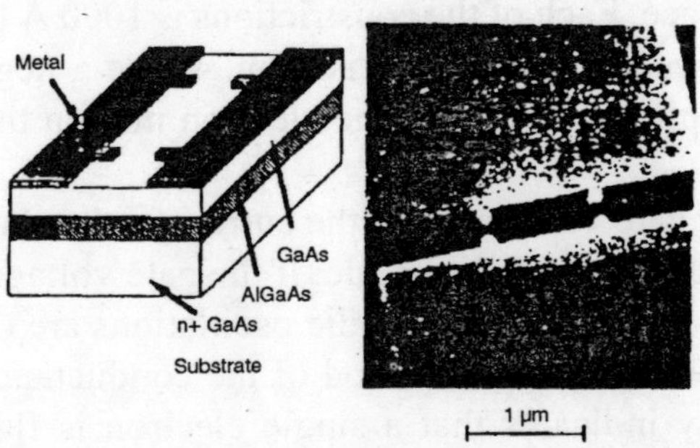

Fig. Schematic drawing of the device structure along with a scanning electron micrograph of one of the double potential barrier samples. An electron gas forms at the top GaAs-AlGaAs interface, with an electron density controlled by the gate voltage V_g. The patterned metal electrodes on top define a narrow channel with two potential barriers.

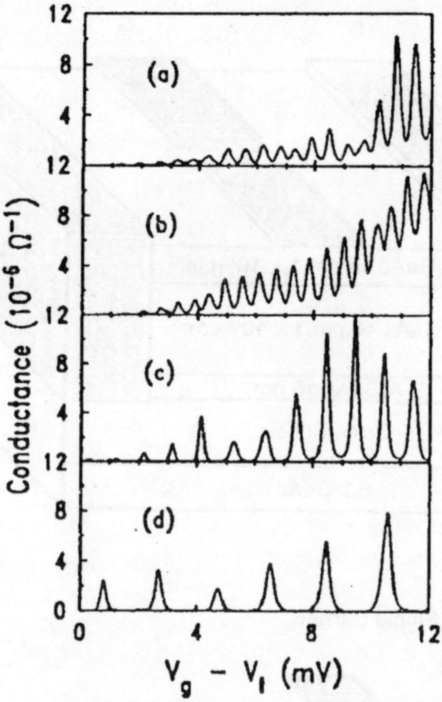

Fig. Periodic oscillations of the conductance vs gate voltage V_g, measured at T ≈ 50mK on a sample-dependent threshold V_t. Traces (a) and (b) are for two samples with the same electrode geometry and hence show the same period. Traces (c) and (d) show data for progressively shorter distances between the two constrictions, with a corresponding increase in period. Each oscillation corresponds to the addition of a single electron between the barriers.

GaAlAs interface. Each of the constrictions is 1000 Å long and the length of the channel between constrictions is 1μm. When a negative bias voltage (V_b ~ –0.5V) is applied to the gate, the electron motion through the gates is constrained.

At a threshold gate voltage of V_t, the current in the channel goes to zero. This is referred to as Coulomb blockade. If the gate voltage is now increased (positively) above V_t, a series of periodic oscillations are observed.

The correlation between the period of the conductance oscillations and the electron density indicates that a single electron is flowing through the double gated structure per oscillation.

The oscillations shows the same periodicity when prepared under the same conditions [as in traces (a) and (b)].

As the length L_0 between the constrictions is reduced below 1μm, the oscillation period gets longer. To verify that they had seen single electron charging behaviour, Meirav et al. fit the experimental lineshape for a single oscillation to the functional form for the conductance

$$G(\mu) \sim \frac{\partial F}{\partial E} \sim \cosh^{-2}\left[\frac{E-\mu}{2k_B T}\right].$$

where μ is the chemical potential, $F(E-\mu, T)$ is the Fermi function and E is the single electron energy in the 2DEG.

Chapter 6

Rutherford Backscattering Spectroscopy

INTRODUCTION TO THE TECHNIQUE

Ions of all energies incident on a solid influence its materials properties. Ions are used in many different ways in research and technology. We here review some of the physics of the interaction of ion beams with solids and some of the uses of ion beams in semiconductors.

We start by noting that the interaction of ion beams with solids depends on the energy of the incident ion. A directed low energy ion (~ 10–100 eV) comes to rest at or near the surface of a solid, possibly growing into a registered epitaxial layer upon annealing. A 1–keV heavy ion beam is the essential component in the sputtering of surfaces. For this application, a large fraction of the incident energy is transferred to the atoms of the solid resulting in the ejection of surface atoms into the vacuum. The surface is left in a disordered state. Sputtering is used for removal of material from a sample surface on an almost layer–by–layer basis. It is used both in semiconductor device fabrication, ion etching or ion milling, and more generally in materials analysis, through depth profiling. The sputtered atoms can also be used as a source for the sputter deposition technique discussed in connection with the growth of superlattices.

At higher energies, ~ 100–300 keV, energetic ions are used as a source of atoms to modify the properties of materials. Low concentrations, $\ll 0.1$ atomic percent, of implanted atoms are used to change and control the electrical properties of semiconductors. The implanted atom comes to rest ~ 1000Å below the surface in a region of disorder created by the passage of the implanted ion. The electrical properties of the implanted layer depend on the species and concentration of impurities, the lattice position of the impurities and the amount of lattice disorder that is created.

Lattice site and lattice disorder as well as epitaxial layer formation can readily be analyzed by the channeling of high energy light ions (such as H^+ and He^+), using the Rutherford backscattering technique. In this case MeV ions are used because they penetrate deeply into the crystal (microns) without substantially perturbing the lattice. This is an attractive ion energy regime because scattering cross sections, flux distributions in the crystal, and the rate of energy loss are quantitatively established. Particle–solid interactions in the range from about 0.1 MeV to 5 MeV are understood and one can use this well–characterized tool for investigations of solid state phenomena. Outside of this range many of the concepts discussed below remain valid: however, the use of MeV ions is favored in solid state characterization applications because of both experimental convenience and the ability to probe both surface and bulk properties.

In § 6.2 and § 6.3, we will review the two last energy regimes, namely ion implantation in the 100 keV region and ion beam analysis (Rutherford backscattering and channeling) by MeV light mass particles (H^+, He^+). We start by describing the most important features of ion implantation. Then we will review a two atom collision in order to introduce the basic atomic scattering concepts needed to describe the slowing down of ions in a solid. Then we will state the results of the LSS theory (J. Lindhard, M. Scharff and H. Schiτtt, Mat.

Fys. Medd. Dan. Vid. Selsk. 33, No. 14 (1963)) which is the most successful basic theory for describing the distribution of ion positions and the radiation–induced disorder in the implanted sample. A number of improvements to this model have been made in the last two decades and are used for current applications. We will conclude this introduction to ion implantation with a discussion of the lattice damage caused by the slowing down of the energetic ions in the solid.

We then go on and describe the use of a beam of energetic (1–2 MeV) light mass particles (H^+ , He^+) to study material properties in the near surface region (< μm) of a solid. We will then see how to use Rutherford backscattering spectrometry (RBS) to determine the stoichiometry of a sample composed of multiple chemical species and the depth distribution of implanted ions. Furthermore, we see that with single–crystal targets, the effect of channeling also allows investigation of the crystalline perfection of the sample as well as the lattice location of the implanted atomic species. Finally, we will review some examples of the modification of material properties by ion implantation, including the use of ion beam analysis in the study of these materials modifications.

ION IMPLANTATION

Ion implantation is an important technique for introducing impurity atoms in a controlled way, thus leading to the synthesis of new classes of materials, including metastable materials. The technique is important in the semiconductor industry for making *p–n* junctions by, for example, implanting *n*–type impurities into a *p*–type host material. From a materials science point of view, ion implantation allows essentially any element of the periodic table to be introduced into the near surface region of essentially any host material, with quantitative control over the depth and composition profile by proper choice of ion energy and fluence. More generally, through ion implantation, materials with increased strength and corrosion resistance or other desirable properties can be synthesized.

A schematic diagram of an ion implanter in this diagram the "target" is the sample that is being implanted. In the implantation process, ions of energy E and beam current i_b are incident on a sample surface and come to rest at some characteristic distance R_p with a Gaussian distribution of half width at half maximum ΔR_p.

Typical values of the implantation parameters are: ion energies $E \sim 100$ keV, beam currents $i_b \sim 50\mu A$, penetration depths $R_p \sim 1000Å$ and half-widths $\Delta R_p \sim 300Å$. In the implanted regions, implant concentrations of 10i3 to 10i5 relative to the host materials are typical. In special cases, local concentrations as high as 20 at.% (atomic percent) of implants have been achieved.

Some characteristic features of ion implantation are the following:

1. The ions characteristically only penetrate the host material to a depth $R_p \leq 1\mu m$. Thus ion implantation is a near surface phenomena. To achieve a large percentage of impurity ions in the host, the host material must be thin (comparable to R_p), and high fluences of implants must be used ($\phi > 10^{16}/cm^2$).
2. The depth profile of the implanted ions (R_p) is controlled by the ion energy. The impurity content is controlled by the ion fluence ϕ.
3. Implantation is a non–equilibrium process. Therefore there are no solubility limits on the introduction of dopants. With ion implantation one can thus introduce high concentrations of dopants, exceeding the normal solubility limits. For this reason ion implantation permits the synthesis of metastable materials.
4. The implantation process is highly directional with little lateral spread. Thus it is possible to implant materials according to prescribed patterns using masks. Implantation proceeds in the regions where the masks are not present. An application of this technology is to the ion implantation of polymers to make photoresists with sharp boundaries. Both positive and negative photoresists can be prepared using ion implantation, depending on the choice of the polymer. These masks are widely used in the semiconductor industry.
5. The diffusion process is commonly used for the introduction of impurities into semiconductors. Efficient diffusion occurs at high temperatures. With ion implantation, impurities can be introduced at much lower temperatures, as for example room temperature, which is a major convenience to the semiconductor industry.

6. The versatility of ion implantation is another important characteristic. With the same implanter, a large number of different implants can be introduced by merely changing the ion source. The technique is readily automated, and thus is amenable for use by technicians in the semiconductor industry. The implanted atoms are introduced in an atomically dispersed fashion, which is also desirable. Furthermore, no oxide or interfacial barriers are formed in the implantation process.
7. The maximum concentration of ions that can be introduced is limited. As the implantation process proceeds, the incident ions participate in both implantation into the bulk and the sputtering of atoms off the surface. Sputtering occurs because the surface atoms receive sufficient energy to escape from the surface during the collision process. The dynamic equilibrium between the sputtering and implantation processes limits the maximum concentration of the implanted species that can be achieved.
   Sputtering causes the surface to recede slowly during implantation.
8. Implantation causes radiation damage. For many applications, this radiation damage is undesirable. To reduce the radiation damage, the implantation can be carried out at elevated temperatures or the materials can be annealed after implantation.
   In practice, the elevated temperatures used for implantation or for post–implantation annealing are much lower than typical temperatures used for the diffusion of impurities into semiconductors.

A variety of techniques are used to characterize the implanted alloy. Ion backscattering of light ions at higher energies (e.g., 2 MeV He^+) is used to determine the composition versus depth with ~ 10nm depth resolution. Depth profiling by sputtering in combination with Auger or secondary ion mass spectroscopy is also used.

Lateral resolution is provided by analytical transmission electron microscopy. Electron microscopy, glancing angle *x*–ray analysis and ion channeling in single crystals provide detailed information on the local atomic structure of the alloys formed. Ion backscattering and channeling.

Basic Scattering Equations

An ion penetrating into a solid will lose its energy through the Coulomb interaction with the atoms in the target. This energy loss will determine the final penetration of the projectile into the solid and the amount of disorder created in the lattice of the sample. When we look more closely into one of these collisions we see that it is a very complicated event in which:

- The two nuclei with masses M_1 and M_2 and charges Z_1 and Z_2, respectively, repel each other by a Coulomb interaction, screened by the respective electron clouds.

- Each electron is attracted by the two nuclei (again with the corresponding screening) and is repelled by all other electrons.

In addition, the target atom is bonded to its neighbors through bonds which usually involve its valence electrons. The collision is thus a many–body event described by a complicated Hamiltonian. However experience has shown that very accurate solutions for the trajectory of the nuclei can be obtained by making some simplifying assumptions. The most important assumption for us, is that the interaction between the two atoms can be separated into two components:

- Ion (projectile)–nucleus (target) interaction
- Ion (projectile)–electron (target) interaction.

Let us now use the very simple example of a collision between two masses M_1 and M_2 to determine the relative importance of these two processes and to introduce the basic atomic scattering concepts required to describe the stopping of ions in a solid. The classical collision between the incident mass M_1 and the target mass M_2 which can be a target atom or a nearly free target electron.

By applying conservation of energy and momentum, the following relations can be derived

- The energy transferred (T) in the collision from the incident projectile M_1 to the target particle M_2 is given by

$$\mathrm{T} = T_{\max} \sin^2\left(\frac{\Theta}{2}\right)$$

where

$$T_{\max} = 4E_0 \frac{M_1 M_2}{(M_1 + M_2)^2}$$

is the maximum possible energy transfer from M_1 to M_2.

- The scattering angle of the projectile in the laboratory system of coordinates is given by

$$\cos\theta = \frac{1-(1+M_2+M_1)(T/2E)}{\sqrt{(1-T/E)}}$$

- The energies of the projectile before (E_0) and after (E_1) scattering are related by

$$E_1 = k^2 E_0$$

where the kinematic factor k is given by

$$k = \left(\frac{M_1\cos\theta \pm (M_2^2 - M_1^2\sin^2\theta)^{1/2}}{M_1 + M_2}\right)$$

These relations are absolutely general no matter how complex the force between the two particles, so long as the force acts along the line joining the particles and the electron is nearly free so that the collision can be taken to be elastic. In reality, the collisions between the projectile and the target electrons are inelastic because of the binding energy of the electrons. The case of inelastic collisions will be considered in what follows.

Using equations and assuming either that M_2 is of the same atomic species as the projectile ($M_2 = M_1$), or that M_2 is a nearly free electron ($M_2 = m_e$) we construct Table.

With the help of the previous example, we can formulate a qualitative picture of the slowing down process of the incident energetic ions. As the incident ions penetrate into the solid, they lose energy. There are two dominant mechanisms for this energy loss:

Table: Maximum energy transfer T_{max} and scattering angle θ for nuclear ($M_2 = M_1$) and electronic (M_2 / m_e) collisions.

	"nuclear" collision	"electronic" collision
T_{max}	T_{max}; E_0	Tmax; $(4m/M_1)$ E0
θ	$0 < \theta \leq \pi/2$	$\theta = 0°$

1. The interaction between the incident ion and the electrons of the host material. This inelastic scattering process gives rise to electronic energy loss.
2. The interaction between the incident ions and the nuclei of the host material. This is an elastic scattering process which gives rise to nuclear energy loss.

The ion–electron interaction induces small losses in the energy of the incoming ion as the electrons in the atom are excited to higher bound states or are ionized.

These interactions do not produce significant deviations in the projectile trajectory. In contrast, the ion–nucleus interaction results in both energy loss and significant deviation in the projectile trajectory. In the ion–nucleus interaction, the atoms of the host are also significantly dislodged from their original positions giving rise to lattice defects, and the deviations in the projectile trajectory will give rise to the lateral spread of the distribution of implanted species.

Let us now further develop our example of a "nuclear" collision. For a given interaction potential $V(r)$, each ion coming into the annular ring of area $2\pi p dp$ with energy E, will be deflected through an angle θ where p is the impact parameter. We define $T = E_0 - E_1$ as the energy transfer from the incoming ion to the host and we define $2\pi p dp = d\sigma$ as the differential cross section. When the ion moves a distance Δx in the host material, it will interact with $N\,\Delta x 2\pi p dp$ atoms where N is the atom density of the host.

The energy ΔE lost by an ion traversing a distance Δx will be

$$\Delta \mathrm{E} = N\Delta x \int T\, 2\pi p dp$$

so that as $\Delta x \rightarrow 0$, we have for the stopping power

$$\frac{dE}{dx} = N \int T d\sigma$$

where σ denotes the cross sectional area. We thus obtain for the stopping cross section E

$$E = \frac{1}{N}\frac{dE}{dx} = \int T d\sigma$$

The total stopping power is due to both electronic and nuclear processes

$$\frac{dE}{dx} = \left(\frac{dE}{dx}\right)_e + \left(\frac{dE}{dx}\right)_n = N(E_e + E_n)$$

where N is the target density and E_e and E_n are the electronic and nuclear stopping cross sections, respectively. Likewise for the stopping cross section E.

$$E = E_e + E_n.$$

Table: Typical Values of E_1, E_2, E3 for silicon.[a]

Ion	$E_1(keV)$	$E_2(keV)$	$E_3(keV)$
B	3	17	3000
P	17	140	~ 3 £10^4
As	73	800	> 10^5
Sb	180	2000	> 10^5

From the energy loss we can obtain the ion range or penetration depth

$$R = \int \frac{dE}{dE/dx}.$$

Since we know the energy transferred to the lattice (including both phonon generation and displacements of the host ions), we can calculate the energy of the incoming ions as a function of distance into the medium $E(x)$.

At low energies of the projectile ion, nuclear stopping is dominant, while electron stopping dominates at high energies as shown in the characteristic stopping power curves. Note the three important energy parameters on the curves: E_1 is the energy where the nuclear stopping power is a maximum, E_3 where the electronic stopping power is a maximum, and E_2 where the electronic and nuclear stopping powers are equal. As the atomic number of the ion increases for a fixed target, the scale of E_1, E_2 and E_3 increases. Also indicated on the diagram is the functional form of the energy dependence of the stopping power in several of the regimes of interest.

Typical values of the parameters E_1, E_2 and E_3 for various ions in silicon. Ion implantation in semiconductors is usually done in the regime where nuclear energy loss is dominant. The region where $(dE/dx) \sim 1/E$ corresponds to the regime where light ions like H^+ and He^+ have incident energies of 1–2 MeV and is therefore the region of interest for Rutherford backscattering and channeling phenomena.

Radiation Damage

The energy transferred from the projectile ion to the target atom is usually sufficient to result in the breaking of a chemical bond and the permanent displacement of the target atom from its original site. The condition for this process is that the energy transfer per collision T is greater than the binding energy E_d.

Because of the high incident energy of the projectile ions, each incident ion can dislodge multiple host ions. The damage profile for low dose implantation gives rise to isolated regions of damage.

As the fluence is increased, these damaged regions coalesce. The damage profile also depends on the mass of the projectile ion, with heavy mass ions of a given energy causing more local lattice damage as the ions come to rest. Since $(dE/dx)_n$ increases as the energy decreases, more damage is caused as the ions are slowed down and come to rest. The damage pattern schematically for light ions (such as boron in silicon) and for heavy ions (such as antimony in silicon). Damage is caused both by the incident ions and by the displaced energetic (knock-on) ions.

The types of defects caused by ion implantation. Here we see the formation of vacancies and interstitials, Frenkel pairs (the pair formed by the Coulomb attraction of a vacancy and an interstitial). The formation of multiple vacancies leads to a depleted zone while multiple interstitials lead to ion crowding.

Applications of Ion Implantation

For the case of semiconductors, ion implantation is dominantly used for doping purposes, to create sharp p-n junctions in the near-surface region. To reduce radiation damage, implantation is sometimes done at elevated temperatures. Post implantation annealing is also used to reduce radiation damage, with elevated temperatures provided by furnaces, lasers or flash lamps. The ion implanted samples are characterized by a variety of experimental techniques for the implant depth profile, the lattice location of the implant, the residual lattice disorder subsequent to implantation and annealing, the electrical properties (and conductivity) and the device performance.

A major limitation of ion implantation for modifying metal surfaces has been the shallow depth of implantation. In addition, the sputtering of atoms from the surface sets a maximum concentration of elements which can be added to a solid, typically ~ 20 to 40 at.%.

To form thicker layers and higher concentrations, combined processes involving ion implantation and film deposition are being investigated. Intense ion beams are directed at the solid to bring about alloying while other elements are simultaneously brought to the surface, for example by sputter deposition, vapour deposition or the introduction of reactive gases.

One process of interest is ion beam mixing, where thin films are deposited onto the surface first and then bombarded with ions. The dense collision cascades of the ions induce atomic–scale mixing between elements. Ion beam mixing is also a valuable tool to study metastable phase formation.

With regard to polymers, ion implantation can enhance the electrical conductivity by many orders of magnitude, as is for example observed for ion implanted polyacrylonitrile (PAN, a graphite fibre precursor). Some of the attendant property changes of polymers due to ion implantation include cross-linking and scission of polymer chains, gas evolution as volatile species are released from polymer chains and free radical formation when vacancies or interstitials are formed.

Implantation produces solubility changes in polymers and therefore can be utilized for the patterning of resists for semiconductor mask applications. For the positive resists, implantation enhances the solubility, while for negative resists, the solubility is reduced. The high spatial resolution of the ion beams makes ion beam lithography a promising technique for sub–micron patterning applications.

For selected polymers (such as PAN), implantation can result in transforming a good insulator into a conducting material with an increase in conductivity by more than 10 orders of magnitude upon irradiation. Thermoelectric power measurements on various implanted polymers show that implantation can yield either *p*–type or *n*–type conductors and in fact a *p*–*n* junction has recently been made in a polymer through ion implantation.

The temperature dependence of the conductivity for many implanted polymers is of the form $\sigma / \sigma_o \exp(T_0/T)^{1/2}$ which is also the relation characteristic of the one–dimensional hopping conductivity model for disordered materials. Also of interest is the long term chemical stability of implanted polymers.

Due to recent developments of high brightness ion sources, focused ion beams to submicron dimensions can now be routinely produced, using ions from a liquid metal source.

Potential applications of this technology are to ion beam lithography, including the possibility of maskless implantation doping of semiconductors. Instruments based on these ideas may be developed in the future. The applications of ion implantation represent a rapidly growing field.

Instruments based on these ideas may be developed in the future. The applications of ion implantation represent a rapidly growing field.

Further discussion of the application of ion implantation to the preparation of metastable materials is presented after the following sections on the characterization of ion implanted materials by ion backscattering and channeling.

ION BACKSCATTERING

In Rutherford backscattering spectrometry (RBS), a beam of mono-energetic (1–2 MeV), collimated light mass ions (H^+, He^+) impinges (usually at near normal incidence) on a target and the number and energy of the particles that are scattered backwards at a certain angle θ are monitored to obtain information about the composition of the target (host species and impurities) as a function of depth. We will review the fundamentals of the RBS analysis.

Particles scattered at the surface of the target will have the highest energy E upon detection. Here the energy of the backscattered ions E is given by the relation.

$$E = k^2 E_0$$

where

$$k = \left(\frac{M_1 \cos\theta \pm (M_2^2 - M_1^2 \sin^2 \theta)^{1/2}}{M_1 + M_2} \right)$$

For a given mass species, the energy E_s of particles scattered from the surface corresponds to the edge of the spectrum. In addition, the scattered energy depends through k on the mass of the scattering atom. Thus different species will appear displaced on the energy scale, thereby allowing for their chemical identification. We next show that the displacement along the energy scale from the surface contribution gives information about the depth where the backscattering took place. Thus the energy scale is effectively a depth scale.

The height H of the RBS spectrum corresponds to the number of detected particles in each energy channel ΔE.

CHANNELING

If the probing beam is aligned nearly parallel to a close-packed row of atoms in a single crystal target, the particles in the beam will be steered by the potential field of the rows of atoms, resulting in an undulatory motion in which the "channeled" ions will not approach the atoms in the row to closer than 0.1–0.2 Å. This is called the channeling effect. Under this channeling condition, the probability of large angle scattering is greatly reduced. As a consequence, there will be a drastic reduction in the scattering yield from a channeled probing beam relative to the yield from a beam incident in a random direction. Two characteristic parameters for channeling are the normalized minimum yield $\chi_{min}/H_A/H$ which is a measure of the crystallinity of the target, and the critical angle for channeling $\psi_1/2$ (the halfwidth at half maximum intensity of the channeling resonance) which determines the degree of alignment required to observed the channeling effect.

The RBS–channeling technique is frequently used to study radiation–induced lattice disorder by measuring the fraction of atom sites where the channel is blocked.

In general, the channeled ions are steered by the rows of atoms in the crystal. However if some portion of the crystal is disordered and lattice atoms are displaced so that they partially block the channels, the ions directed along nominal channeling directions can now have a close collision with these displaced atoms, so that the resulting scattered yield will be increased above that for an undisturbed channel.

Furthermore, since the displaced atoms are of equal mass to those of the surrounding lattice, the increase in the yield occurs at a position in the yield vs. energy spectrum corresponding to the depth at which the displaced atoms are located. The increase in the backscattering yield from a given depth will depend upon the number of displaced atoms, so the depth (or equivalently, the backscattering energy E) dependence of the yield, reflects the depth dependence of displaced atoms, and integrations over the whole spectrum will give a measure of the total number of displaced atoms.

Another very useful application of the RBS–channeling technique is in the determination of the location of foreign atoms in a host lattice. Since channeled ions cannot approach the rows of atoms which form the channel closer than ~ 0.1Å, we can think of a "for-bidden region", as a cylindrical region along each row of atoms with radius ~ 0.1Å, such that there are no collisions between the channeled particles and atoms located within the forbidden zone. In particular, if an impurity is located in a forbidden region it will not be detected by the channeled probing beam. On the other hand, any target particle can be detected by (i.e., will scatter off) probing particles from a beam which impinges in a random direction. Thus, by comparing the impurity peak observed for channeling and random alignments, the fraction of impurities sitting in the forbidden region of a particular channel (high symmetry crystallographic axis) can be determined. Repeating the procedure for other crystallographic directions allows the identification of the lattice location of the impurity atom in many cases.

Rutherford backscattering will not always reveal impurities embedded in a host matrix, in particular if the mass of the impurities is smaller than the mass of the host atoms. In such cases, ion induced x–rays and ion induced nuclear reactions are used as signatures for the presence of the impurities inside the crystal, and the lattice location is derived from the changes in yield of these processes for random and channeled impingement of the probing beam.

Ion markers consist of a very thin layer of a guest atomic species are embedded in an otherwise uniform host material of a different species to establish reference distances. Backscattering spectra are taken before and after introduction of the marker. The RBS spectrum taken after insertion of the marker can be used as a reference for various applications. Some examples where marker references are useful include:

- Estimation of surface sputtering by ion implantation. In this case, recession of the surface from the reference position set by the marker can be measured by RBS and can be analyzed to yield the implantation–induced surface sputtering.
- Estimation of surface material vapourized through laser annealing, rapid thermal annealing or laser melting of a surface.
- Estimation of the extent of ion beam mixing.

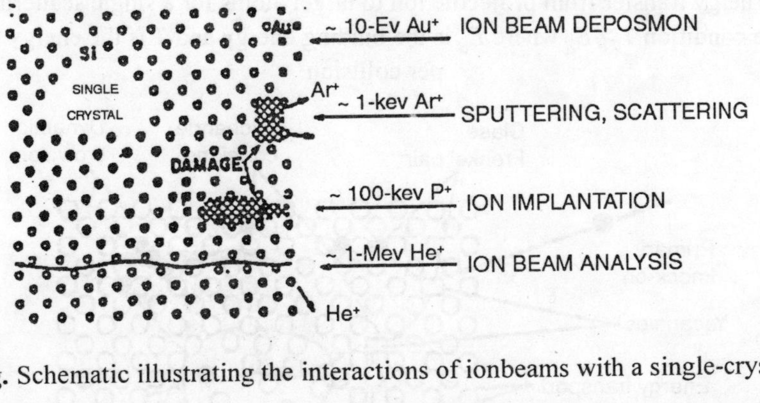

Fig. Schematic illustrating the interactions of ionbeams with a single-crystal solid. Directed beams of ~10 eV are used for film deposition and epitaxial formation. Ion beams of energy ! 1 keV are employed in sputtering applications; ~ 100 keV ions are used in ion implantation. Both the sputtering and implantation processes damage and disorder the crystal. Higher energy light ions are used for ion beam analysis.

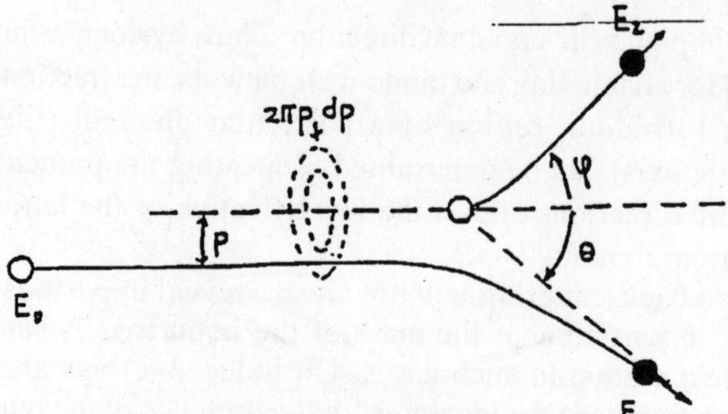

Fig. Classical model for the collision of a projectile of energy E with a target at rest. The open circles denote the initial state of the projectile and target atoms, and the full circles denote the two atoms after the collision.

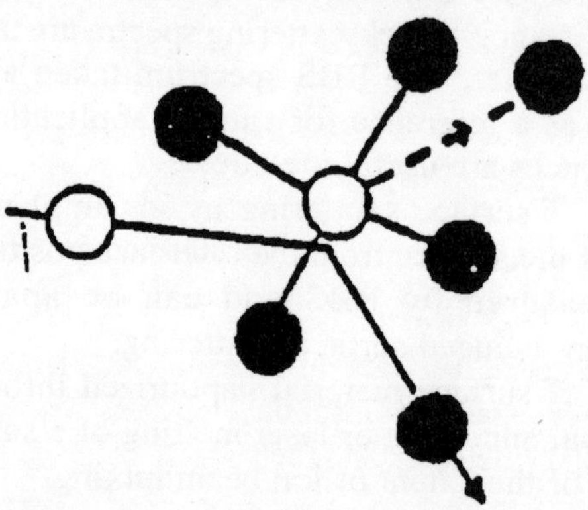

Fig. Energy transfer from projectile ion to target atoms for a single scattering event for the condition $T > E_d$ where E_d is the binding energy and T is the energy transfer per collision.

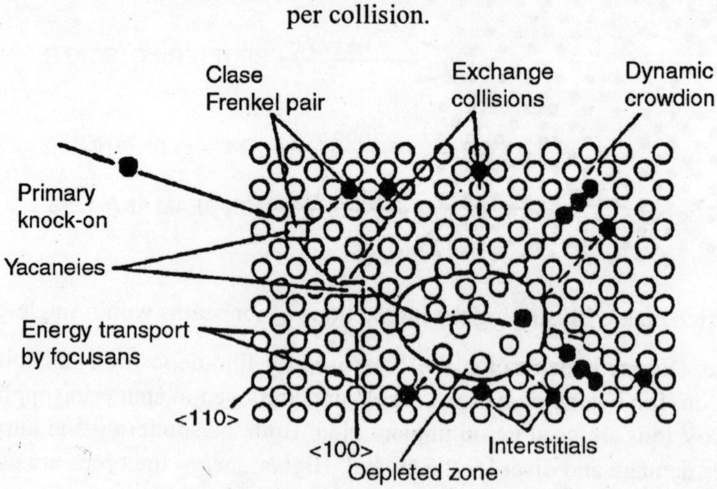

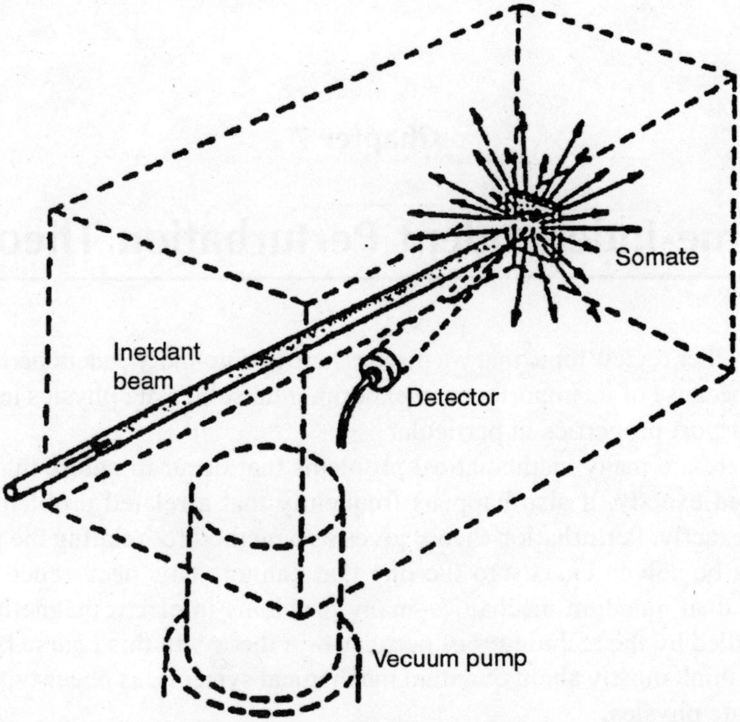

Fig. In the RBS experiment, the scattering chamber where the analysis/experiment is actually performed contains the essential elements: the sample, the beam, the detector, and the vacuum pump.

Chapter 7

Time-Independent Perturbation Theory

Another review topic that we discuss here is time-independent perturbation theory because of its importance in experimental solid state physics in general and transport properties in particular.

There are many mathematical problems that occur in nature that cannot be solved exactly. It also happens frequently that a related problem can be solved exactly. Perturbation theory gives us a method for relating the problem that can be solved exactly to the one that cannot. This occurrence is more general than quantum mechanics–many problems in electromagnetic theory are handled by the techniques of perturbation theory. In this course however, we will think mostly about quantum mechanical systems, as occur typically in solid state physics.

Suppose that the Hamiltonian for our system can be written as

$$H = H_0 + H'$$

where H_0 is the part that we can solve exactly and H^0 is the part that we cannot solve. Provided that $H'^0 \ll H_0$ we can use perturbation theory; that is, we consider the solution of the unperturbed Hamiltonian H_0 and then calculate the effect of the perturbation Hamiltonian H'. For example, we can solve the hydrogen atom energy levels exactly, but when we apply an electric or a magnetic field we can no longer solve the problem exactly. For this reason, we treat the effect of the external field s as a perturbation, provided that the energy associated with these field s is small:

$$H = \frac{p^2}{2m} - \frac{e^2}{r} - e\vec{r}\cdot\vec{E} = H_0 + H'$$

where

$$H_0 = \frac{p^2}{2m} - \frac{e^2}{r}$$

and

$$H' = -e\vec{r}\cdot\vec{E}\cdot$$

As another illustration of an application of perturbation theory, consider a weak periodic potential in a solid. We can calculate the free electron energy levels (empty lattice) exactly. We would like to relate the weak potential situation to the empty lattice problem, and this can be done by considering the weak periodic potential as a perturbation.

NON-DEGENERATE PERTURBATION THEORY

In non-degenerate perturbation theory we want to solve SchrÄodinger's equation

$$H\psi_n = E_n\psi_n$$

where

$$H = H_0 + H'$$

and

$$H' \ll H_0.$$

It is then assumed that the solutions to the unperturbed problem

$$H_0\psi_n^0 = E_n^0\psi_n^0$$

are known, in which we have labeled the unperturbed energy by E_n^0 and the unperturbed wave function by ψ_n^0. By non-degenerate we mean that there is only one eigenfunction ψ_n^0 associated with each eigenvalue E_n^0. The wave functions ψ_n^0 form a complete orthonormal set

$$\int \psi_n^{*0}\psi_m^0 d^3r = \left\langle \psi_n^0 \psi_m^0 \right\rangle = \delta_{nm}.$$

Since H' is small, the wave functions for the total problem ψ_n do not differ greatly from the wave functions ψ_n^0 for the unperturbed problem. So we expand ψ_n^0 in terms of the complete set of ψ_n^0 functions

$$\psi_{n'} = \sum_n a_n \psi_n^0.$$

Such an expansion can always be made; that is no approximation. We then substitute the expansion of equation A.10 into SchrÄodinger's equation to obtain

$$H\psi_{n'} = \sum_n a_n (H_0 + H')\psi_n^0 = \sum_n a_n (E_n^0 + H')\psi_n^0 = E_{n'} \sum_n a_n \psi_n^0$$

and therefore we can write

$$\sum_n a_n (E_{n'} + E_n^0)\psi_n^0 = \sum_n a_n H' \psi_n^0.$$

If we are looking for the perturbation to the level m, then we multiply equation from the left by ψ_m^{0*} and integrate over all space. On the left hand side of equation we get $\left\langle \psi_m^0 \mid \psi_n^0 \mid \right\rangle = \delta_{mn}$ while on the right hand side we have the matrix element of the perturbation Hamiltonian taken between the unperturbed states:

$$a_m(E_{n'} - E_m^0) = \sum_n a_n \langle \psi_m^0 | H' | \psi_n^0 | \rangle \equiv \sum_n a_n H'_{mn}$$

where we have written the indicated matrix element as H'_{mn}. Equation A.13 is an iterative equation on the a_n coefficients where each a_m coefficient is related to a complete set of a_n coefficients by the relation

$$a_m = \frac{1}{E_{n'} - E_m^0} \sum_n a_n \langle \psi_m^0 | H' | \psi_n^0 \rangle = \frac{1}{E_{n'} - E_m^0} \sum_n a_n H'_{mn}$$

in which the summation includes the $n = n'$ and m terms. We can rewrite equation to involve terms in the sum $n \neq m$

$$a_m(E_{n'} - E_m^0) = a_m H'_{mm} + \sum_{n \neq m} a_n H'_{mn}$$

so that the coefficient a_m is related to all the other a_n coefficients by:

$$a_m = \frac{1}{E_{n'} - E_m^0 - H'_{mm}} \sum_{n \neq m} a_n H'_{mn}$$

where n' is an index denoting the energy of the state we are seeking. The equation (A.16) written as

$$a_m(E_{n'} - E_m^0 - H'_{mm}) = \sum_{n \neq m} a_n H'_{mn}$$

is an identity in the a_n coefficients. If the perturbation is small then $E_{n'}$ is very close to E_m^0 and the first order corrections are found by setting the coefficient on the right hand side equal to zero and $n' = m$. The next order of approximation is found by substituting for of an on the right hand side of equation and substituting for an the expression

$$a_n = \frac{1}{E_{n'} - E_m^0 - H'_{mm}} \sum_{n \neq m} a_{n''} H'_{mn''}$$

which is obtained from equation by the transcription $m \to n$ and $n \to n''$. In the above, the energy level $E_{n'} = E_m$ is the level for which we are calculating the perturbation. We now look for the a_m term in the sum $\sum_{n'' \neq m} a_{n''} H'_{mn''}$ of equation and bring it to the left hand side of equation. If we are satisfied with our solutions, we end the perturbation calculation at this point. If we are not satisfied, we substitute for the $a_{n''}$ coefficients in equation using the same basic equation to obtain a triple sum. We then select out the a_m term, bring it to the left hand side of equation etc. This procedure gives us an easy recipe to find the energy to any order of perturbation theory. We now write these iterations down more explicitly for first and second order perturbation theory.

1st ORDER PERTURBATION THEORY

In this case, no iterations are needed and the sum $\sum_{n\neq m} a_n H'_{mn}$ on the right hand side of equation is neglected, for the reason that if the perturbation is small, $\psi_{n'} \sim \psi^0_{n'}$. Hence only a_m contributes significantly. We merely write $E_{n'} = E_m$ to obtain:

$$a_m(E_n - E^0_m - H'_{mm})$$

Since the a_m coefficients are arbitrary coefficients, this relation must hold for all a_m so that

$$(E_m - E^0_m - H'_{mm}) = 0$$

or

$$E_m = E^0_m + H'_{mm}.$$

We write even more explicitly so that the energy for state m for the perturbed problem E_m is related to the unperturbed energy E^0_m by

$$E_m = E^0_m + \langle \psi^0_m | H' | \psi^0_m \rangle$$

where the indicated diagonal matrix element of H' can be integrated as the average of the perturbation in the state ψ^0_m. The wave functions to lowest order are not changed

$$\psi_m = \psi^0_m.$$

2nd ORDER PERTURBATION THEORY

If we carry out the perturbation theory to the next order of approximation, one further iteration is required:

$$a_m(E_m - E^0_m - H'_{mm}) = \sum_{n\neq m} \frac{1}{E_m - E^0_n - H'_{nn}} \sum_{n''\neq m} a_{n''} H'_{nn''} H'_{mn}$$

in which we have substituted for the a_n coefficient in equation using the iteration relation given. We now pick out the term on the right hand side of equation for which $n'' = m$ and bring that term to the left hand side of equation. If no further iteration is to be done, we throw away what is left on the right hand side of equation and get an expression for the arbitrary a_m coefficients

$$a_m \left[(E_m - E^0_m - H'_{mm}) - \sum_{n\neq m} \frac{H'_{nm} H'_{mn}}{E_m - E^0_n - H'_{nn}} \right].$$

Since a_m is arbitrary, the term in square brackets in Equation vanishes and the second order correction to the energy results:

$$E_m = E^0_m + H'_{mm}) + \sum_{n\neq m} \frac{|H'_{mn}|^2}{E_m - E^0_n - H'_{nn}}$$

in which the sum on states $n \neq m$ represents the 2nd order correction.

To this order in perturbation theory we must also consider corrections to the wave function

$$\psi_m = \sum_n a_n \psi_n^0 = \psi_m^0 + \sum_{n \neq m} a_n \psi_n^0$$

in which ψ_m^0 is the large term and the correction terms appear as a sum over all the other states $n \neq m$. In handling the correction term, we look for the a_n coefficients, which from Equation are given by

$$a_n = \frac{1}{E_n' - E_n^0 - H_{nn}'} \sum_{n'' \neq n} a_{n''} H_{nn''}'$$

If we only wish to include the lowest order correction terms, we will take only the most important term, i.e., $n'' = m$, and we will also use the relation $a_m = 1$ in this order of approximation. Again using the identification $n' = m$, we obtain

$$a_n = \frac{H_{nm}'}{E_m - E_n^0 - H_{nn}'}$$

and

$$\psi_m = \psi_m^0 + \sum_{n \neq m} \frac{H_{nm}' \psi_n^0}{E_m - E_n^0 - H_{nn}'}$$

For homework, you should do the next iteration to get 3rd order perturbation theory, in order to see if you really have mastered the technique (this will be an optional homework problem).

Now look at the results for the energy E_m and the wave function ψ_m for the 2nd order perturbation theory and observe that these solutions are implicit solutions. That is, the correction terms are themselves dependent on E_m. To obtain an explicit solution, we can do one of two things at this point.

1. We can ignore the fact that the energies differ from their unperturbed values in calculating the correction terms. This is known as Raleigh-SchrÄodinger perturbation theory. This is the usual perturbation theory given in Quantum Mechanics texts and for homework you may review the proof as given in these texts.

2. We can take account of the fact that E_m differs from E^0 m by calculating the correction

terms by an iteration procedure; the first time around, you substitute for E_m the value that comes out of 1st order perturbation theory. We then calculate the second order correction to get E_m. We next take this E_m value to compute the new second order correction term etc. until a convergent value for E_m is reached. This iterative procedure is what is used in Brillouin–Wigner perturbation theory and is a better approximation than Raleigh-SchrÄodinger perturbation theory to both the wave function and the energy eigenvalue for the same order in perturbation theory.

The Brillouin–Wigner method is often used for practical problems in solids. For example, if you have a 2-level system, the Brillouin–Wigner perturbation theory to second order gives an exact result, whereas Rayleigh–SchrÄodinger perturbation theory must be carried out to infinite order.

Let us summarize these ideas. If you have to compute only a small correction by perturbation theory, then it is advantageous to use Rayleigh-Schroodinger perturbation theory because it is much easier to use, since no iteration is needed. If one wants to do a more convergent perturbation theory (i.e., obtain a better answer to the same order in perturbation theory), then it is advantageous to use Brillouin–Wigner perturbation theory. There are other types of perturbation theory that are even more convergent and harder to use than Brillouin–Wigner perturbation theory. But these two types are the most important methods used in solid state physics today.

For your convenience we summarize here the results of the second–order non–degenerate Rayleigh-Schrödinger perturbation theory:

$$E_m = E'_m + H'_{mm} + \sum_n{}' \frac{|H'_{nm}|^2}{E_m^0 - E_n^0} + \cdots$$

$$\psi_m = \psi_m^0 + \sum_n{}' \frac{H'_{nm}\psi_n^0}{E_m^0 - E_n^0} +$$

where the sums in equations denoted by primes exclude the m = n term. Thus,

Brillouin–Wigner perturbation theory contains contributions in second order which occur in higher order in the Rayleigh-Schrödinger form. In practice, Brillouin– Wigner perturbation theory is useful when the perturbation term is too large to be handled conveniently by Rayleigh–Schrödinger perturbation theory but still small enough for perturbation theory to work insofar as the perturbation expansion forms a convergent series.

DEGENERATE PERTURBATION THEORY

It often happens that a number of quantum mechanical levels have the same or nearly the same energy. If they have exactly the same energy, we know that we can make any linear combination of these states that we like and get a new eigenstate also with the same energy. In the case of degenerate states, we have to do perturbation theory a little differently.

Suppose that we have an f-fold degeneracy (or near-degeneracy) of energy levels

$$\underbrace{\psi_1^0, \psi_2^0, \ldots \psi_f^0}_{\text{States with the same or nearly the same energy}} \qquad \underbrace{\psi_{f+1}^0, \psi_{f+2}^0, \cdots}_{\text{States with quite different energy}}$$

We will call the set of states with the same (or approximately the same) energy a "nearly degenerate set" (NDS). In the case of degenerate sets, the iterative Equation still holds. The only difference is that for the degenerate case we solve for the perturbed energies by a different technique, as described below.

Starting with Equation, we now bring to the left-hand side of the iterative equation all terms involving the f energy levels that are in the NDS. If we wish to calculate an energy within the NDS in the presence of a perturbation, we consider all the a_n's within the NDS as large, and those outside the set as small. To first order in perturbation theory, we ignore the coupling to terms outside the NDS and we get f linear homogeneous equations in the an's where $n = 1, 2, \ldots f$. We thus obtain the following equations from Equation:

$$\begin{array}{lllll} a_1(E_1^0 + H'_{11} - E) & +a_2 H'_{12} & +\cdots & +a_f H'_1 f & = 0 \\ a_1 H'_{21} & +a_2(E_2^0 + H'_{22} - E) & +\ldots & +a_f H'_2 f & = 0 \\ \vdots & \vdots & \ddots & \vdots & \\ a_1 H'_{f1} & +a_2 H'_{f2} & +\ldots & +a_f(E_f^0 + H'_{ff} - E) & = 0 \end{array}$$

In order to have a solution of these f linear equations, we demand that the coefficient determinant vanish:

$$\begin{vmatrix} (E_1^0 + H'_{11} - E) & H'_{12} & H'_{13} & \cdots & H'_{1f} \\ H'_{21} & (E_2^0 + H'_{22} - E) & H'_{23} & \cdots & H'_{2f} \\ \vdots & \vdots & \vdots & \ddots & \vdots \\ H'_{1f} & H'_{f2} & \cdots & \cdots & (E_f^0 + H'_{ff} - E) \end{vmatrix} = 0$$

The f eigenvalues that we are looking for are the eigenvalues of the matrix in Equation and the set of orthogonal states are the corresponding eigenvectors. Remember that the matrix elements H'_{ij} that occur in the above determinant are taken between the unperturbed states in the NDS.

The generalization to second order degenerate perturbation theory is immediate. In this case, equations have additional terms. For example, the first relation in equation would then become

$$a_1(E_1^0 + H'_{11} - E) + a_2 H'_{12} + a_2 H'_{13} + \ldots + a_f H'_{1f} = -\sum_{n \neq N\,DS} a_n H'_{1n}$$

and for the a_n in the sum in equation, which are now small (because they are outside the NDS), we would use our iterative form

$$a_n = \frac{1}{E - E_n^0 - H'_{nn}} \sum_{m \neq n} a_n H'_{nm}.$$

But we must only consider the terms in the above sum which are large; these terms are all in the NDS. This argument shows that every term on the left side of equation will have a correction term. For example the correction term to a general coefficient a_i will look as follows:

$$a_i H'_{1i} + a_i \sum_{n \neq N\,DS} \frac{H'_{1n} H'_{ni}}{E - E_n^0 - H'_{nn}}$$

where the first term is the original term from 1st order degenerate perturbation theory and the term from states outside the NDS gives the 2nd order correction terms. So, if we are doing higher order degenerate perturbation theory, we write for each entry in the secular equation the appropriate correction terms that are obtained from these iterations. For example, in 2nd order degenerate perturbation theory, the (1,1) entry to the matrix in Equation would be

$$E_1^0 + H'_{11} + \sum_{n \neq N\,DS} \frac{|H'_{1n}|^2}{E - E_n^0 - H'_{nn}} - E.$$

As a further illustration let us write down the (1,2) entry:

$$H'_{12} + \sum_{n \neq N\,DS} \frac{H'_{1n} H'_{n2}}{E - E_n^0 - H'_{nn}}$$

Again we have an implicit dependence of the 2nd order term in equations on the energy eigenvalue that we are looking for. To do 2nd order degenerate perturbation we again have two options. If we take the energy E in equation as the unperturbed energy in computing the correction terms, we have 2nd order degenerate Rayleigh-SchrÄodinger perturbation theory. On the other hand, if we iterate to get the best correction term, then we call it Brillouin–Wigner perturbation theory.

How do we know in an actual problem when to use degenerate 1st or degenerate 2nd order perturbation theory? If the matrix elements $H^{0\,ij}$ coupling members of the NDS vanish, then we must go to 2nd order. Generally speaking, the first order terms will be much larger than the 2nd order terms, provided that there is no symmetry reason for the first order terms to vanish.

Let us explain this further. By the matrix element H'_{12} we mean $(\psi_1^0|H'|\psi_2^0)$. Suppose the perturbation Hamiltonian H' under consideration is due to an electric field $\vec{E}$

$$H' = -e\vec{r} \cdot \vec{E}$$

where $-e\vec{r}$ is the dipole moment of our system. If now we consider the effect of inversion on H', we see that $\vec{r}$ changes sign under inversion $(x, y, z) \rightarrow -(x, y, z)$, i.e., $\vec{r}$ is an odd function.

Suppose that we are considering the energy levels of the hydrogen atom in the presence of an electric field. We have s states (even), p states (odd), d states (even), etc. The electric dipole moment will only couple an even state to an odd state because of the oddness of the dipole moment under inversion.

Hence there is no effect in 1st order non-degenerate perturbation theory. For the $n = 1$ level, there is an effect due to the electric field in second order so that the correction to the energy level goes as the square of the electric field, i.e., $|\vec{E}|^2$.

For the $n/2$ levels, we treat them in degenerate perturbation theory because the 2s and 2p states are degenerate in the simple treatment of the hydrogen atom. Here, first order terms only appear in entries coupling s and *p* states. To get corrections which split the *p* levels among themselves, we must go to 2nd order degenerate perturbation theory.

Chapter 8

s and Electron-Phonon Interaction

S

The solution of the problem in quantum mechanics using raising and lowering operators. This is aimed at providing a quick review as background for the lecture on phonon scattering processes and other topics in this course.

The Hamiltonian for the in one-dimension is written as:

$$H = \frac{p^2}{2m} + \frac{1}{2}\kappa x^2 .$$

We know classically that the frequency of oscillation is given by $\omega = \sqrt{\kappa / m}$ so that

$$H = \frac{p^2}{2m} + \frac{1}{2}m\omega^2 x^2$$

Define the lowering and raising operators a and $a^\dagger$ respectively by

$$a = \frac{p - im\omega x}{\sqrt{2\hbar m\omega}}$$

$$a^\dagger = \frac{p + im\omega x}{\sqrt{2\hbar m\omega}}$$

Since $[p, x] = \hbar/i$, then $[a, a^\dagger] = 1$ so that

$$H = \frac{1}{2m}[(p + i\omega mx)(p - i\omega mx) + m\hbar\omega)]$$
$$= \hbar\omega[a^\dagger a + 1/2].$$

Let $N = a^\dagger a$ denote the number operator and its eigenstates $|n\rangle$ so that $N|n\rangle = n|n\rangle$ where n is any real number. However

$$\langle n|N|n\rangle = \langle n|a^\dagger a|n\rangle = \langle y|y\rangle = n \geq 0$$

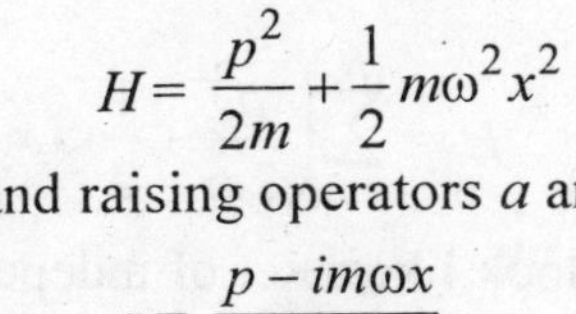

Fig. Simple with Single Spring.

where $|y| = a|n\rangle$ and the absolute value square of the eigenvector cannot be negative. Hence n is a positive number or zero.

$$Na \mid n\rangle = a^{\dagger} aa \mid n\rangle = (aa^{\dagger} - 1)a \mid n\rangle = (n-1)a \mid n\rangle$$

Hence $a|n\rangle = c|n-1\rangle$ and $\langle n|a^{\dagger}a|n\rangle = |c|^2$. However from equation $\langle n|a^{\dagger}a|n\rangle$ $= n$ so that $c = \sqrt{n}$ and $a|n\rangle = \sqrt{n}\,|\,n-1\rangle$. Since the operator a lowers the quantum number of the state by unity, a is called the annihilation operator. Therefore n also has to be an integer, so that the null state is eventually reached by applying operator a for a sufficient number of times.

$$Na^{\dagger} \mid n\rangle = a^{\dagger} aa^{\dagger} \mid n\rangle = a^{\dagger}(1 + a^{\dagger}a) \mid n\rangle = (n+1)a^{\dagger} \mid n\rangle$$

Hence $a^{\dagger} \mid n\rangle = \sqrt{n+1} \mid n+1\rangle$ so that $a^{\dagger}$ is called a raising operator or a creation operator. Finally,

$$H|n\rangle = \hbar\omega[N + 1/2]\ |n\rangle = \hbar\omega(n + 1/2)|n\rangle$$

so the eigenvalues become

$$E = \hbar\omega(n + 1/2), \quad n = 0, 1, 2, ...$$

PHONONS

The lattice vibrations to s and identify the quanta of the lattice vibrations with phonons. Consider the 1-D model of atoms connected by springs. The Hamiltonian for this case is written as:

$$H = \sum_{s=1}^{N}\left(\frac{p_s^2}{2m} + \frac{1}{2}\kappa(x_{s+1} - x_s)^2\right)$$

This equation doesn't look like a set of independent s since x_s and x_s+1 are coupled. Let

$$x_s = 1/\sqrt{N}\sum\nolimits_k Q_k e^{iksa}$$

$$p_s = 1/\sqrt{N}\sum\nolimits_k P_k e^{iksa}$$

These Q_k, P_k's are called phonon coordinates. It can be verified that the commutation relation for momentum and coordinate implies a commutation relation between P_k and $Q_{k'}$

$$[p_s, x_{s'}] = \frac{\hbar}{i}\delta_{ss'} \Rightarrow [P_k, Q_{k'}] = \frac{\hbar}{i}\delta_{kk'}$$

The Hamiltonian in phonon coordinates is:

$$H = \sum_{k}\left(\frac{1}{2m}P_k^{\dagger}P_k + \frac{1}{2}m\omega_k^2 Q_k^{\dagger}Q_k\right)$$

with the dispersion relation given by

$$\omega_k = \sqrt{2k}\,(1-\cos ka)$$

This is all in Kittel ISSP, pp. 611-613. Again let

$$a_k = \frac{iP_k^{\dagger} + m\omega_k Q_k}{\sqrt{2\hbar m\omega_k}},$$

$$a_k^{\dagger} = \frac{-iP_k + m\omega_k Q_k^{\dagger}}{\sqrt{2\hbar m\omega_k}}$$

so that the Hamiltonian is written as:

$$H = \sum_k \hbar\omega_k (a_k^{\dagger}a_k + 1/2) \Rightarrow E = \sum_k (n_k + 1/2)\hbar\omega_k$$

The quantum of energy $\hbar\omega_k$ is called a phonon. The state vector of a system of phonons is written as $|n_1, n_2, ..., n_k, ...\rangle$, upon which the raising and lowering operator can act:

$$a_k|n_1, n_2, ..., n_k, ...\rangle = \sqrt{n_k}\,|n_1, n_2, ..., n_k - 1, ...\rangle$$

$$a_k^{\dagger}|n_1, n_2, ..., n_k, ...\rangle = \sqrt{n_k + 1}\,|n_1, n_2, ..., n_k + 1, ...\rangle$$

From equation it follows that the probability of annihilating a phonon of mode k is the absolute value squared of the diagonal matrix element or n_k.

ELECTRON-PHONON INTERACTION

The basic Hamiltonian for the electron-lattice system is

$$H = \sum_k \frac{p_k^2}{2m} + \frac{1}{2}\sum_{kk'}{}' \frac{e^2}{|\vec{r}_k - \vec{r}_{k'}|} + \sum_i \frac{P_i^2}{2M} + \frac{1}{2}\sum_{ii'}{}' V_{\text{ion}}(\vec{R}_i - \vec{R}_{i'}) + \sum_{k,i} V_{\text{el-ion}}(\vec{r}_k - \vec{R}_i)$$

where the first two terms constitute H_{electron}, the third and fourth terms are denoted by Hion and the last term is $H_{\text{electron-ion}}$. The electron-ion interaction term can be separated into two parts: the interaction of electrons with ions in their equilibrium positions, and an additional term due to lattice vibrations:

$$H_{\text{el-ion}} = H^0_{\text{el-ion}} + H_{\text{el-ph}}$$

$$\sum_{k,i} V_{\text{el-ion}}(\vec{r}_k - \vec{R}_i) = \sum_{k,i} V_{\text{el-ion}}[\vec{r}_k - \vec{R}_i^0 + \vec{s}_i)$$

where $\vec{R}^0$ is the equilibrium lattice site position and $\vec{s}_i$ is the displacement of the atoms from their equilibrium positions in a lattice vibration so that

$$H^0_{\text{el-ion}} = \sum_{k,i} V_{\text{el-ion}}(\vec{r}_k - \vec{R}_i^0)$$

and

$$H_{\text{el-ph}} = \sum_{k,i} \vec{s}_i \cdot \vec{\nabla} V_{\text{el-ion}} (\vec{r}_k - \vec{R}_i^0).$$

In solving the Hamiltonian we use an adiabatic approximation, which solves the electronic part of the Hamiltonian by

$$(H_{\text{electron}} + H^0_{\text{el-ion}})\psi = \text{E}_{\text{el}\psi}$$

and seeks a solution of the total problem as

$$\Psi = \psi(\vec{r}_1, \vec{r}_2, \cdots \vec{R}_1, \vec{R}_2, \cdots)\varphi(\vec{R}_1, \vec{R}_2, \cdots)$$

such that $H\Psi = E\Psi$. Here Ψ is the wave function for the electron-lattice system. Plugging this into the equation , we find

$$E\Psi = \text{H}\Psi = \psi(\text{H}_{\text{ion}} + E_{\text{el}})\varphi - \sum_i \frac{\hbar^2}{2M_i}\left(\varphi\nabla_i^2\psi + 2\vec{\nabla}_{i\varphi} \cdot \vec{\nabla}_{i\psi}\right)$$

Neglecting the last term, which is small, we have

$$H_{\text{ion}}\varphi = (E - E_{\text{el}})\varphi$$

Hence we have decoupled the electron-lattice system.

$$(H_{\text{electron}} + H^0_{\text{el-ion}})\psi = E_{\text{el}}\psi$$

which gives us the energy band structure and ψ satisfies Bloch's theorem while φ is the wave function for the ions

$$H_{\text{ion}}\varphi = E_{\text{ion}}\varphi$$

which gives us phonon spectra and like wave functions, as we have already seen in §B.2.

The discussion has thus far left out the electron-phonon interaction $H_{\text{el-ph}}$

$$\text{H}_{\text{el-ph}} = \sum_{k,i} \vec{s}_i \cdot \vec{\nabla} V_{\text{el-ion}} (\vec{r}_k - \vec{R}_i^0)$$

which is then treated as a perturbation. Since the displacement vector can be written in terms of the normal coordinates $Q_{\vec{q},j}$

$$\vec{s}_i = \frac{1}{\sqrt{NM}} \sum_{\vec{q},j} Q_{\vec{q},j} e^{i\vec{q}\cdot\vec{R}_i^0} \hat{e}_j$$

where j denotes the polarization index, N is the total number of ions and M is the ion mass. Hence

$$\text{H}_{\text{el-ph}} = \sum_{k,i} \frac{1}{\sqrt{NM}} \sum_{\vec{q},j} Q_{\vec{q},j} e^{i\vec{q}\cdot\vec{R}_i^0} \hat{e}_j \cdot \vec{\nabla} V_{\text{el-ion}} (\vec{r}_k - \vec{R}_i^0)$$

where the normal coordinate can be expressed in terms of the lowering and raising operators

$$Q_{\vec{q},j} = \left(\frac{\hbar}{2\omega_{\vec{q},j}}\right)^{\frac{1}{2}} (a_{\vec{q},j} + a^{\dagger}_{-\vec{q},j})$$

Writing out the time dependence explicitly,

$$a_{\vec{q},j}(t) = a_{\vec{q},j} e^{-i\omega\vec{q},jt}$$

$$a^{\dagger}_{\vec{q},j}(t) = a^{\dagger}_{\vec{q},j} e^{i\omega_{\vec{q},j}t}$$

we obtain

$$H_{\text{el-ph}} = -\sum_{\vec{q},j}\left(\frac{\hbar}{2NM\omega_{\vec{q},j}}\right)^{\frac{1}{2}} (a_{\vec{q},j} e^{-i\omega_{\vec{q},j}t} + a^{\dagger}_{\vec{q},j} e^{i\omega_{\vec{q},j}t})$$

$$\times \sum_{k,i} (e^{i\vec{q}} \cdot \vec{R}_i^0 + e^{-i\vec{q}} \cdot \vec{R}_i^0)\hat{e}_j \vec{\nabla} V_{\text{el-ion}}(\vec{r}_k - \vec{R}_i^0)$$

$$= -\sum_{\vec{q},j}\left(\frac{\hbar}{2NM\omega_{\vec{q},j}}\right)^{\frac{1}{2}} \left(a_{\vec{q},j} \sum_{k,i} \hat{e}_j\, e^{i(\vec{q}\vec{R}_i^0 - \omega_{\vec{q},j}t)} \cdot \nabla V_{\text{el-ion}}(\vec{r}_k - \vec{R}_i^0) \right.$$

$$\left. + \, c.c. \right)$$

If we are only interested in the interaction between one electron and a phonon on a particular branch, say the longitudinal acoustic (LA) branch, then we drop the summation over j and k

$$H_{\text{el-ph}} = -\left(\frac{\hbar}{2NM\omega_{\vec{q}}}\right)^{\frac{1}{2}} \left(a_{\vec{q}} \sum_i \hat{e}\, e^{i(\vec{q}\cdot\vec{R}_i^0 - \omega_{\vec{q}}t)} \cdot \nabla V_{\text{el-ion}}(\vec{r}_k - \vec{R}_i^0) + c.c \right)$$

where the first term in the bracket corresponds to the phonon absorption and the c.c. term corresponds to the phonon emission.

With $H_{\text{el-ph}}$ at hand, we can solve transport problems (e.g., τ due to phonon scattering) and optical problems (e.g., indirect transitions) exactly since all of these problems involve the matrix element $\langle f \,|\, H_{\text{el-ph}} \,|\, i \rangle$ of $H_{\text{el-ph}}$ linking states $|i\rangle$ and $|j\rangle$.

Chapter 9

Artificial Atoms and Superconductivity

The charge and energy of a sufficiently small particle of metal or semiconductor are quantized just like those of an atom. The current through such a quantum dot or oneelectron transistor reveals atom-like features in a spectacular way.

The wizardry of modern semiconductor technology makes it possible to fabricate particles of metal or "pools" of electrons in a semiconductor that are only a few hundred angstroms in size. Electrons in these structures can display astounding behaviour. Such structures, coupled to electrical leads through tunnel junctions, have been given various names: single electron transistors, quantum dots, zero-dimensional electron gases and Coulomb islands. Artificial atoms-atoms whose effective nuclear charge is controlled by metallic electrodes.

Like natural atoms, these small electronic systems contain a discrete number of electrons and have a discrete spectrum of energy levels. Artificial atoms, however, have a unique and spectacular property: The current through such an atom or the capacitance between its leads can vary by many orders of magnitude when its charge is changed by a single electron. Why this is so, and how we can use this property to measure the level spectrum of an Artificial atom, is the subject of this article.

To understand Artificial atoms it is helpful to know how to make them. One way to confine electrons in a small region is by employing material boundaries by surrounding a metal particle with insulator, for example. Alternatively, one can use electric field s to confine electrons to a small region within a semiconductor.

Either method requires fabricating very small structures. This is accomplished by the techniques of electron and x-ray lithography. Instead of explaining in detail how Artificial atoms are actually fabricated.

Two kinds of what is sometimes called, for reasons that will soon become clear, a single-electron transistor. In the first type, the all-metal Artificial atom, electrons are confined to a metal particle with typical dimensions of a few thousand angstroms or less. The particle is separated from the leads by thin insulators, through which electrons must tunnel to get from one side to the other. The leads are labeled "source" and "drain" because the electrons enter through the former and leave through the latter the same way the leads are labeled for conventional field effect transistors, such as those in the memory of your personal computer. The entire structure sits near a large, well-insulated metal electrode, called the gate.

A structures that is conceptually similar to the all-metal atom but in which the confinement is accomplished with electric field s in gallium arsenide. Like the all-metal atom, it has a metal gate on the bottom with an insulator above it; in this type of atom the insulator is AlGaAs. When a positive voltage V_g is applied to the gate, electrons accumulate in the layer of GaAs above the AlGaAs. Because of the strong electric field at the

AlGaAs-GaAs interface, the electrons' energy for motion perpendicular to the interface is quantized, and at low temperatures the electrons move only in the two dimensions parallel to the -interface. The special feature that makes this an Artificial atom is the pair of electrodes on the top surface of the GaAs. When a negative voltage is applied between these and the source or drain, the electrons are repelled and cannot accumulate underneath them.

Consequently the electrons are confinedin a narrow channel between the two electrodes. Constrictions sticking but into the channel repel the electrons and create potential barriers at either end of the channel. An electron to travel from the source to the drain it must tunnel through the barriers. The "pool" of electrons that accumulates between the two constrictions plays the same role that the small particle plays in the all-metal atom, and the potential barriers from the constrictions play the role of the thin insulators. Because one can control the height of these barriers by varying the voltage on the electrodes, This type of Artificial atom the controlled-barrier atom. Controlled- barrier atoms in which the heights of the two potential barriers can be varied independently have also been fabricated. In addition, there are structures that behave like controlled-barrier atoms but in which the barriers are caused by charged impurities or grain boundaries.

The electrons in a layer of GaAs are sandwiched between two layers of insulating AlGaAs. One or both of these insulators acts as a tunnel barrier. If both barriers are thin, electrons can tunnel through them, and the structure is analogous to the single-electron transistor without the gate. Such structures, usually called quantum dots, have been studied extensively. The cylinder can be made by etching away unwanted regions of the layer structure, or a metal electrode on the surface, can be used to repel electrons everywhere except in a small circular section of GaAs. Although a gate electrode can be added to this kind of structure, most of the experiments have bee done without one, so we call this the two-probe atom.

CHARGE QUANTIZATION

One way to learn about natural atoms is to measure the energy required to add or remove electrons. This usually done by photoelectron spectroscopy. For example the minimum photon energy needed to remove an electron is the ionization potential, and the maximum energy (photons emitted when an atom captures an electron is the electron affinity. To learn about Artificial atoms we also measure the energy needed to add or subtract electron.

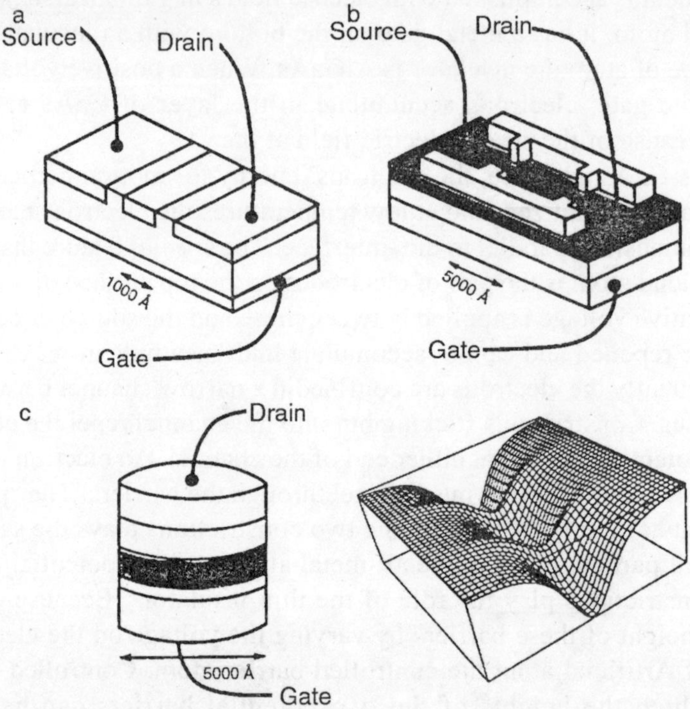

Fig. The many forms of Artificial atoms include the all-metal atom (a), the controlled-barrier atom (b) and the two-probe atom, or "quantum dot" (c). A potential similar to the one in the controlled-barrier atom, plotted as a function of position at the semiconductor-insulator interface. The electrons must tunnel through potential barriers caused by the two constrictions. For capacitance measurements with a two-probe atom, only the source barrier is made thin enough for tunneling, but for current measurements both source and drain barriers are thin.

However, we do it by measuring the current through the Artificial atom.

The current through a controller barrier atom as a function of the voltage V_g between the gate and the atom. One obtains this plot by applying very small voltage between the source and drain, just large enough to measure the tunneling conductance between them.

The results are astounding. The conductance(displays sharp resonances that are almost periodic in V_g By calculating the capacitance between the Artificial atom and the gate we can show[2];8 that the period is the voltage necessary to add one electron to the confinedpool of electrons. That is why we sometimes call the controller barrier atom a single-electron transistor: Whereas the transistors in your personal computer turn on only on(when many electrons are added to them, the Artificial atom turns on and off again every time a single electron added to it.

A simple theory, the Coulomb blockade model, explains the periodic conductance resonances. This model is quantitatively correct for the all-metal atom and qualitatively correct for the controlled-barrier atom. To understand the model, think about how an electron in the all-metal atom tunnels from one lead onto the metal particle and then onto the other lead.

Suppose the particle is neutral to begin with. To add a charge Q to the particle requires energy $Q^2/2C$, where C is the total capacitance between the particle and the rest of the system; since you cannot add less than one electron the flow of current requires a Coulomb energy $e^2/2C$. This energy barrier is called the Coulomb blockade. A fancier way to say this is that charge quantization leads to an energy gap in the spectrum of states for tunneling: For an electron to tunnel onto the particle, its energy must exceed the Fermi energy of the contact by $e^2/2C$, and for a hole to tunnel, its energy must be below the Fermi energy by the same amount. Consequently the energy gap has width e^2/C. If the temperature is low enough that $kT < e^2/2C$, neither electrons nor holes can flow from one lead to the other.

The gap in the tunneling spectrum is the difference between the "ionization potential" and the "electron affinity" of the Artificial atom. For a hydrogen atom the ionization potential is 13.6 eV, but the electron affinity, the binding energy of H^-, is only 0.75 eV. This large difference arises from the strong repulsive interaction between the two electrons bound to the same proton. Just as for natural atoms like hydrogen, the difference between the ionization potential and electron affinity for artificial atoms arises from the electronelectron interactions; the difference, however, is much smaller for Artificial atoms because they are much bigger than natural ones.

By changing the gate voltage V_g one can alter the energy required to add charge to the particle. V_g is applied between the gate and the source, but if the drain-source voltage is very small, the source, drain and particle will all be at almost the same potential. With V_g applied, the electrostatic energy of a charge Q on the particle is

$$E = QV_g + Q^2/2C$$

For negative charge Q, the first term is the attractive interaction between Q and the positively charged gate electrode, and the second term is the repulsive interaction among the bits of charge on the particle. Equation C.1 shows that the energy as a function of Q is a parabola with its minimum at $Q = -CV_g$.

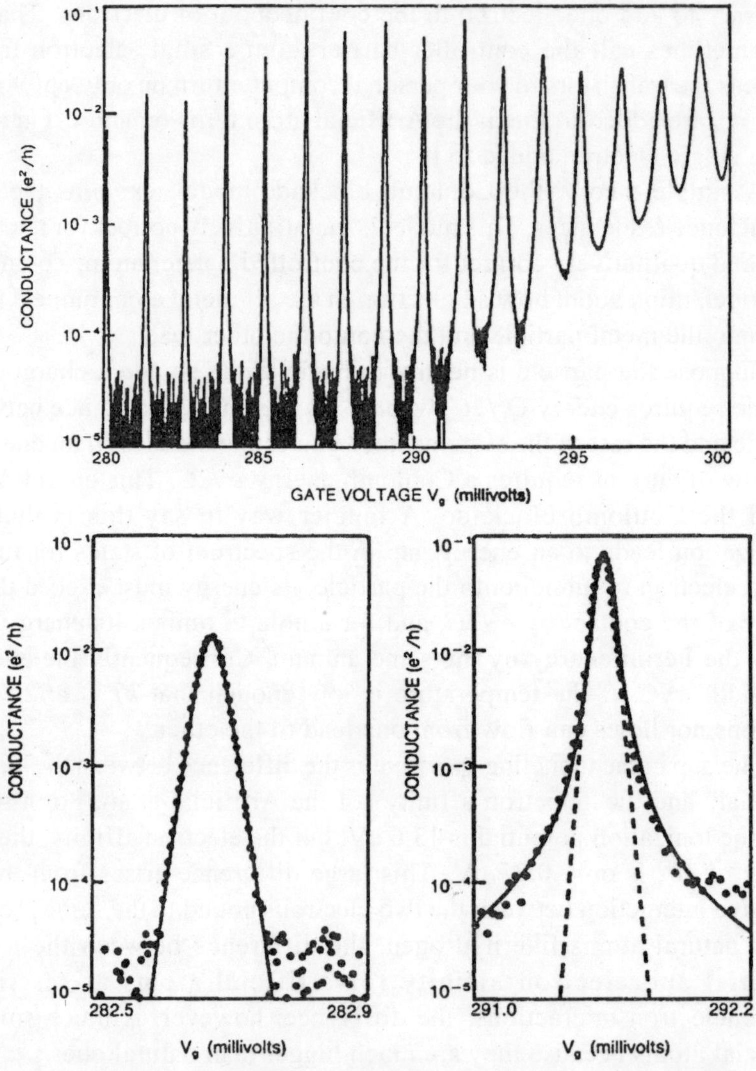

Fig. Conductance of a controlled-barrier atom as a function of the voltage V_g on the gate at a temperature of 60 mK. At low V_g (solid curve) the shape of the resonance is given by the thermal distribution of electrons in the source that are tunneling onto the atom, but at high V_g a thermally broadened Lorentzian is a better description than the thermal distribution alone (dashed curve).

For simplicity the gate is the only electrode that contributes to C, in reality, there are other contributions.

By varying V_g we can choose any value of Q_0, the charge that would minimize the energy in equation C.1 if charge were not quantized. However, because the real charge is quantized, only discrete values of the energy E are possible.

When $Q_0 = -Ne$, an integral number N of electrons minimizes E, and the Coulomb interaction results in the same energy difference $e^2/2C$ for increasing or decreasing N by 1. For all other values of Q_0 except $Q_0 = -(N + 1/2)e$ there is a smaller, but nonzero, energy for either adding or subtracting an electron. Under such circumstances no current can flow at low temperature.

However, if $Q_0 = -(N + 1/2)$e the state with $Q = -Ne$ and that with $Q = (N + 1)$e are degenerate, and the charge fluctuates between the two values even at zero temperature. Consequently the energy gap in the tunneling spectrum disappears, and current can flow. The peaks in conductance are therefore periodic, occurring whenever $CV_g = Q_0 = -(N + 1/2)e$, spaced in gate voltage by e/C.

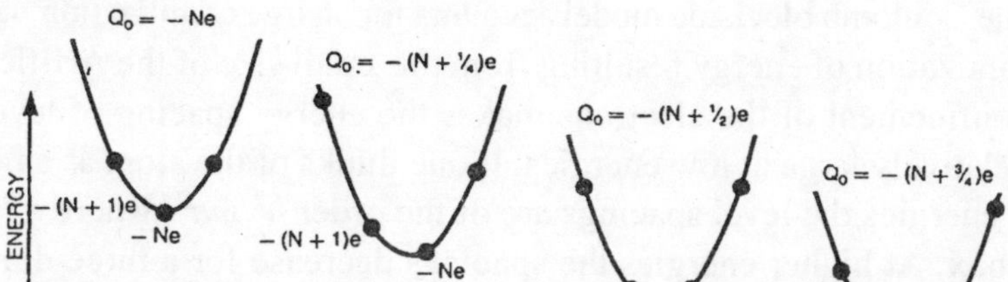

Fig. Total energy (top) and tunneling energies (bottom) for an Artificial atom. As voltage is increased the charge Q_0 for which the energy is minimized changes from $-Ne$ to $-(N + 1/4)e$. Only the points corresponding to discrete numbers of electrons on the atom are allowed (dots on upper curves). Lines in the lower diagram indicate energies needed for electrons or holes to tunnel onto the atom. When $Q_0 = -(N + 1/2)e$ the gap in tunneling energies vanishes and current can flow.

There is a gap in the tunneling spectrum for all values of V_g except the charge-degeneracy points. The more closely spaced discrete levels shown outside this gap are due to excited states of the electrons present on the Artificial atom. As V_g is increased continuously, the gap is pulled down relative to the Fermi energy until a charge degeneracy point is reached. On moving through this point there is a discontinuous change in the tunneling spectrum: The gap collapses and then reappears shifted up by e^2/C. Simultaneously the charge on the Artificial atom increases by 1 and the process starts over again. A charge-degeneracy point and a conductance peak are reached every time the voltage is increased by e/C, the amount necessary to add one electron to the Artificial atom. Increasing the gate voltage of an Artificial atom is therefore analogous to moving through the periodic table for natural atoms by increasing the nuclear charge.

The quantization of charge on a natural atom is something we take for granted. However, if atoms were larger, the energy needed to add or remove electrons would be smaller, and the number of electrons on them would fluctuate except at very low temperature. The quantization of charge is just one of the properties that Artificial atoms have in common with natural ones.

ENERGY QUANTIZATION

The Coulomb blockade model accounts for charge quantization but ignores the quantization of energy resulting from the small size of the Artificial atom. This confinement of the electrons makes the energy spacing of levels in the atom relatively large at low energies. If one thinks of the atom as a box, at the lowest energies the level spacings are of the order $\hbar^2/ma^2$, where a is the size of the box. At higher energies the spacings decrease for a three-dimensional atom because of the large number of standing electron waves possible for a given energy. If there are many electrons in the atom, they fill up many levels, and the level spacing at the Fermi energy becomes small. The all-metal atom has so many electrons (about 10^7) that the level spectrum is effectively continuous. Because of this, many experts do not regard such devices as "atoms," it is helpful to think of them as being atoms in the limit in which the number of electrons is large. In the controlled-barrier atom, however, there are only about 30–60 electrons, similar to the number in natural atoms like krypton through xenon.

Two-probe atoms sometimes have only one or two electrons. (There are actually many more electrons that are tightly bound to the ion cores of the semiconductor, but those are unimportant because they cannot move.) For most cases, therefore, the spectrum of energies for adding an extra electron to the atom is discrete, just as it is for natural atoms.

One can measure the energy level spectrum directly by observing the tunneling current at fixed V_g as a function of the voltage V_{ds} between drain and source. Suppose we adjust V_g so that, for example, $Q_0 = -(N + 1/4)e$ and then begin to increase V_{ds}. The Fermi level in the source rises in proportion to V_{ds} relative to the drain, so it also rises relative to the energy levels of the Artificial atom. Current begins to flow when the Fermi energy of the source is raised just above the first quantized energy level of the atom. As the Fermi energy is raised further, higher energy levels in the atom fall below it, and more current flows because there are additional channels for electrons to use for tunneling onto the Artificial atom. We measure an energy level by measuring the voltage at which the current increases or, equivalently, the voltage at which there is a peak in the derivative of the current, dI/dV_{ds}. (We need to correct for the increase in the energy of the atom with V_{ds}, but this is a small effect.) Many beautiful tunneling spectra of this kind have been measured[5] for two-terminal atoms.

Increasing the gate voltage lowers all the energy levels in the atom by eV_g, so that the entire tunneling spectrum shifts with V_g. One can observe this effect by plotting the values of V_{ds} at which peaks appear in dI/dV_{ds}. As V_g increases you can see the gap in the tunneling spectrum shift lower and then disappear at the charge-degeneracy point, just as the Coulomb blockade model predicts.

The discrete energy levels of the Artificial atom. For the range of V_s the voltage is only large enough to add or remove one electron from the atom; the discrete levels above the gap are the excited states of the atom with one extra electron, and those below the gap are the excited states of the atom with one electron missing (one hole). At still higher voltages one observes levels for two extra electrons or holes and so forth. The charge-degeneracy points are the values of V_g for which one of the energy levels of the Artificial atom is degenerate with the Fermi energy in the leads when $V_{ds} = 0$, because only then can the charge of the atom fluctuate.

In a natural atom one has little control over the spectrum of energies for adding or removing electrons. There the electrons interact with the fixed potential of the nucleus and with each other, and these two kinds of interaction determine the spectrum. In an Artificial atom, however, one can change this spectrum completely by altering the atom's geometry and composition. For the all-metal atom, which has a high density of electrons, the energy spacing between the discrete levels is so small that it can be ignored. The high density of electrons also results in a short screening length for external electric field s, so electrons added to the atom reside on its surface. Because of this, the electron-electron interaction is always e^2/C (where C is the classical geometrical capacitance), independent of the number of electrons added. This is exactly the case for which the Coulomb blockade model was invented, and it works well: The conductance peaks are perfectly periodic in the gate voltage. The difference between the "ionization potential" and the "electron affinity" is e^2/C, independent of the number of electrons on the atom.

In the controlled-barrier atom, the level spacing is one or two tenths of the energy gap. The conductance peaks are not perfectly periodic in gate voltage, and the difference between ionization potential and electron affinity has a quantum mechanical contribution.

In the two-probe atom the electron-electron interaction can be made very small, so that

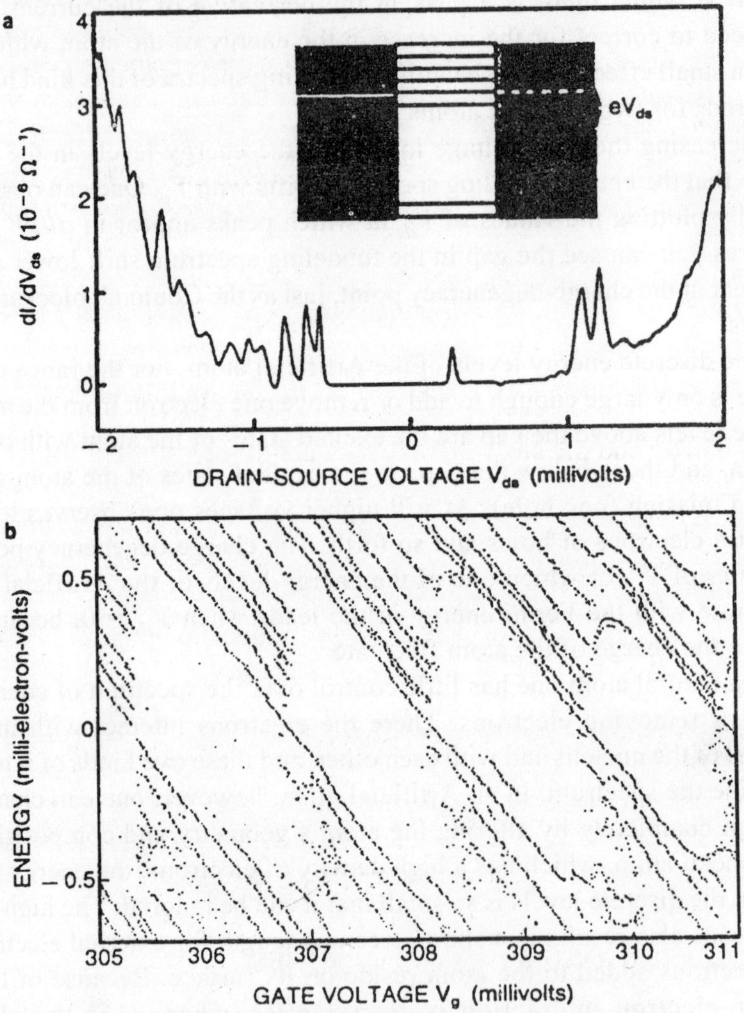

Fig. Discrete energy levels of an Artificial atom can be detected by varying the drain-source voltage.

When a large enough V_{ds} is applied, electrons overcome the energy gap and tunnel from the source to the Artificial atom. a: Every time a new discrete state is accessible the tunneling current increases, giving a peak in dI/dV_{ds}. The Coulomb blockade gap is the region between about –0.5 mV and + 0.3

mV where there are no peaks. *b*: Plotting the positions of these peaks at various gate voltages gives the level spectrum.

Note how the levels and the gap move downward as V_g increases. One can in principle reach the limit opposite to that of the all-metal atom. One can find the energy levels of a two-probe atom by measuring the capacitance between its two leads as a function of the voltage between them. When no tunneling occurs, this capacitance is the series combination of the source-atom and atom-drain capacitances. For capacitance measurements, two-probe atoms are made with the insulating layer between the drain and atom so thick that current cannot flow under any circumstances. Whenever the Fermi level in the source lines up with one of the energy levels of the atom, however, electrons can tunnel freely back and forth between the atom and the source. This causes the total capacitance to increase, because the source-atom capacitor is effectively shorted by the tunneling current. The amazing thing about this experiment is that a peak occurs in the capacitance every time a single electron is added to the atom. The voltages at which the peaks occur give the energies for adding electrons to the atom, just as the voltages for peaks in dI/dV_{ds} do for the controlled-barrier atom or for a two probe atom in which both the source-atom barrier and the atom-drain barrier are thin enough for tunneling.

The energies for adding electrons to a two-probe atom vary with a magnetic field perpendicular to the GaAs layer. In an all-metal atom the levels would be equally spaced, by e^2/C , and would be independent of magnetic field because the electron-electron interaction completely determines the energy. By contrast, the levels of the two-probe atom are irregularly spaced and depend on the magnetic field in a systematic way. For the two-probe atom the fixed potential determines the energies at zero field. The level spacings are irregular because the potential is not highly symmetric and varies at random inside the atom because of charged impurities in the GaAs and AlGaAs. It is clear that the electron-electron interactions that are the source of the Coulomb blockade are not always so important in the two-probe atom as in the all-metal and controlled-barrier atoms. Their relative importance depends in detail on the geometry.

ARTIFICIAL ATOMS IN A MAGNETIC FIELD

Level spectra for natural atoms can be calculated theoretically with great accuracy, and it would be nice to be able to do the same for Artificial atoms. No one has yet calculated. an entire spectrum. However, for a simple geometry we can now predict the charge-degeneracy points, the values of V_g corresponding to conductance peaks. From the earlier discussion it should be clear that in such a calculation one must take into account the electron's interactions with both the fixed potential and the other electrons.

The simplest way to do this is with an extension of the Coulomb blockade model. It is assumed, as before, that the contribution to the gap in the tunneling spectrum from the Coulomb interaction is e^2/C no matter how many electrons are added to the atom.

To account for the discrete levels one pretends that once on the atom, each electron interacts independently with the fixed potential. All one has to do is solve for the energy levels of a single electron in the fixed potential that creates the Artificial atom and then fill those levels in accordance with the Pauli exclusion principle. Because the electron-electron interaction is assumed always to be e^2/C , this is called the constant-interaction model.

Now think about what happens when one adds electrons to a controlled-barrier atom by increasing the gate voltage while keeping V_{ds} just large enough so one can measure the conductance. When there are $N-1$ electrons on the atom the $N-1$ lowest energy levels are filled. The next conductance peak occurs when the gate voltage pulls the energy of the atom down enough that the Fermi level in the source and drain becomes degenerate with the Nth level. Only when an energy level is degenerate with the Fermi energy can current flow; this is the condition for a conductance peak.

When V_g is increased further and the next conductance peak is reached, there are N electrons on the atom, and the Fermi level is degenerate with the $(N+1)$-th level. Therefore to get from one peak to the next the Fermi energy must be raised by $e^2/C + (E_{N+1} - E_N)$, where $E_{N'}$ is the energy of the Nth level of the atom. If the energy levels are closely spaced the Coulomb blockade result is recovered, but in general the level spacing contributes to the energy between successive conductance peaks.

It turns out that we can test the results of this kind of calculation best if a magnetic field is applied perpendicular to the GaAs layer. For free electrons in two dimensions, applying the magnetic field results in the spectrum of Landau levels with energies $(n + 1{=}/2)\hbar\omega_c$ where the cyclotron frequency is $\omega_c = eB/m^*c$, and m^* is the effective mass of the electrons. In the controlled barrier atom and the two-probe atom, we expect levels that behave like Landau levels at high field s, with energies that increase linearly in B. This behaviour occurs because when the field is large enough the cyclotron radius is much smaller than the size of the electrostatic potential well that confines the electrons, and the electrons act as if they were free. Levels shifting proportionally to B , as expected, are seen experimentally.

To calculate the level spectrum we need to model the fixed potential, the analog of the potential from the nucleus of a natural atom. The simplest choice is a potential, and this turns out to be a good approximation for the controlled-barrier atom.

The calculated level spectrum as a function of magnetic field for noninteracting electrons in a two dimensional potential. At low field s the energy levels dance around wildly with magnetic field. This occurs because some states have large angular momentum and the resulting magnetic moment causes their energies to shift up or down strongly with magnetic field.

As the field is increased, however, things settle down. For most of the field range shown there are four families of levels, two moving up, the other two down. At the highest field s there are only two families, corresponding to the two possible spin states of the electron.

Suppose we measure, in an experiment like the one whose results, the gate voltage at which a specific peak occurs as a function of magnetic field. This value of V_g is the voltage at which the Nth energy level is degenerate with the Fermi energy in the source and drain. A shift in the energy of the level will cause a shift in the peak position.

The calculated energy of the 39th level, so it gives the prediction of the constant-interaction model for the position of the 39th conductance peak. As the magnetic field increases, levels moving up in energy cross those moving down, but the number of electrons is fixed, so electrons jump from upward-moving filled levels to downward-moving empty ones. The peak always follows the 39th level, so it moves up and down in gate voltage.

A measurement of V_g for one conductance maximum, as a function of B. The behaviour is qualitatively similar to that predicted by the constant-interaction model. The peak mover up and down with increasing B , and the frequency of level crossings changes at the field where only the last two families of levels remain. However, at high B the frequency is predicted to be much lower than what is observed experimentally. While the constant-interaction model is in qualitative agreement with experiment, it is not quantitatively correct.

To anyone who has studied atomic physics, the constant-interaction model seems quite crude. Even the simplest models used to calculate energies of many electron atoms determine the charge density and potential self-consistently. One begins by calculating the charge density that would result from noninteracting electrons in the fixed potential, and then one calculates the effective potential an electron sees because of the fixed potential and the potential resulting from this charge density. Then one calculates the charge density again. One does this repeatedly until the charge density and potential are self-consistent.

The constant-interaction model fails because it is not self- consistent. The results of a self-consistent calculation for the controlled-barrier atom. It is in good agreement with experiment-much better agreement than the constant-interaction model gives.

CONDUCTANCE LINE SHAPES

In atomic physics, the next step after predicting energy levels is to explore how an atom interacts with the electromagnetic field , because the absorption and emission of photons teaches us the most about atoms. For Artificial atoms, absorption and emission of electrons plays this role, so we had better understand how this process works.

Think about what happens when the gate voltage in the controlled-barrier atom is set at a conductance peak, and an electron is tunneling back and forth between the atom and the leads. Since the electron spends only a finite time τ on the atom, the uncertainty principle tells us that the energy level of the electron has a width $\hbar/\tau$. Furthermore, since the probability of finding the electron on the atom decays as $e^{t/\tau}$, the level will have a Lorentzian line shape.

This line shape can be measured from the transmission probability spectrum $T(E)$ of electrons with energy E incident on the Artificial atom from the source. The spectrum is given by

$$T(E)=\frac{\Gamma^2}{\Gamma^2+(E-E_N)^2}$$

where Γ is approximately $\hbar/\tau$ and E_N is the energy of the Nth level. The probability that electrons are transmitted from the source to the drain is approximately,proportional to the conductance G.

In fact, G; $(e^2/h)T$, where e^2/h is the quantum of conductance. It is easy to show that one must have $G < e^2 = h$ for each of the barriers separately to observe conductance resonances. This condition is equivalent to requiring that the separation of the levels is greater than their width Γ.

Like any spectroscopy, our electron spectroscopy of Artificial atoms has a finite resolution. The resolution is determined by the energy spread of the electrons in the source, which are trying to tunnel into the Artificial atom. These electrons are distributed according to the Fermi-Dirac function,

$$f(E)=\frac{1}{\exp[(E-E_F)/kT]+1}$$

where E_F is the Fermi energy. The tunneling current is given by

$$I=\int\frac{e}{h}T(E)[f(E)-f(E-eV_{ds})]dE.$$

Equation says that the net current is proportional to the probability $f(E)T(E)$ that there is an electron in the source with energy E and that the electron can tunnel between the source and drain minus the equivalent probability for electrons going from drain to source.

The best resolution is achieved by making $V_{ds} \ll kT$. Then $[f(E) - f(E - eV_{ds})]$; $eV_{ds}\,(df/dE)$, and I is proportional to V_{ds}, so the conductance is I/V_{ds}.

The experiments well: At low V_g, where Γ is much less than kT, the shape of the conductance resonance is given by the resolution function df/dE. But at higher V_g one sees the Lorentzian tails of the natural line shape quite clearly.

The width Γ depends exponentially on the height and width of the potential barrier, as is usual for tunneling. The height of the tunnel barrier decreases with *Vg*, which is why the peaks become broader with increasing *Vg*. Just as we have control over the level spacing in Artificial atoms, we also can control the coupling to the leads and therefore the level widths.

It is clear why the present generation of Artificial atoms show unusual behaviour only at low temperatures: When kT becomes comparable to the energy separation between resonances, the peaks overlap and the features disappear.

APPLICATIONS

The behaviour of Artificial atoms is so unusual that it is natural to ask whether they will be useful for applications to electronics. Some clever things can be done., Because of the electron-electron interaction, only one electron at a time can pass through the atom.

With devices like the "turnstile" device shown on the cover of this issue the two tunnel barriers can be raised and lowered independently. Suppose the two barriers are raised and lowered sequentially at a radio or microwave frequency ν.

Then, with a small source-drain voltage applied, an electron will tunnel onto the atom when the source-atom barrier is low and off it when the atom-drain barrier is low. One electron will pass in each time interval ν^{-1}, producing a current eν.

Other applications, such as sensitive electrometers, can be imagined. However, the most interesting applications may involve devices in which several Artificial atoms are coupled together to form Artificial molecules or in which many are coupled to form Artificial solids. Because the coupling between the Artificial atoms can be controlled, new physics as well as new applications may emerge. The age of Artificial atoms has only just begun.

SUPERCONDUCTIVITY

Before we start with the theoretical treatment of superconducitvity, we review some of the charactistic experimental facts, in order to gain an overall picture of this striking manifestation of quantum mechanics on a macrosocpic scale.

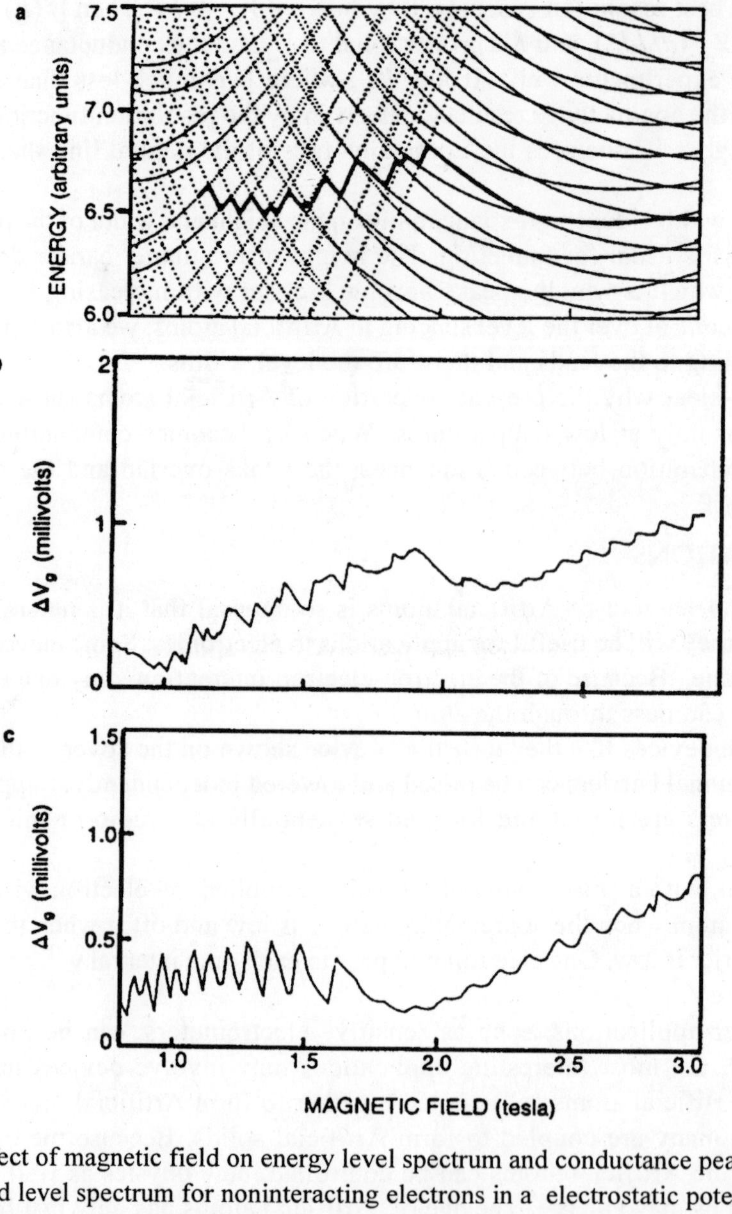

Fig. Effect of magnetic field on energy level spectrum and conductance peaks. a: Calculated level spectrum for noninteracting electrons in a electrostatic potential as a function of magnetic field. The prediction that the constant interaction model gives for the gate voltage for the 39th conductance peak. b: Measured position of a conductance peak in a controlled-barrier atom as a function of field. c: Position of the 39th conductance peak versus field , calculated self-consistently. The scale in c does not match that in b because parameters in the calculation were not precisely matched to the experimental conditions.

SUPERCONDUCTIVITY

We start with an experimental result that gives the name of this phenomenon, namely with the rather abrupt change in resistivity at a temperature $T = T_c$ called the critical temperature. At temperatures above T_c, the metal is in the normal state,

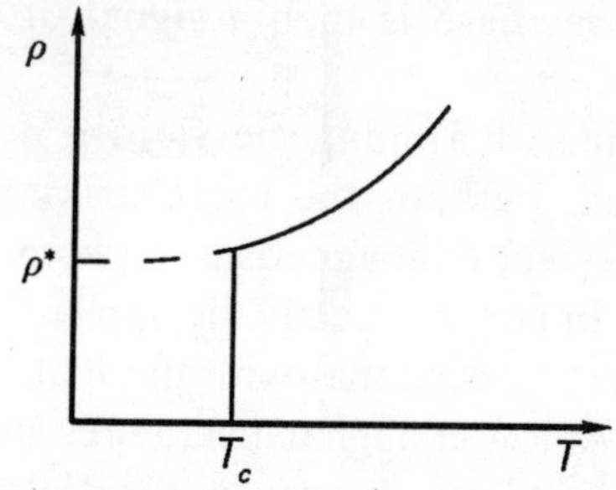

Fig. Schematic representation of the resistivity of a metal with a transition to a superconducting phase at T_c.

where $\rho(T) \sim T^2$ at low enough temperatures in the case of a Fermi liquid. In the absence of the transition, the resistivity extrapolates in general to a finite value as $T \to 0$, the residual resisitivity, that arises due to the presence of impurities, the residual resistivity is denoted ρ^*. In the case of a superconductor, however, the resisitivity vanishes at $T = T_c$ and no dissipation is present anymore for $T < T_c$. Accompanying this change in the resistivity, there is also a feature in the specific heat c_V that is common to phase transitions.

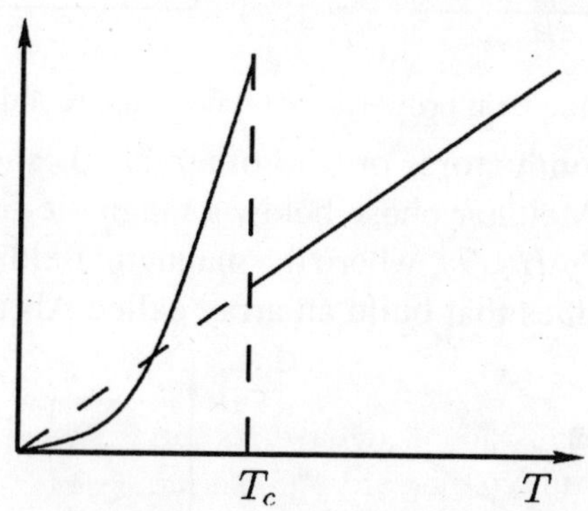

Fig. Schematic representation of the specific heat of a metal with a transition to a superconducting phase at T_c.

At temperatures $T > T_c$ the specific heat (actually the electronic contribution to it) decreases linearly with decreasing temperature, as expected for Fermi liquids. It can be easily seen in the case of the Fermi gas, that this is a consequence of the Fermi-Dirac statistics. At T_c, however, there is a singularity in c_V that manifests experimentally as a jump. At low temperatures the specific heat decreases exponentially,

$$c_V \sim \exp\left(-\frac{a}{T}\right).$$

Since the specific heat is associated with the density of states at the Fermi energy, an exponential decrease is then a signal of an energy gap in the electronic spectrum.

Also in the presence of a magnetic field a number of remarkable phenomena are observed. Perhaps the most striking one is the fact that superconductors show perfect diamagnetism, a phomenon known under the name of Meißner effect. In normal metals the applied magnetic field and the total magnetic field B are proportional with the magnetic permeability as a porportionality constant. At low enough temperatures and aplied magnetic field H in the superconducting phase, however, the field B inside the sample vanishes. In the H, T plane, the phase diagram shows a line $H_c(T)$ separating the superconducting and the normal phases. Both features correspond to so-called type I superconductors.

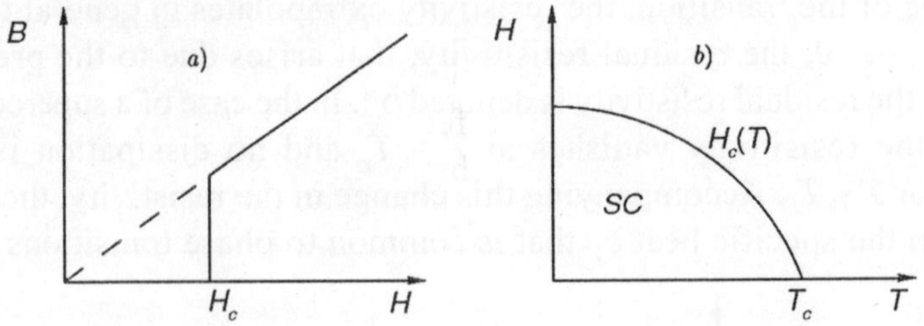

Fig. Magnetic properties of type I superconductors.

Type II superconductors, on the other hand, show a more complex behaviour, with the Meißner phase below a magnetic field $H_{c1}(T)$ and a new phase for $H_{c1}(T) < H < H_{c2}(T)$, where the magnetic field penetrates the sample in the form of flux tubes that build an array called Abrikosov lattice.

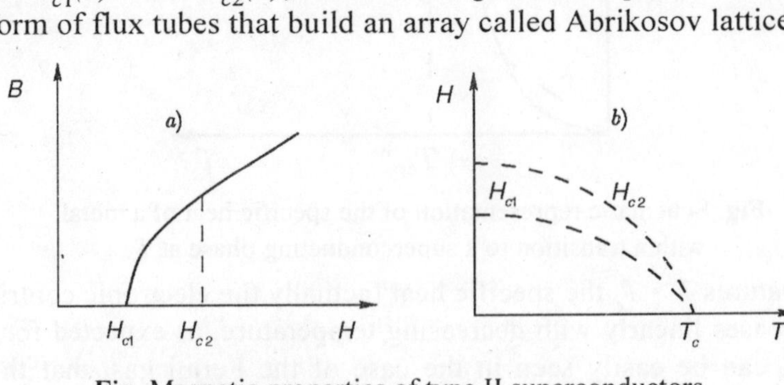

Fig. Magnetic properties of type II superconductors.

We will discuss first a phenomenological theory due to Ginzburg and Landau, that starting with fro phase transitions will be able to describe consistently many of the features discussed above. Later, we discuss electron-phonon coupling and the BCS-theory due to Bardeen, Cooper and Schrieûer, that will give us a microscopic insight into the mechanism of superconductivity.

THEORY

This phenomelogical theory is a special case of the general theory for phase transitions developed by Landau.

In this frame, we consider the F as a functional of a field that should describe the possible phases of the system. This order parameter ψ has the property of being zero in the high temeprature phase, where the new order is still not established, and $\psi \neq 0$ below T_c, where the new phase appears. The order parameter should be such that it contains information relevant to the ordered phase.

In the case of superconductivity, we consider that the new phase can be described by a wave function, and hence it will have in general a real and an imaginary part. We say then, that the order parameter has two components. In the general frame of the theory of phase transitions, the number of components of the order parameter will determine in general the major features of the transition. For example, in the case of a ferromagnet, if the system is isotropic in spin space, the order parameter will be a vector with three components.

WITHOUT MAGNETIC FIELD

We consider first the form that the is supposed to have without taking into account a magnetic field. We assume that close to the phase transition, the should be a functional of Ψ with Ψ small so that an expansion in powers of it can be performed. Since the is a real quantity, it will depend only on the modulus of the order parameter. We assume the following form for the .

$$F[\psi] = F_n(T) + \frac{1}{V}\int d^3x \left\{ a\frac{T - T_c}{T_c}|\psi(x)|^2 + \frac{b}{2}|\psi(x)|^2 + \cdots \right.$$
$$\left. + \frac{\hbar^2}{2m}|\nabla\psi(x)|^2 + \cdots \right\}$$

where $F_n(T)$ is the contribution for the normal state, that we assume to be continuous and analytical across the transition. Furthermore, $\cdots$ represent higher order terms that in principle could be included. However, nowadays we know through renormalization group arguments that these terms are irrelevant for the critical properties of the phase transition.

In order to have a physical understanding of the parameters entering the , let us consider the case where $| \psi (x) |$ is homogeneous, i.e. it does not depend on position. Then, we can regard F as a potential depending only on $| \psi |$. For $| \psi |$ very small we need only to consider the term of order $O(| \psi |^2)$. Taking a > 0, we have a close to $| \psi | = 0$.

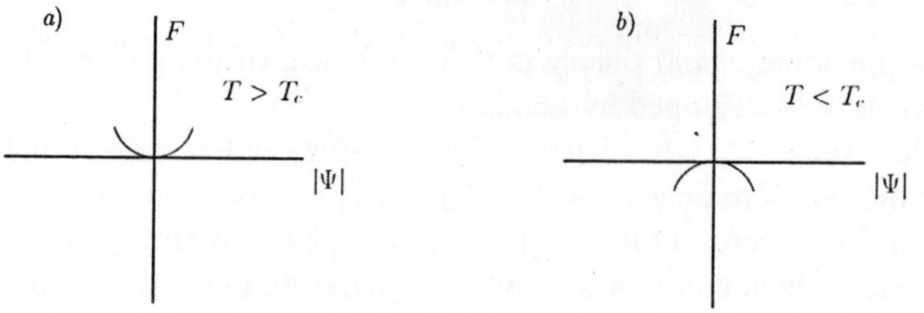

Fig. close to $| \psi | = 0$ for $a > 0$.

Since the physical state corresponds to the minimum of the , the choice a > 0, the phase at $T > T_c$ corresponds to a vanishing order paramter, whereas for $T < T_c$ such a state is unstable. If the expansion in powers of $| \psi (x) |$ is cut at the fourth order, then, we should have $b > 0$ in order to have a stable system.

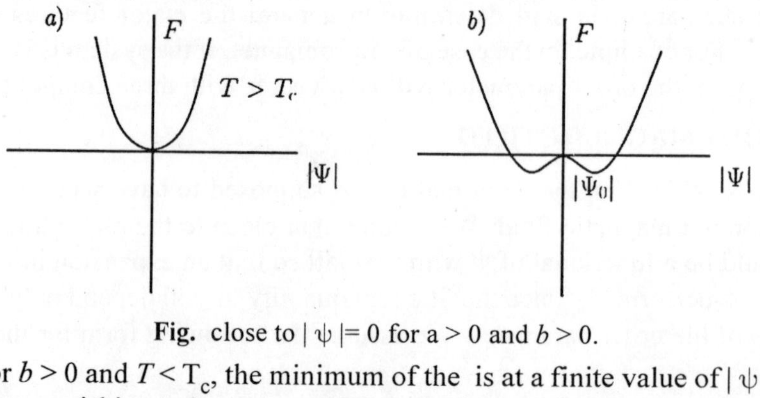

Fig. close to $| \psi | = 0$ for a > 0 and $b > 0$.

For $b > 0$ and $T < T_c$, the minimum of the is at a finite value of $| \psi |$, such that a non-vanishing order parameter characterizes the low temperature region. Finally, the gradient term in (2) allows in principle for non-homogenous solutions but since the constants in front of it are positive, the homogeneous solutions will lead to a lower and will be therefore prefered.

Since, as we stated previously, the physical state is the one for which the is at its minumum, and since with the argument above, the homogeneous case leads to a lower , we only need to minimize in order to have a quantitative estimate of the order parameter.

$$F[\psi] \rightarrow a\frac{T - T_c}{T_c}|\psi|^2 + \frac{b}{2}|\psi|^4,$$

such that the value of the order parameter is determined by the equation

$$\frac{\partial F}{\partial|\psi|} = 2|\psi|\left[a\frac{T-T_c}{T_c} + b|\psi|^2\right] = 0$$

Since we already fixed a, $b > 0$, for $T > 0$ the minimum is at $|\psi| = 0$, where the continuation of the picture to the negative axis is only meant to recall the fact that the order parameter has actually a phase, such that in the twodimensiona! space of the order parameter, the has revolution symmetry. Then, the order parameter is zero and the is given by the the normal state contribution in (2). On the other hand, for $T < T_c$, we have the solution

$$|\psi| = \sqrt{\frac{aT - T_c}{bT_c}}$$

showing that the order parameter can be expressed as

$$|\psi| \sim (T_c - T)^{\beta}$$

with β a critical exponent that in the theory has the value $\beta = 1/2$.

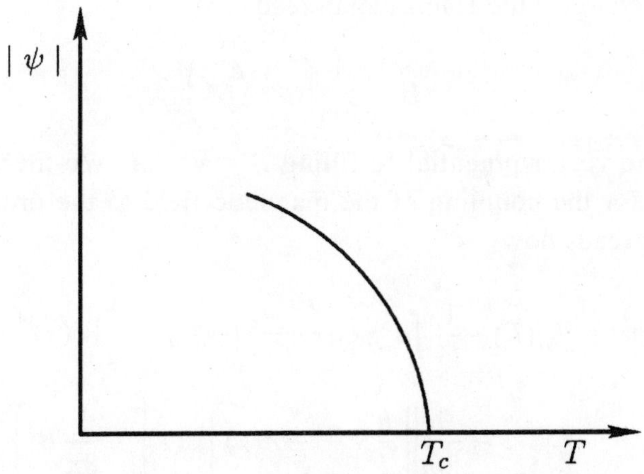

Fig. Order parameter for temperatures close to T_c.

Once the is obtained, we can calculate the specific heat, as

$$c_V = -T\left(\frac{\partial^2 F}{\partial T^2}\right).$$

Inserting (4) into (2), and carrying out the derivatives, we have

$$\left.\frac{c_V^{sc} - C_V^{n}}{T}\right|_{T_c} = \frac{a^2}{bT_c^2},$$

where the superscript sc refers to the superconducting phase, whereas n to the normal one. Thus, the phenomenological theory leads to a jump of the specific heat, as observed experimentally.

Actually, in phase transitions of second order as the one to a superconducting state, the specific heat shows either a divergence or a cusp at T_c. The result obtained here is the one correponding to so-called mean-field theories. In superconductors, the critical region, i.e. the region where divergencies in different qunatities are observed is very small of the order of $| T - T_c |/T_c \sim 10^{-5}$, such that in conventional experiments only the mean-field behaviour is observed. However, in magnetic systems also presenting phase transitions, the deviations from mean-field behaviour can be clearly seen.

WITH MAGNETIC FIELD

In the case a magnetic field is applied, we have to postulate the form in which it couples to the order parameter. It is useful in this case, to recall how a magnetic field couples to a charged particle. From the lectures in mechanics and electrodynamics, we know that in the presence of a magnetic field B, a particle with charge e the Hamiltonian reads

$$H = \frac{1}{2m}\left(p - \frac{e}{c}A\right)^2,$$

where A is the vector potential fulfilling $B = \nabla \times A$. We therefore adopt a similar form for the coupling of the magnetic field to the order parameter, such that the reads now

$$\text{F}\,[\psi] = F_n(T) + \frac{1}{V}\int d^3x \left\{ a\frac{T - T_c}{T_c}|\psi(x)|^2 + \frac{b}{2}|\psi(x)|^4 + \cdots \right.$$

$$\left. + \frac{1}{2m}\left|\left[\frac{\hbar}{i}\nabla - \frac{e^*}{c}A(x)\right]\psi(x)\right|^2 + \frac{1}{8\pi}B^3 \right\}$$

The first line of the equation above contains those parts that were already taken into account in the absence of a magnetic field, for a homogeneous order parameter.

The second line gives the modification of the term in (2) that took into account the contributions due to inhomogeneities of the order parameter. It looks like the momentum for a quantum mechanical particle of mass m and charge e^*, in the presence of a magnetic field. This term is clearly gauge invariant, since a change in A of the form

$$A\,(x) \rightarrow A\,(x) + \nabla\Lambda(x)$$

is compensated by a corresponding change in the phase of the order parameter

$$\psi(x) \to \psi(x) \exp\left[i\frac{e^*}{\hbar c}\Lambda(x)\right].$$

The third line in (10) takes into account the energy due to the magnetic field. Once we have the , we have to determine the values of the fields by minimizing it. Since in this case we are dealing with functions, we have to perform a variation of functions, as done in mechanics when deducing the Euler-Lagrange equations of motions. This can be done by carrying out a functional derivative. Since this was possibly not shown in other lectures, let us give a definition, which can be directly applied. For a functional $F[\psi]$, the funtional derivative is defined as

$$\int d^d x \frac{\delta F}{\delta\psi(x)} f(x) = \lim_{\in \to 0} \frac{1}{\in}\{F[\psi + \in] - F[\psi]\},$$

where $f(x)$ is an arbitrary function. The (10) has to be varied with respect to A, ψ, and ψ^*. Let us consider first the variation with respect to A, with $i = x$, y, or z, of the last term in (10).

$$\int d^3 x \frac{\delta F}{\delta\psi(x)} f(x) \to \lim_{\in \to 0} \frac{1}{\in}\left\{\int d^3 x B^2[A + \in f] - \int d^3 x B^2[A]\right\}$$

where we introduced an arbitrary vector f. Since

$$B[A + \in f] = \nabla \times A + \in \nabla \times f,$$

we have

$$= 2\int d^3 x B \cdot \nabla \times f = 2\int d^3 x B^i \varepsilon^{ijk} \frac{\partial f^k}{\partial x^j}$$

$$= -2\varepsilon^{ijk} \int d^3 x \frac{\partial B^i}{\partial x^j} f^k = 2\varepsilon^{kji} \int d^3 x \frac{\partial B^i}{\partial x^j} f^k$$

$$= \int d^3 x 2\nabla \times B \cdot f.$$

In a similar way we can perform the other functional derivatives, leading to

$$\frac{1}{4\pi}\nabla \times B = \frac{\hbar e^*}{2mc}\left[\psi^*\left(\frac{1}{i}\nabla - \frac{e^*}{\hbar c}A\right)\psi + \psi\left(-\frac{1}{i}\nabla - \frac{e^*}{\hbar c}A\right)\psi^*\right].$$

Since in general

$$\nabla \times B = \frac{4\pi}{c} j,$$

we see that a magnetic field induces a current due to the order parameter, such that

$$j_s = \frac{\hbar e^*}{2m}\left[\psi^*\left(\frac{1}{i}\nabla - \frac{e^*}{\hbar c}A\right)\psi + \psi\left(-\frac{1}{i}\nabla - \frac{e^*}{\hbar c}A\right)\psi^*\right].$$

is called the supercurrent density.

By considering the variation of the with respect to the order parameter, or its complex conjugate, another equation is obtained,

$$\frac{1}{2m}\left(\frac{\hbar}{i}\nabla - \frac{e^*}{\hbar c}A\right)^2\psi + a\frac{T-T_c}{T_c}\psi + b|\psi(x)|^2 = 0.$$

Equation and constitute the equations for a superconductor. In the following we consider some of the consequences of the GinzburgLandau equations.

Superfluid Density, Superfluid Velocity, and Critical Current

Consider a thin superconductor, such that the space dependence needs only to be assumed one-dimensional:

$$\psi = \psi_0 e^{iqx}.$$

Assume also that no magnetic field is present ($A = 0$). Then, eq. (20) becomes

$$\frac{\hbar^2 q^2}{2m}\psi_0 + a\frac{T-T_c}{T_c}\psi_0 + b\psi_0^3 = 0.$$

There are again two solution, the trivial one $\psi_0 = 0$ for $T > T_c$ and

$$\psi_0 = \sqrt{\frac{a(T_c - T)}{bT_c} - \frac{\hbar^2 q^2}{2mb}}$$

for $T < T_c$. On the other hand, from (19) we can calculate the supercurrent density in this case

$$j_s = e^*\frac{\hbar q}{m}\psi_0^2 \equiv e^* v_s \rho_s$$

where we have defined the superfluid velocity $v_s = \hbar q/m$ and the superfluid density $\rho_s = \psi_0^2$.

Since ψ_0 can be only real, equation shows that the supercurrent density can only reach a maximal value called the critical current j_c. Beyond this value, the metal has only normal conduction.

Meißner Effect and Penetration Depth

Let us consider a homogeneous superconductor, such that the superfluid density canbe considered constant in space. The whole space dependence is the in the phase of the order parameter,

$$\psi = \psi_0 \exp[i\varphi(x)].$$

If a magnetic fiels is present, then the supercurrent density is given by

$$j_s = \frac{\hbar e^*}{m}\rho_s \nabla\varphi(x) - \frac{e^{*2}}{mc}\rho_s A(x).$$

Since the curl of a scalar is zero, we have

$$\nabla \times j_s = -\frac{e^{*2}}{mc}\rho_s B.$$

This equation is called London equation. It was formulated by London empirically, in order to explain the Meißner effect. In fact, once we arrive at it, since on the other hand equation also relates B and j_s, and $\nabla \cdot B = 0$, we have

$$\nabla \times (\nabla \times B) = \nabla(\nabla \cdot B) - \nabla^2 B = \nabla^2 B$$

$$= \frac{4\pi}{c}\nabla \times j_s = -\frac{4\pi e^{*2}}{mc^2}\rho_s B.$$

From here we see that a length scale is present in the problem, namely

$$\lambda_L = \sqrt{\frac{mc^2}{4\pi e^{*2}\ \rho_s}},$$

called the London penetration depth. The name becomes clear by solving the equation

$$\nabla^2 B = \frac{1}{\lambda_L^2}B.$$

Let us assume that the superconductor is occupying the half-space $x > 0$, and the magnetic field is parallel to the surface of the superconsuctor. Then, the differential equation becomes a one-dimensional one with a solution

$$B(x) = B_0 e^{-x/\lambda L}$$

where B_0 is the magnetic field at the surface of the superconductor. Here we see that when a magnetic field is present, supercurrents are induced that shield the magnetic field in the interior of a superconductor. In fact, London equation shows that these currents produce a magnetic field opposed to the applied one.

FLUX QUANTIZATION

We consider here a superconducting ring

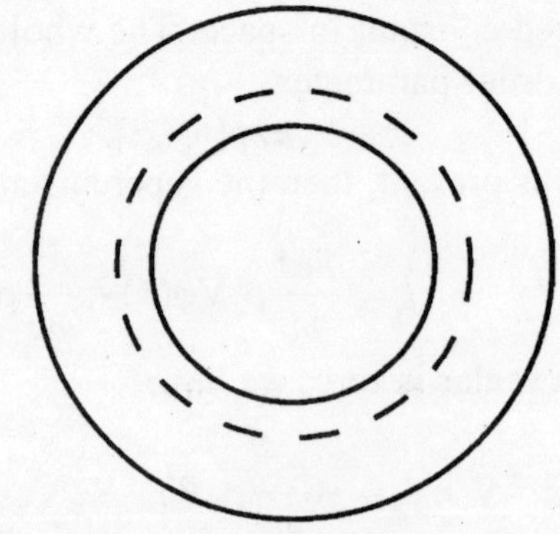

Fig. Superconducting ring with a contour (dashed line) deep in the interior of the superconductor.

Deep in the interior of the superconductor, the magnetic field is completely screened out, and therefore, supercurrents should be absent there. This means that on the dashed line, $j_s = 0$, or equivalently,

$$\oint_C j_s \cdot d\ell = 0$$

Using, we have

$$0 = \frac{\hbar e^*}{m}\rho_s \oint_C \nabla_\varphi \cdot d\ell - \frac{e^{*2}}{m}\rho_s \oint_C A \cdot d\ell$$

but on the one hand, we ask the wavefunction to be single-valued, and hence

$$\oint_C \nabla_\varphi \cdot d\ell = 2\pi n,$$

with n integer. On the other hand,

$$\oint_C A \cdot d\ell = \int_S B \cdot dS = \Phi_S\ ,$$

is the magnetic flux through the ring. We see then, that the magnetic flux is quantized

$$\Phi_S = n\frac{2\pi\hbar c}{e^*} = n\Phi_0$$

in units of the flux quantum

$$\Phi_0 = \frac{hc}{e^*}.$$

ELECTRON-PHONON COUPLING

The previous discussion of the phenomenon of superconductivity did not make any reference to the mechanism that leads to its appearance. In order to come closer to a microscopic understanding of superconductivity we should mention a decisive experimental finding, namely the isotope effect, where the critical temperature varies as

$$Tc \sim M^{-\alpha},$$

with M the mass of the ions, and $\alpha \sim 1/2$. Such a relationship, points to the fact that phonons play an important role for a system that becomes superconducting.

Here we will consider a Hamiltonian introduced by Fr¨olich, who actually predicted the isotope effect before its observation.

$$H_F = H_{\text{el}} + H_{\text{ph}} + H_{\text{el-ph}},$$

where

$$H_{\text{el}} = \sum_k \in (k) f_k^\dagger f_k$$

describes the electronic part. Here it is assumed that the role of Coulomb interaction is not important for the phenomenon of superconductivity, such that $\in (k)$ describes some band-structure, that takes into account the Coulomb interaction in the frame of a Fermi liquid theory.

The phonons are described by

$$H_{\text{ph}} = \sum_k \hbar\omega_q \left(b_q^\dagger b_q + \frac{1}{2} \right)$$

i.e. s with a dispersion ω_q.

For the coupling of electrons and phonons we can take the following simple picture. First we consider acoustic phonons since they are allway present, such that our discussion remains general. We consider further those phonons that produce a local change of the ionic density, and hence, directly couple to the density of electrons. Let us describe such a change in ionic potential with a field $D(x)$, such that

$$H_{\text{el-ph}} = \int d^3x \psi_\alpha^\dagger (x)\, D(x)\, \hat{\psi}_\alpha (x).$$

Changes in the ionic density correspond to displacement fields u (x) with

$$D(x) = \nabla \cdot \mathrm{u}(x),$$

corresponding to local dilatations and contractions. This means that we need only to consider longitudinal phonons, i.e. with $u \sim q$, where q is the wavevector of the phonon. From the relation between creation and annihilation of phonon quanta and the displacement fields, we can obtain the following re!ationship.

$$u(x) = \frac{1}{\sqrt{N}} \sum_q \frac{q}{|q|} \left(\frac{\hbar}{2M\omega_q} \right)^{\frac{1}{2}} \left[b_q \exp(iq \cdot x) + b_q^\dagger \exp(-iq \cdot x) \right]$$

From here we can then calculate the field D using (43), and the fact that for an acoustic phonon, $\omega q = v_s q$, with vs the sound velocity.

$$D(x) = \frac{i}{\sqrt{N}} \sum_q \left(\frac{\hbar q}{2Mv_s} \right)^{\frac{1}{2}} \left[b_q \exp(iq \cdot x) - b_q^\dagger \exp(-iq \cdot x) \right]$$

For the electronic field operators, we use an expansion in Bloch states

$$\hat{\psi}_\alpha(x) = \sum_k f_{k\alpha} u_k(x) \exp(ik \cdot x),$$

$$\hat{\psi}_\beta^\dagger(x) = \sum_k f_{k\beta}^\dagger u_k^*(x) \exp(-ik \cdot x).$$

Then, the electron-phonon interaction looks as follows.

$$H_{\text{el-ph}} = \frac{i}{\sqrt{N}} \sum_q \left(\frac{\hbar q}{2Mv_s} \right)^{\frac{1}{2}} \Big\{ b_q \sum_{kk'} f_{k\alpha}^\dagger f_{k'\alpha}$$

$$\times \int d^3x\, u_k^*(x)\, u_{k'}(x) \exp\left[i(q - k + k') \cdot x \right] - h.c. \Big\}$$

$$= i \sum_{k,q} D_q \left(b_q f_{k+q,\alpha}^\dagger f_{k,\alpha} - b_{k-q,\alpha}^\dagger f_{k,\alpha} \right),$$

where in going to the last line, we restricted ourselves to a model on the lattice, where Wannier functions are taken, such that no further k-dependence appears in *Dq*.

Finally, we can write the full Hamiltonian as

$$H_F = \sum_{k,q} \in(k) f_k^\dagger f_k + \sum_q \hbar\omega_q \left(b_q^\dagger b_q + \frac{1}{2} \right)$$

$$+ i \sum_{k,q} D_q \left(b_q \rho_q^\dagger - b_q^\dagger \rho_q \right)$$

where

$$\rho_q = \sum_k f_{k-q,\alpha}^\dagger f_{k,\alpha} = \rho_{-q}^\dagger$$

is an electron-density operator.

EFFECTIVE ELECTRON-ELECTRON INTERACTION

In the following we will see that it is possible to extract an effective electron-electron interaction from the electron-phonon interaction. This can be normally achieved, when the different degrees of freedom appear in bilinear forms and the coupling between different kinds of excitations is linear. This can be easily shown in a pathintegral formalism. Here, we will restrict ourselves to work in the frame of second quantization and use a canonical transformation. A canonical transformation one in which the canonical commutation relations of operators is preserved.

We start with an antiunitary operator S, i.e. $S^\dagger = -S$, such that exp S is unitary. By performing a unitary transformation, all physical quantities remain unchanged. Applied on the Hamiltonian, we transform it as follows

$$\tilde{H}F = e^{-S}H_F e^S$$
$$= H_F + [H_F, S] + \frac{1}{2!}[[H_F, S], S] + \frac{1}{3!}[[[H_F, S], S], S] + \cdots$$

For the choice of S, we look at the diagramatic representation of the electron-phonon interaction in (48). There, a phonon is created or annihilated in a scattering process

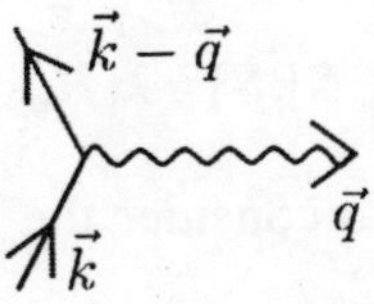

Fig. Diagramatic representation of the electron-phonon coupling in eq. (48).

with an electron. However, after a phonon is e.g. created, it can be annihilated by another scattering process with an electron. Thus, a phonon is exchanged by two electrons giving rise to an effective interaction.

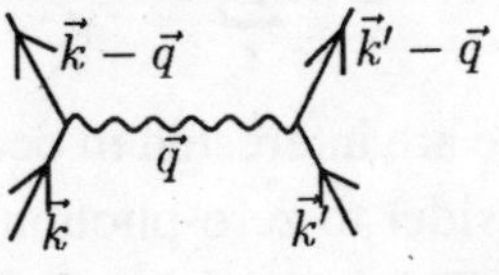

Fig. Effective electron-electron interaction throught the exchange of a phonon.

Such a process can occur only in second order in the electron-phonon interaction or in higher orders. Therefore, the canonical transformation should be chosen in such a way that contributions from the electron-phonon interaction are cancelled in first order. Let us write H_F as

$$H_F = H_0 + \lambda H_{\text{el-ph}},$$

such that

$$\tilde{H}_F = e^{-S} H_F e^S$$
$$= H_0 + \lambda H_{\text{el-ph}} + [H_0, S] + \lambda [H_{\text{el-ph}}, S]$$
$$+ \frac{1}{2!}[[H_0 + \lambda H_{\text{el-ph}}, S], S] + \cdots$$

In the case that

$$\lambda \mathrm{H}_{\text{el-ph}} + [H_0, S] = 0,$$

$S \sim O(\lambda)$, implying thatHF has no term linear in λ. Let us consider now the˜ possible matrix elements of the condition above for states $|m>$ that are eingenstates of H_0.

$$\begin{aligned} \lambda < n| H_{\text{el-ph}} \mid m > &= < n \mid [S, H_0] \mid m > \\ &= (E_m - E_n) < n \mid S \mid m >, \end{aligned}$$

leading to

$$< n \mid S \mid m > = \lambda \frac{< n|H_{\text{el-ph}} \mid m >|}{(E_m - E_n)},$$

such that we have S in the basis that diagonalizes H_0. Once we have S, we can consider the first non-vanishing correction to H_0. In the next order in λ we have

$$\lambda[H_{\text{el-ph}}, S] + \frac{1}{2}[[H_0, S], S] = \lambda[H_{\text{el-ph}}, S] - \frac{\lambda}{2}[H_{\text{el-ph}}, S]$$

Therefore, we need now to consider the matrix elements of $[H_{\text{el-ph}}, S]$. Let us look first at $H_{\text{el-ph}}S$.

$$\begin{aligned} < n \mid H_{\text{el-ph}} S \mid m > &= \sum_{\ell} < n|H_{\text{el-ph}}|\ell >< \ell \mid S \mid m > \\ &= \lambda \sum_{\ell} \frac{< n|H_{\text{el-ph}}|\ell >< \ell \mid H_{\text{el-ph}} \mid m >}{(E_m - E_\ell)}. \end{aligned}$$

Since the phenomena we are interested in occur at very low temperatures, we restrict the states we consider to zero-phonon states, the one-phonon state being only reached virtually. This means that in the expression above we have for a particular phonon mode,

$$\begin{aligned} &\rightarrow \frac{< 0|H_{\text{el-ph}}|1_q >< 1_q \mid H_{\text{el-ph}} \mid 0 >}{(E_m - E_\ell)} \\ &= D_q^2 \sum_{k,k'} f_{k'}^\dagger f_{k'-q} f_{k-q}^\dagger f_k \frac{1}{\epsilon_k - \epsilon_{k-q} - \hbar\omega q}. \end{aligned}$$

Now we consider $SH_{\text{el-ph}}$. In this case we have

$$< \mathrm{n} \mid SH_{\text{el-ph}} \mid m > = \sum_{\ell} < n|S|\ell >< \ell \mid H_{\text{el-ph}} \mid m >$$

$$= \lambda \sum_{\ell} \frac{< n | H_{\text{el-ph}} | \ell >< \ell | H_{\text{el-ph}} | m >}{(E_{\ell} - E_n)}$$

$$\rightarrow D_q^2 \sum_{k,k'} f_{k'}^{\dagger} f_{k'-q} f_{k-q}^{\dagger} f_k \frac{1}{\in_k - \in_{k-q} + \hbar\omega q}.$$

Putting all the contributions together, we obtain an effective electron-electron interaction of the form

$$H_{\text{el-el}} = \sum_{q,k,k'} W_{kq} f_{k'}^{\dagger} f_{k'+q} f_{k-q}^{\dagger} f_k$$

with

$$W_{kq} = \frac{D_q^2 \hbar\omega q}{(\in_k - \in_{k+q}) - \hbar^2 \omega_q^2}$$

This means that for a small region around the Fermi energy with $|\in_k - \in_{k+q}| < \hbar\omega q$, an attractive interaction among electrons arises mediated by phonons. One could then roughly argue that a new bound state consisting of two fermions could appear, such that the bound pair would have bosonic character, and hence, similarly to 4 He, a superfluid phase could appear. Since in this case the pair would be charged, superconductivity would take place. That such a bound state can occur even in the presence of many electrons, is the subject of the theory by Bardeen, Cooper ans Schrieffer.

THE BCS THEORY

After we have seen that the electron-phonon coupling gives rise to an effective attractive interaction between electrons, the idea that pairs will form becomes natural. The theory developed by Bardeen, Cooper and Schrieffer (BCS), allows for a calculation of several features related to the transition to superconductivity like the isotope effect, and also features related to the ground-state like the opening of a gap in the one-particle excitation spectrum. Although the full developement of the consequences of the BCS-theory lie beyond the scope of these lectures, we should mention that this theory is one of the most predictive theories in solid state physics.

The following features should be taken into account, when discussing pairing.

(i) As shown by (61), there is an attractive interaction for states with energy $\in_k - \in_{k+q} | < \delta \sim \hbar\omega_D$, where ω_D is the Debye frequency, that gives the characteristic phonon frequency of the system.

Fig. Shell around the Fermi surface with an attractive interaction.

(ii) Pairs with total momentum zero and total spin zero. The states would be $| k > = | k, \sigma >, | ek > = |-k, -\sigma >$. Such states can be realized on the whole Fermi sphere and have therefore a high density of states. Since electrons are $S - \frac{1}{2}$ particles, a pair can be in a state with total spin $S = S_1 + S_2 = 0$ or 1. If $S = 0$ one speaks of singlet pairing and if $S = 1$ triplet pairing.

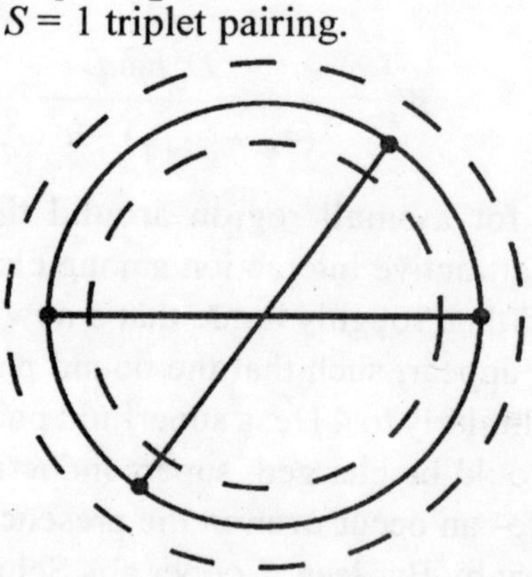

Fig. All states diametrically opposed can form time-reversal invariant pairs

The Hamiltonian introduced by Bardeen, Cooper, and Schrieffer (BCS) is restricted to the conditions mentioned above.

$$H_{BCS} = \sum_{k,\sigma} \in (k) f_{k,\sigma}^{\dagger} f_{k,\sigma} + \frac{1}{2} \sum_{k,k'} W_{kk'} f_k^{\dagger} f_{-k}^{\dagger} f_{-k'} f_k$$

where

$$W_{kk'} = \begin{cases} < 0 & \text{for } |\in(k) - E_F| < \frac{\delta}{2} \text{ and } |\in(k') - E_F| < \frac{\delta}{2} \\ 0 & \text{otherwise} \end{cases}$$

This is still an interacting Hamiltonian and, hence in general diffcult to solve.

However, the following equalities will proove to be useful.

$$f_{-k} f_k = < f_{-k} f_k > + (f_{-k} f_k^{-} - < f_{-k} f_k >)$$

$$f_k^\dagger f_{-k}^\dagger = < f_k^\dagger f_{-k}^\dagger > + (f_k^\dagger f_{-k}^\dagger - < f_k^\dagger f_{-k}^\dagger >)$$

Such a decomposition of operators, makes explicit the picture that when a certain type of order is established, the operators can be viewed as a mean-field (first terms in eqs. (64)) and fluctuations around it. This mean-field is in the case of superconductivity of a rather strange nature. In the normal state, $< f_k^\dagger f_{-k}^\dagger > = < f_k^\dagger f_{-k}^\dagger > = 0$, since it is an expectation number of operators that change the number of particles in the system. On the other hand, if it happens to be non-zero, we have $< f_k^\dagger f_{-k}^\dagger > = < f_{-k} f_k >^*$, and hence, this field is complex. From these characteristics, we can identify it with the order parameter we had in the phenomenological treatment above. However, if $< f_{-k} f_k > \neq 0$, then the new state is one without a definite number of particles. Although we do not show this here, let us mention that this is directly related to the fact that gauge invariance is broken.

After the remarks above, let us use the relations (64) to express the interaction term, but neglecting terms quadratic in fluctuations (second terms in eqs. We have then,

$$\begin{aligned} f_k^\dagger f_{-k}^\dagger f_{-k'} f_{-k'} &= < f_k^\dagger f_{-k}^\dagger >< f_{-k'} f_{-k'} > \\ &+ < f_k^\dagger f_{-k}^\dagger > (f_{-k'} f_{-k'} - < f_{-k'} f_{k'} >) \\ &+ (_k^\dagger f_{-k}^\dagger - <_k^\dagger f_{-k}^\dagger >) < f_{-k'} f_{-k'} > \\ &= f_k^\dagger f_{-k}^\dagger \psi_{k'} + \psi_k^* f_{-k'} f_{-k'} - \psi_k^* \psi_{k'}, \end{aligned}$$

where we defined

$$\psi_k \equiv < f_{-k} f_k >.$$

After such an approximation, and extending it by addition of the chemical potential, the BCS-Hamiltonian becomes

$$\begin{aligned} H_{BCS} \to\ & H_{BCS} - \mu \hat{N} \\ \simeq & \sum_k [\in(k) - \mu] f_k^\dagger k_k \\ & + \frac{1}{2} \sum_{kk'} W_{kk'} (f_k^\dagger f_{-k}^\dagger \psi_{k'} + \psi_k^* f_{-k'} f_{k'} - \psi_k^* \psi_{k'}) \end{aligned}$$

With such an approximation (mean-field approximation), the originally interacting system is approximated by non-interacting fermions under the influence of a field that has to be determined self-consistently. On the other hand, since now the Hamiltonian is bilinear in the fermionic operators, it is possible to diagonalize it by means of a Bogoliubov-transformation, where new operators are introduced as follows

$$f_k = u_k\gamma_k + v_k\gamma^\dagger_{-k}$$

$$\gamma^\dagger_{-k} = -u_k\gamma_k + u_k\gamma^\dagger_{-k}$$

with the properties of the coeffcients

$$u_k = u_{-k},\ v_k = -v_{-k},\ u_k^2 + v_k^2 = 1,$$

such that the new operators obey canonical anticommutation relations for fermions,

$$\left\{\gamma_k, \gamma^\dagger_{k'}\right\} = \delta_{kk'}, \quad \left\{\gamma_k, \gamma_{k'}\right\} = \left\{\gamma^\dagger_k, \gamma^\dagger_{k'}\right\} = 0\,,$$

and hence, we have a canonical transformation. This leads to the different products

$$\begin{aligned} f^\dagger_k f_k &= (-v_{-k}\gamma_{-k} + u_{-k}\gamma^\dagger_k)(u_k\gamma_k + v_k\gamma^\dagger_{-k}) \\ &= u_{-k}u_k\gamma^\dagger_k\gamma_k + v_{-k}u_k\ \gamma^\dagger_{-k}\gamma_{-k} \\ &\quad -v_{-k}u_k\gamma_{-k}\gamma_k + u_{-k}u_k\ \gamma^\dagger_k\gamma^\dagger_{-k} - v_{-k}v_k \end{aligned}$$

$$\begin{aligned} f^\dagger_k f^\dagger_{-k} &= (-v_{-k}\gamma_{-k} + u_{-k}\gamma^\dagger_k)(-v_k\gamma_k + u_k\gamma^\dagger_{-k}) \\ &= v_{-k}u_k\gamma^\dagger_{-k}\gamma_{-k} - u_{-k}v_k\ \gamma^\dagger_k\gamma_k \\ &\quad +v_{-k}v_k\gamma_{-k}\gamma_k + u_{-k}u_k\ \gamma^\dagger_k\gamma^\dagger_{-k} - v_{-k}u_k. \end{aligned}$$

Inserting these products into, we have

$$\begin{aligned} &= \sum_k \left\{[\in(k) - \mu]\left(u_k^2 - v_k^2\right) - 2\sum_{k'} W_{kk'}\ \psi_{k'}u_k v_k\right\}\gamma^\dagger_k\psi_k \\ &\quad + \sum_k \left\{[\in(k) - \mu]\, u_k v_k + \frac{1}{2}\sum_{k'} W_{kk'}\ \psi_{k'}\left(u_k^2 - v_k^2\right)\right\} \\ &\qquad \times\left(\gamma^\dagger_k\gamma^\dagger_{-k} + \gamma_{-k}\gamma_k\right) \end{aligned}$$

We can get rid of the non-diagonal terms by requiring

$$2[\in(k) - \mu]u_k v_k = -\left(u_k^2 - v_k^2\right)\sum_{k'} W_{kk'}\ \psi_{k'}k \equiv \left(u_k^2 - v_k^2\right)\Delta_k\,,$$

where we defined

$$\Delta_k = -\sum_{k'} W_{kk'}\ \psi_{k'}.$$

The minus sign was taken into account due to the fact that $W_{kk'} < 0$, as stated in. Equation together with the condition $u_k^2 + v_k^2 = 1$ constitute the equations that determine the coeffcients u_k and v_k of the canonical transformation.

The solution is easily found by setting

$$u_k = \cos\phi_k \qquad v_k = \sin\phi_k,$$

such that goes over into

$$[\in(k) - \mu]\sin 2\phi_k = \Delta_k \cos 2\phi_k \Rightarrow \tan 2\phi_k = \frac{\Delta_k}{\in(k) - \mu}.$$

Using the relations

$$1 + \tan^2 x = \frac{1}{\cos^2 x},\ 1 + \frac{1}{\tan^2 x} = \frac{1}{\sin^2 x},$$

we have

$$\cos 2\phi_k = u_k^2 - v_k^2 = \pm\frac{\in(k) - \mu}{E_k}$$

$$\sin 2\phi_k = \pm\frac{\Delta_k}{E_k},$$

where we defined

$$E_k \equiv \sqrt{[\in(k) - \mu]^2 + \Delta_k^2}.$$

Replacing the results above into, we have the Hamiltonian $\tilde{H}$ after the canonical transformation

$$\tilde{H} = \sum_k E_k \gamma_k^\dagger \gamma_k,$$

where we have taken the positive root, since a meaningfull Hamiltonian should have a ground state. This shows that the spectrum of the quasiparticles has now a gap given by Δ_k. The ground-state of the system is given by the vacuum for the new operators

$$\gamma_k \mid 0> = 0, \qquad <0 \mid \gamma_k^\dagger = 0,$$

such that in the ground-state,

$$\psi_k = <f_{-k} f_k> = u_k v_k = \frac{\Delta_k}{2E_k},$$

according to and. With this result relating the order parameter to energy gap, we finally obtain the gap-equation, by recalling the definition

$$\Delta_k = -\frac{1}{2}\sum_{k'} W_{kk'} \frac{\Delta_{k'}}{E_{k'}}.$$

Since this is a homogeneous equation, the trivial case $\Delta_k = 0$ is allways a solution.

SOLUTION OF THE GAP-EQUATION

In the following we consider the solution of the gap-equation for the case where $W_{kk'} = -W$, is a constant in the shell given by δ in. Then, we have

$$\Delta_k = \frac{W}{2}\sum_{k'}\frac{\Delta_{k'}}{E_{k'}},$$

showing that Δ should also be momentum indenpendent. Then, the gap-equation reduces to

$$1 = \frac{W}{2}\sum_{k'}\frac{1}{\sqrt{[\in(k)-\mu]^2+\Delta^2}}.$$

At this point, it can be clearly seen that we can have a non-trivial solution only when $W > 0$, i.e. when an attractive interaction among electrons is present. Repulsive interactions can only lead to superconductivity, when the order parameter has a non-trivial momentum dependence. This is the case e.g., in high temperature superconductors, where the experimentally measured properties of the order parameter are consistent with a so-called d-wave symmetry, with

$$\Delta_k \sim \cos k_x a - \cos k_y a,$$

a being the lattice constant. In the following we will concentrate us on the simple case that is generally found in ordinary superconductors, where the gap, and isotope effect result in a natural form from the BCS-theory.

As we have already stated at the begining of Sec. 3, the attractive present for $|\in_k - \in_{k+q}| < \hbar\omega_D$, where ωD is the Debye frequency. We write

$$W_{kk} = -\frac{\lambda}{V}\theta(\hbar\omega_D - |\xi_k|\theta(\hbar\omega_D - |\xi_{k'}|),$$

where we introduced the short notation

$$\xi_k \equiv \in(k) - \mu.$$

Introducing this into the gap-equation (85), we have

$$\Delta_k = \frac{\lambda}{2V}\sum_{k'}\theta(\hbar\omega_D - |\xi_k|)\theta(\hbar\omega_D - |\xi_{k'}|)\frac{\Delta_{k'}}{E_{k'}}$$

with a solution $\Delta_k = \Delta\theta\,(\hbar\omega_D - |\,\xi_k\,|)$ that leads to the equation

$$1 = \frac{\lambda}{2V}\sum_k\frac{\theta(\hbar\omega_D - |\xi_k|)}{\sqrt{\Delta^2+\xi_k^2}}.$$

It is convenient to go from a summation over momenta over to energies introducing thus the density of states:

$$\frac{1}{V}\sum_k \to \int N(\xi)d\xi \simeq N(0)\int d\xi,$$

where we have assumed that $\hbar\omega_D \ll E_F$. Then, we have for,

$$1 = \frac{\lambda N(0)}{2}\int_{-\hbar\omega_D}^{\hbar\omega_D}\frac{d\xi}{\sqrt{\Delta^2+\xi^2}} = \lambda N(0)\int_0^{\hbar\omega_D}\frac{d\xi}{\sqrt{\Delta^2+\xi^2}}$$

$$\to 1 = \lambda N(0)\ln\left[\frac{\hbar\omega_D+\sqrt{(\hbar\omega_D)^2+\Delta^2}}{\Delta}\right]$$

$$\simeq \lambda N(0)\ln\frac{2\hbar\omega_D}{\Delta},$$

leading finally to

$$\Delta = 2\hbar\omega_D \exp\left[-\frac{1}{\lambda N(0)}\right],$$

where the gap depends in a non-analytic fashion on the coupling constant, showing that such a result cannot be obtained in the frame of a perturbation theory.

Once we obtained the gap, we can go back to the coeffcients u_k and v_k. Using the condition $u_k^2+v_k^2=1$, and, we have

$$u_k^2 = \frac{1}{2}\left[1+\frac{\xi_k}{\sqrt{\Delta^2+\xi_k^2}}\right],$$

$$v_k^2 = \frac{1}{2}\left[1-\frac{\xi_k}{\sqrt{\Delta^2+\xi_k^2}}\right]$$

Apart from having the full solution, we notice that if we take the ground-state expectation value of, we have

$$<f_k^\dagger f_k> = v_k^2,$$

i.e. we have the occupation number of the original fermions in the new ground state. In fact, if $\Delta = 0$, then we have

$$v_k^2 = \frac{1}{2}\left[1-\frac{\xi_k}{|\xi_k|}\right] = \theta[\in(k)-\mu],$$

as expected for free fermions.

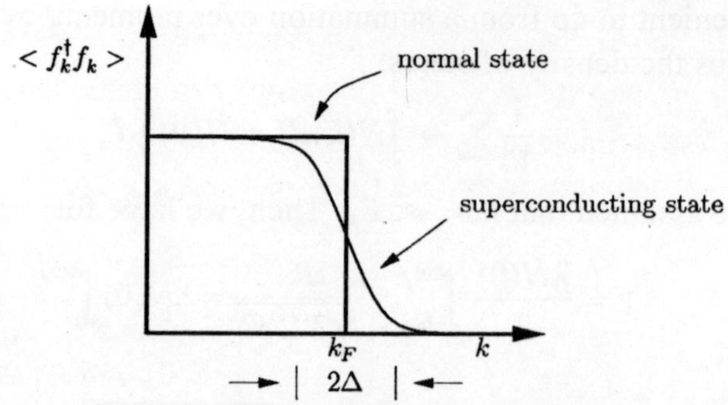

Fig. Momentum distribution function for the normal and superconducting state (red line) with a gap Δ, both at $T = 0$.

For the superconducting state with $\Delta \neq 0$, a gap opens destroying the Fermi surface, such that the momentum distribution function shows a smooth decrease across the Fermi energy.

In order to find the critical temperature, we should perform an analogous calculation at finite temerpatures. Without going through the calculation, we know that the temperature scale should be essentially given by the value of the gap at $T = 0$. In fact, the result is showing that $T_c \sim \nabla\omega_D \sim M^{-1/2}$, explaining thus, the isotope effect.

$$T_c \simeq 1.114\hbar\omega_D \exp\left[-\frac{1}{\lambda N(0)}\right]$$

The exponent of the mass can vary from system to system, since the strength of the Coulomb interaction, that was not accounted for in this treatment may vary. A whole theoretical development was devoted after the BCS-theory appeared to describe the behaviour of superconductors beyond the limitations of the BCS-theory and they can be seen in elective lectures.

We close by noticing that the success of the BCS-theory is not only limited to ordinary superconductors, but it was also applied to understand the pairing states in He, and nuclear matter.

THE

The derivation of the , does not account for that the velocities of most of the electrons are only weakly perturbed by the magnetic field. The more accurate treatment of the Boltzmann equation, leads to the result that

$$\bar{\bar{\rho}} = \frac{1}{\sigma_0}\begin{pmatrix} 1 & -RB\sigma_0 & 0 \\ RB\sigma_0 & 1 & 0 \\ 0 & 0 & 1 \end{pmatrix}, \quad \sigma_0 = \frac{ne^2\pi}{m^*}$$

in the case of B = (0, 0, B). The resistivity tensor $\overline{\overline{\rho}} = (\overline{\overline{\sigma}})^{-1}1$ is defined by $E = \overline{\overline{\rho}}j$, see equation. In order to get this result, the band mass tensor has to be isotropic (equals to m" times the unit matrix).

Inverting the resistivity tensor one finds

$$\overline{\overline{\rho}} = \frac{\sigma_0}{1+(RB\sigma_0)^2}\begin{pmatrix} 1 & RB\sigma_0 & 0 \\ -RB\sigma_0 & 1 & 0 \\ 0 & 0 & 1+(RB\sigma_0)^2 \end{pmatrix}$$

In the limit of $|RB\sigma_0| \gg 1$ then $\overline{\overline{\sigma}}$ has the same B dependence as given by, but R should be placed in the denominators, rather than in the numerators, of the different matrix elements in:

$$\overline{\overline{\sigma}} = \begin{pmatrix} \frac{1}{R^2B^2\sigma_0} & \frac{1}{RB} & 0 \\ -\frac{1}{RB} & \frac{1}{R^2B^2\sigma_0} & 0 \\ 0 & 0 & \sigma_0 \end{pmatrix}, \quad |RB\sigma_0| \gg 1$$

THE DIPOLE MOMENT

A displacement of a point charge q from the point charge "q by a vector x gives rise to an electric dipole moment $p = q\,x$ (where $|x|$ has to be small compared with any other distances considered). A certain dipole at r, within the volume &!, contributes to the polarization $P(r)$ as determined by

$$P(r)\Omega = \int^t J(t')dt' = \int^t q\dot{x}(t')dt' = qx = p(r)$$

and the (averaged) polarization in the volume V is

$$P = \frac{1}{V}\sum_i p(r_i) = \frac{N}{V}p = np$$

if all dipole moments are equal $p(r_i) = p$. The scalar potential at r due to a dipole at origo is

$$\phi(r) = \frac{q}{r_+} - \frac{q}{r_-} = \frac{p \cdot r}{r^3} \quad \Rightarrow \quad E(r) = -\nabla\phi(r) = 3\frac{(p \cdot r)r}{r^5} - \frac{p}{r^3}$$

as obtained, when $x \ll r$, from $r_\pm = \mathrm{r} \mp (x/2)\cos\theta$ and pr $\cos\theta = p \cdot r$. In the presence of an electric field E, the energy of a dipole moment at the position r is

$$U = q\phi\left(r+\frac{1}{2}x\right)-q\phi\left(r-\frac{1}{2}x\right)$$
$$= q\left(\phi(r)+\frac{1}{2}x\cdot\nabla f(r)-q(\phi(r)-\frac{1}{2}x\cdot\nabla\phi(r)\right)$$
$$= qx\cdot\nabla f(r) = -p(r)\cdot E(r)$$

The combination of and determines the field energy of two dipoles p_1 (r_1) and $p_2(r_2)$ to be

$$U_{12} = -3\frac{(p_1\cdot r_{12})(p_2\cdot r_{12})}{r_{12}^5}+\frac{p_1\cdot p_2}{r_{12}^3},\quad r_{12}=r_1-r_2$$

We shall consider an ellipsoidal sample of a homogeneous dielectric material. Classical electrostatics shows that the polarization P, and thus also the electric field E inside the ellipsoid, is uniform in the presence of an applied uniform electric field E_0. We shall assume this to be the situation, in which case the total field E_{cell} acting on a certain dipole in the system is the same for all sites and equal to the field E_{cell} (0) acting on the dipole placed in the centre of the sample. The dipole-dipole interaction is of long range.

It declines proportionally with r^{-3}, but the number of dipoles at the distance r increases as r^2 and, as a consequence, the distribution of dipoles on the surface affects the dipole in the centre no matter how far the distance is to the surface. In order to handle this enormous span of length scales, we make a cut between the microscopic world of a discrete lattice of dipoles and the macroscopic continuous material with its uniform polarization. The cut to be used is the surface of a sphere centered at the dipole considered. The radius of this sphere R_C should be large compared with the lattice spacing, but smaller than the shortest distance to the surface (or smaller than the size of crystallites in a polycrystalline sample or the size of the domains in the case of an ordered material). The total field is the sum of the applied field and the field due to all other dipoles in the sample, i.e. the energy of the dipole at $r = 0$ is

$$U(0) = -p\cdot E_{cell}(0) = -p\cdot E_0 - \sum_{r_i\neq 0}\left[\frac{(p_1\cdot r_i)^2}{r_i^5}-\frac{p^2}{r_i^3}\right]$$

Introducing the cut and applying a continuous approximation for the macroscopic contribution of the dipoles outside the sphere, we get

$$U(0) = -p\cdot E_0 - \sum_{r_i<R_C}\left[3\frac{(p\cdot r_i)^2}{r_i^5}-\frac{p^2}{r_i^3}\right]-\int_{\text{sphere}}^{\text{sample}}\left[3\frac{(p\cdot r)^2}{r^5}-\frac{p^2}{r^3}\right]\frac{dr}{\Omega}$$

Ω is the volume per dipole moment, and defining z to be along the direction of the dipole moments of length $p = P\Omega$, the lattice sum may be written

$$\sum_{r_i<R_C}\left[3\frac{(p\cdot r_i)^2}{r_i^5}-\frac{p^2}{r_i^3}\right] = pP\Omega\sum_{r_i<R_C}\frac{3z_i^2-x_i^2-y_i^2-z_i^2}{(x_i^2+y_i^2+z_i^2)^{5/2}} = pP\xi$$

The field $P\xi$ due to the lattice sum vanishes by symmetry in the case of a cubic system. This is true no matter the direction of the z axis compared to the lattice vectors, but it is only true when the cutting surface is that of a sphere (the reason for making the cut of this shape). If the system is not cubic, the lattice sum may be calculated straightforwardly by numerical methods as it converges quite rapidly. The macroscopic contribution is

$$\int_{\text{sphere}}^{\text{sample}}\left[3\frac{(p\cdot r_i)^2}{r^5}-\frac{p^2}{r^3}\right]\frac{dr}{\Omega} = -pP\int_{\text{sphere}}^{\text{sample}}\nabla\cdot\left(\frac{z}{r^3}\right)dr$$

$$= \int_{\text{sphere}}\frac{z\cdot dS}{r^3} - pP\int_{\text{sample}}\frac{z\cdot dS}{r^3} - pP\left(\frac{4\pi}{3}-N_{zz}\right)$$

The first surface integral (over the sphere) is calculated straightforwardly using $z \cdot dS = r\cos\theta(2\pi r^2 \sin\theta d\theta)\cos\theta$ in spherical coordinates. The last integral over the surface of the sample depends on the shape of the ellipsoid. It is $4\pi/3$ in case it is a sphere, but, in general, it is a number between 0 (long thin cigar along the *z* axis) and 4π (flat pancake). Combining the results above, the total field is found to be

$$E_{\text{cell}} = E_0 + \left(\overline{\overline{\xi}} + \frac{4\pi}{3} - \overline{\overline{N}}\right)P = E + \left(\overline{\overline{\xi}} + \frac{4\pi}{3}\right)P$$

The second expression derives from that the internal electric field is $E = E_0 - \overline{\overline{N}}P$, and this way of writing the result is of more general applicability, since it also applies when the surface charges are not determined alone by the homogeneous polarization of the bulk material. Introducing $D = E + 4\pi P = \overline{\overline{\epsilon}}E$, then the permittivity $\overline{\overline{\epsilon}}$ is a material parameter, which is independent of the shape of the system. Introducing the polarization coeffcient α by $p = \overline{\overline{\alpha}}E_{\text{cell}}$ and assuming all tensors $\overline{\overline{\epsilon}}$, ξ, and $\overline{\overline{\alpha}}$ to be diagonal, we get the Clausius-Mossotti relation (valid for each of the three diagonal components)

$$\epsilon = \frac{3+(8\pi-3\xi)n\alpha}{3-(4\pi+3\xi)n\alpha}$$

MAGNETIC ENERGY AND DOMAINS

In order to derive the magnetic energy of a macroscopic sample of microscopic dipoles or of supercurrent loops in a superconductor, we start out with the classical description of the fields. Introducing the magnetic H field, the Maxwell equations to be considered are

$$\nabla \times H = \frac{4\pi}{c} j_{\text{ext}} + \frac{1}{c}\frac{\partial D}{\partial t}, \qquad \nabla \cdot B = 0$$

where

$$B = H + 4\pi M, \qquad \nabla \cdot H = -4\pi \nabla \cdot M$$

The sample is assumed to be placed in a uniform applied field $B_0 = H_0$, and the sources of the fields (the external currents) are assumed to lie outside the reach of the fields B_d and H_d produced by the sample. The total fields are

$$H = H_0 + H_d, \qquad B = B_0 + B_d, \; B_d = H_d + 4\pi M$$

The magnetic energy of the sample is the total magnetic energy minus the magnetic energy of the space without the sample:

$$U = \frac{1}{4\pi} \int_{\text{all space}} \left[\int H \cdot \delta B - \frac{1}{2} H_0^2 \right] dr$$

As shown in the appendix, $\int_{\text{all space}} H_0 \cdot B_d dr = \int_{\text{all space}} H_0 \cdot B_d dr = 0$ and the magnetic energy reduces to

$$U = \int_{\text{sample}} \left[-M \cdot H_0 - \frac{1}{2} M \cdot H_d + \int H \cdot \delta M \right] dr = - \int_{\text{sample}} [M \cdot \delta H_0] dr$$

Two things are worth to notice. First, the energy only involves an integration over the space occupied by the sample (where M is non-zero). Second, the applied field H_0 is the independent variable in the final expression. This is in accordance with the experimental situation and is in contrast with the original expression, which involves B as the independent variable. The magnetic energy U is equal the magnetic contribution F_M to the of the sample, and

$$f_M(r) = \int_0^{H_0} M(r, H_0') \cdot dH_0', \qquad F_M = \int_{\text{sample}} f_M(r) dr$$

is of general applicability.

If a homogeneous sample in the shape of an ellipsoid is placed in a uniform field, then the magnetization M is constant throughout the sample. The Maxwell equations imply that the magnetic field within the sample, i.e. the internal field, is reduced from the applied field by the "demagnetization field $\bar{\bar{N}}M$, or that

$$H_d = \bar{\bar{N}}M,$$

or

$$H = H_i = H_0 - \bar{\bar{N}}M, \; B = B_i = H_0 = - \bar{\bar{N}}M + 4\pi M$$

inside the sample. The demagnetization tensor $\bar{\bar{N}}$ is the same one as introduced in the case of an electric polarization, and it is diagonal with respect to the main axes of an ellipsoid.
Assuming the field and thus also the magnetization along a main axis z, then the demagnetization field is determined by N_{zz} , which is a number lying between 0 (long thin needle along z) and 4π (thin disk perpendicular to z). In the latter case the internal magnetic induction is the same as the applied one $B = B_0$ reflecting that all the magnetic field lines of each individual magnetic moment are looping back through the sample.

Returning to the first expression in for the magnetic energy, then the integration of H with respect to M requires a knowledge of the relation between the magnetization of the sample and the internal magnetic field. Introducing the characteristic material parameter, the magnetic susceptibility tensor (at zero field)

$$\chi_{\alpha\beta} = \left.\frac{\partial M_\alpha}{\partial H_\beta}\right|_{H=0}$$

when assuming the field and magnetization along z (as long as $M = \chi_{zz}H$).

The magnetization of a uniform sample of microscopic dipoles is

$$M = \frac{1}{V}\sum_i \vec{\mu}_i$$

If these magnetic dipoles are only coupled mutually by the classical dipole field, then we may calculate the energy directly in the same way as in the case of electric dipoles (the P and M fields are equivalent and E corresponds to H not to B). The energy of a dipole in the middle of the sample is:

$$U(r=0) = -\vec{\mu}\cdot H_{\text{cell}} = -\vec{\mu}\cdot\left(H_0 + \sum_i \frac{3r_i(\vec{\mu}_i\cdot r_i) - \vec{\mu}_i r_i^2}{r_i^5}\right)$$

and using the procedure explained on "the dipole moment", we get

$$H_{\text{cell}} = H_0 + \left(\bar{\bar{\xi}} + \frac{4\pi}{3}\right)M - \bar{\bar{N}}\langle M\rangle = H + \left(\bar{\bar{\xi}} + \frac{4\pi}{3}\right)M$$

In the paramagnetic phase there is only one domain, and defining a "non-interacting" susceptibility χ_0 (the susceptibility when the dipole field is neglected) by $M = \chi_0 H_{\text{cell}}$, we get

$$\chi = \frac{M}{H} = \frac{\chi_0}{1-\left(\frac{4\pi}{3}+\xi\right)}, \qquad H = \frac{H_0}{1+N\chi}$$

Systems exist where this susceptibility diverges at a suffciently low temperature TC , i.e. where $[(4\pi/3)+\chi]\chi_0 \rightarrow 1$, in which case the system is a ferromagnet below T_C , an ordering produced exclusively by the classical dipole-dipole in-teraction (one example is $LiHoF_4$ where $T_C = 1.5$ K).

In the other extreme case of a ferromagnet, $|\vec{\mu}|$ and hence $|M| = M$ is a constant. Introducing this in nd summing over all sites, the average energy density is found to be

$$u = -\langle M_z\rangle H_0 + \frac{1}{2}N_{zz}\langle M_z\rangle^2 - \frac{1}{2}\left(\frac{4\pi}{3}+\xi\right)M^2$$

Here $\langle M_z\rangle$ is the magnetization, along the direction of the applied field, averaged over the sample, which average is equal M if there is only one domain.

By introducing this average, the expression is of more general validity, since it is the dipolar energy density of the sample also in the presence of many different domains (the cutting sphere is assumed to lie within one domain with the magnetization $|M| = M$). The minimization of (13) with respect to $\langle M_z\rangle$ leads to

$$\langle M_z\rangle \frac{H_0}{N_{zz}} \quad \text{and} \quad H = 0, \quad \text{when} \quad H_0 < N_{zz}M$$

and

$$\langle M_z\rangle = M \quad \text{and} \quad H = 0 - H_0 - N_{zz}M, \quad \text{when} \quad H_0 > N_{zz}M$$

The magnetization rises linearly with H_0 and attains the saturation value M when $H_0 = H_d = N_{zz}M$. The internal field is the most natural choice for the field variable when investigating the material parameters of the sample, see the definition of χ. In terms of H the magnetization is a step function, $\langle M_z\rangle = M$ as soon as H is non-zero.

In this account we have neglected the energy cost due to the domain walls between the different domains. This energy cost should be included in a realistic model together with a more direct evaluation of the demagnetization field, which would change from one domain to the next. However, if the domain walls are easily created as is the case if the magnetic anisotropy is weak, the simple averaging of the domain effects is acceptable. In the case of hard magnetic systems (large magnetic anisotropy) the situation is complicated and irreversible hysteresis phenomena become important (permanent magnets).

In most magnetic material the classical dipole-dipole interaction is weak compared to other "two-ion interactions" (exchange interactions), and the difference between χ_0 and χ in (12) may safely be neglected. However, the long range nature of the coupling implies that it is important, in most cases, to account for the demagnetization field, i.e. to realise that the internal field is

$H = H_0 - N_{zz} \langle M_z \rangle$, and that the system may possibly divide itself into different domains. The approximative inclusion of the dipole-dipole interaction, where χ; χ_0, corresponds to the use of H_{cell} H in, and thus that the energy of the dipole moment $\bar{\mu}$ is $U(r=0) \simeq -\bar{\mu} \cdot H$. Hence, when focussing on the material parameters or the principal behaviour of a certain system, we normally assume that we are left with only one magnetic domain and that the magnetic field variable is the internal field H. The thermodynamic quantities are then derived from the partition function Z according to

$$F = -k_B T \ln Z, \qquad S = -\frac{\partial F}{\partial T}, \qquad U = F + TS$$

and

$$M = -\frac{1}{V}\frac{\partial F}{\partial H}, \quad \chi_{\alpha\beta} = -\frac{1}{V}\frac{\partial^2 F}{\partial H_\alpha H_\beta}, \quad C = \frac{\partial U}{\partial T} = T\frac{\partial S}{\partial T}$$

where the expressions for M and χ are the results derived from and, when approximating H_0 by H in (6). The thermodynamic magnetization derived from the is actually the one, which determines what are the microscopic dipoles in.

In the mean-field approximation the interactions between the magnetic moments are replaced by effective fields acting on the single moments and $H = \Sigma_l H_l$ as for non-interacting moments. In this case, the partition function $Z = \Pi Zl$ ($= Z^N$ for a uniform system)and all thermodynamic quantities are derived from the partition functions of the single moments. Determining the n eigenstates and eigenvalues of the Hamiltonian for the lth site we may introduce the population factor p_i of the ith level

$$H_i \mid i\rangle = \varepsilon_i \mid i\rangle, \quad p_i = \frac{e^{-\beta\varepsilon_i}}{Z_l}, \quad \sum_{i=1}^{n} p_i = 1 \quad (k_B T = 1/\beta)$$

where the last identity follows from $Z_l = \Sigma_i^{-\beta\varepsilon i}$. The ith population factor is the probability that the ith level is occupied. Immediate consequences of this interpretation of p_i are

$$U = N\langle H_l \rangle = N\sum_{i=1}^{n} \varepsilon_i p_i$$

and

$$M_\alpha = \frac{N}{V}\langle \mu_\alpha \rangle = \frac{N}{V}\sum_{i=1}^{n} \langle i | \hat{\mu}_\alpha | i \rangle\, p_i$$

which results agree with those obtained from the thermodynamic definitions and relations (see the solution to HS's problem 4). The Maxwell equations are

$$\nabla \cdot D = 4\pi\rho_{ext} \quad , \quad \nabla \cdot B = 0$$

$$\nabla \times E = -\frac{1}{c}\frac{\partial B}{\partial t} \quad , \quad \nabla \times H = 0\frac{4\pi}{c} j_{ext} + \frac{1}{c}\frac{\partial D}{\partial t}$$

where

$$D = E + 4\pi P \quad , \quad B = H + 4\pi M$$

$$\nabla \cdot P = -\rho_{int} \quad , \quad c\nabla \times M + \frac{\partial P}{\partial t} = j_{int}$$

The work done by the "surroundings" on the system (a certain volume in space) derives exclusively from the current density j_{ext} applied in the outside world, and the energy gain of the system is

$$\delta U = -\int dr(j_{ext} \cdot E)\delta t = -\int dr\left(\frac{c}{4\pi}\nabla \times H - \frac{1}{4\pi}\frac{\partial D}{\partial t}\right) \cdot E\delta t$$

The second term is the electric part of the energy (when $\rho_{ext} = 0$):

$$\delta U_E = \frac{1}{4\pi}\int dr \frac{\partial D}{\partial t} \cdot E\,\delta t = \frac{1}{4\pi}\int dr\, E \cdot \delta t$$

whereas the first term in is the magnetic energy part δU_M. Using the identity $\Delta \cdot (a \times b) = b \cdot \nabla \times a - a \cdot \nabla \times b$, we get

$$\delta U_M = -\int dr \frac{c}{4\pi}[H \cdot \nabla \times E + \nabla \cdot (H \times E)]\delta t$$

Applying the divergence theorem of Gauss, the second integral may be written as an integral of $H \times E$ over the surface of the system. Assuming that the system extents to "infinity", where all fields vanish, this integral is zero, and

$$\delta U_M = -\int dr\, H \cdot \nabla \times E\delta t = \frac{1}{4\pi}\int dr\, H \cdot \frac{\partial B}{\partial t}\delta t = \frac{1}{4\pi}\int dr\, H \cdot \delta B$$

The magnetic fields H_d and $B_d = H_d + 4\pi M$, are the fields produced by a magnetic sample, equation.

The fields within the sample are given by eq. (in the homogeneous case). Outside the sample $M = 0$ and thus $B_d = H_d$, but this field has not been calculated.

It is clear from Maxwell equations that $\nabla \cdot B_d = 0$ and also $\nabla \times H_d = 0$, since $\nabla \times H_0$ accounts for all the external currents. Because $\nabla \times H_d = 0$ it is possible to determine a potential field ψ so that $H_d = -\nabla\psi$. Hence

$$\int_{\text{all space}} H_d \cdot B_d \, dr = -\int \nabla\psi \cdot B_d dr = -\int [\nabla.(\psi B_d) - \psi \nabla \cdot B_d] dr$$

$$= -\int \nabla \cdot (\psi B_d) \, dr = -\lim_{r \to \infty} \int_{\text{surface}} \psi B_d \cdot dS = 0$$

because ψB_d must become vanishing small suffciently far away from the sample ($\psi B_d \propto r^{-5}$ for a dipole field).

By the same arguments, the integral where H_d is replaced by H_0 is also found to be vanishing small. Now $\nabla \times H_0$ is not zero, but it is sensible to assume that the applied field is generated by currents j_{ext} "outside" the space where B_d is of any importance, or that the surface integral has vanished before $\nabla \times H_0 = 0$ is violated.

Introducing $\delta B = \delta(H + 4\pi M)$ in, then

$$U_M = \frac{1}{4\pi} \int dr \int H \cdot \delta(H + 4\pi M) = \frac{1}{8\pi} \int dr \, H^2 + \int dr \int H \cdot \delta M$$

where

$$\frac{1}{8\pi} \int dr H^2 = \frac{1}{8\pi} \int dr \left[H_0^2 + 2H_0 \cdot H_d + H_d^2 \right]$$

$$= \frac{1}{8\pi} \int dr \left[H_0^2 + 2H_0 \cdot (B_d - 4\pi M) + H_d \cdot (B_d - 4\pi M) \right]$$

$$= \frac{1}{8\pi} \int dr H_0^2 - \int dr \left(M \cdot H_0 + \frac{1}{2} M \cdot H_d \right)$$

introducing this result in, we get the expression for the magnetic energy of the sample given by eq., when subtracting the background energy.

Chapter 10

Quantum and Statistical Mechanics

STATES AND OPERATORS

A quantum mechanical system is defined by a Hilbert space, H, whose vectors, $|\psi\rangle$ are associated with the states of the system. A state of the system is represented by the set of vectors $e^{i\alpha}|\psi\rangle$. There are linear operators, O_i which act on this Hilbert space. These operators correspond to physical observables. Finally, there is an inner product, which assigns a complex number, $\langle\chi|\psi\rangle$, to any pair of states, $|\psi\rangle$, $|\chi\rangle$. A state vector, $|\psi\rangle$ gives a complete description of a system through the expectation values, $\langle\psi|O_i|\psi\rangle$ (assuming that $|\psi\rangle$ is normalized so that $\langle\psi|\psi\rangle = 1$), which would be the average values of the corresponding physical observables if we could measure them on an infinite collection of identical systems each in the state

The adjoint, $O^\dagger$, of an operator is defined according to

$$\langle\chi|(O|\psi\rangle) = (\langle\chi|O^\dagger)|\psi\rangle$$

In other words, the inner product between $|\chi\rangle$ and $O|\psi\rangle$ is the same as that between $O^\dagger|\psi\rangle$ and $|\psi\rangle$. An Hermitian operator satisfies

$$O = O^\dagger$$

while a unitary operator satisfies

$$OO^\dagger = O^\dagger O = 1$$

If O is Hermitian, then

$$eiO$$

is unitary. Given an Hermitian operator, O, its eigenstates are orthogonal,

$$\langle\lambda'|O|\lambda\rangle = \lambda\langle\lambda'|\lambda\rangle = \lambda'\langle\lambda'|\lambda\rangle$$

For $\lambda \neq \lambda'$,

$$\langle\lambda'|\lambda\rangle = 0$$

If there are n states with the same eigenvalue, then, within the subspace spanned by these states, we can pick a set of n mutually orthogonal states. Hence, we can use the eigenstates $|\lambda\rangle$ as a basis for Hilbert space. Any state $|\psi\rangle$ can be expanded in the basis given by the eigenstates of O:

$$|\psi\rangle = \sum_{\lambda} c_{\lambda} |\lambda\rangle$$

with

$$c_{\lambda} = \langle \lambda | \psi \rangle$$

A particularly important operator is the Hamiltonian, or the total energy, which we will denote by H. Schrodinge's equation tells us that H determines how a state of the system will evolve in time.

$$i\hbar \frac{\partial}{\partial t} |\psi\rangle = H |\psi\rangle$$

If the Hamiltonian is independent of time, then we can define energy eigenstates,

$$H | E \rangle = E | \psi \rangle$$

which evolve in time according to:

$$| E(t) \rangle = e^{-i\frac{Et}{\hbar}} | E(0) \rangle$$

An arbitrary state can be expanded in the basis of energy eigenstates:

$$|\psi\rangle = \sum_{i} c_i | E_i \rangle$$

It will evolve according to:

$$|\psi(t)\rangle = \sum_{j} c_j e^{-i\frac{E_j t}{\hbar}} | E_j \rangle$$

For example, consider a particle in 1D. The Hilbert space consists of all continuous complex-valued functions, $\psi(x)$. The position operator, $\hat{x}$, and momentum operator, $\hat{p}$ are defined by:

$$\hat{x} \cdot \psi(x) \equiv x\psi(x)$$

$$\hat{p} \cdot \psi(x) \equiv -i\hbar \frac{\partial}{\partial x} \psi(x)$$

The position eigenfunctions,

$$x\, \delta(x - a) = a\delta(x - a)$$

are Dirac delta functions, which are not continuous functions, but can be defined as the limit of continuous functions:

$$\delta(x) = \lim_{a \to 0} \frac{1}{a\sqrt{\pi}} e^{-\frac{x^2}{a^2}}$$

The momentum eigenfunctions are plane waves:

$$-i\hbar \frac{\partial}{\partial x} e^{ikx} = \hbar k\, e^{ikx}$$

Expanding a state in the basis of momentum eigenstates is the same as taking its Fourier transform:

$$\psi(x) = \int_{-\infty}^{\infty} dk\tilde{\psi}(k)\frac{1}{\sqrt{2\pi}}e^{ikx}$$

where the Fourier coefficients are given by:

$$\tilde{\psi}(k) = \frac{1}{\sqrt{2\pi}}\int_{-\infty}^{\infty} dx\psi(x)\, e^{-ikx}$$

If the particle is free,

$$H = -\frac{\hbar^2}{2m}\frac{\partial^2}{\partial x^2}$$

then momentum eigenstates are also energy eigenstates:

$$\hat{H}e^{ikx} = \frac{\hbar^2 k^2}{2m}e^{ikx}$$

If a particle is in a Gaussian wavepacket at the origin at time $t = 0$,

$$\psi(x,0) = \frac{1}{a\sqrt{\pi}}e^{-\frac{x^2}{a^2}}$$

Then, at time t, it will be in the state:

$$\psi(x,t) = \frac{1}{a\sqrt{\pi}}\int_{-\infty}^{\infty} dk\frac{a}{\sqrt{\pi}}e^{-i\frac{\hbar k^2 t}{2m}}e^{-\frac{1}{2}k^2a^2}\;e^{ikx}$$

DENSITY AND CURRENT

Multiplying the free-particle Schrodinger equation by ψ^*,

$$\psi^* i\hbar\frac{\partial}{\partial t}\psi = -\frac{\hbar^2}{2m}\psi^*\frac{\partial^2}{\nabla^2}\psi$$

and subtracting the complex conjugate of this equation, we find

$$\frac{\partial}{\partial t}(\psi^*\psi) = \frac{i\hbar}{2m}\vec{\nabla}\cdot(\psi^*\vec{\nabla}\psi - (\vec{\nabla}\psi^*)\psi)$$

This is in the form of a continuity equation,

$$\frac{\partial}{\partial t} = \vec{\nabla}\cdot\vec{j}$$

The density and current are given by:

$$\rho = \psi^*\psi$$

$$\vec{j} = \frac{i\hbar}{2m}(\psi^*\vec{\nabla}\psi - (\vec{\nabla}\psi^*)\psi)$$

The current carried by a plane-wave state is:

$$\vec{j} = \frac{\hbar}{2m}\vec{k}\frac{1}{(2\pi)^3}$$

δ-FUNCTION SCATTERER

$$H = -\frac{\hbar^2}{2m}\frac{\partial^2}{\partial x^2} + V\delta(x)$$

$$\psi(x) = \begin{cases} e^{ikx} + \text{Re}^{-ikx} & \text{if } x < 0 \\ Te^{ikx} & \text{if } x > 0 \end{cases}$$

$$T = \frac{1}{1 - \frac{mV}{\hbar^2 k}i}$$

$$R = \frac{\frac{mV}{\hbar^2 k}i}{1 - \frac{mV}{\hbar^2 k}i}$$

There is a bound state at:

$$ik = \frac{mV}{\hbar^2}$$

PARTICLE IN A BOX

Particle in a $1D$ region of length L:

$$H = -\frac{\hbar^2}{2m}\frac{\partial^2}{\partial x^2}$$

$$\psi(x) = Ae^{ikx} + Be^{-ikx}$$

has energy $E = \hbar^2k^2/2m$. $\psi(0) = \psi(L) = 0$. Therefore,

$$\psi(x) = A \sin\left(\frac{n\pi}{L}x\right)$$

for integer n. Allowed energies

$$E_n = \frac{\hbar^2\pi^2 n^2}{2mL^2}$$

In a $3D$ box of side L, the energy eigenfunctions are:

$$\psi(x) = A\sin\left(\frac{n_x\pi}{L}x\right)\sin\left(\frac{n_y\pi}{L}y\right)\sin\left(\frac{n_z\pi}{L}z\right)$$

and the allowed energies are:

$$E_n = \frac{\hbar^2\pi^2}{2mL^2}\left(n_x^2 + n_y^2 + n_z^2\right)$$

$$H = -\frac{\hbar^2}{2m}\frac{\partial^2}{\partial x^2} + \frac{1}{2}kx^2$$

Writing $\omega = \sqrt{k/m}$, $\tilde{p} = p/(km)^{1/4}, \tilde{x} = x(km)^{1/4}$,

$$H = \frac{1}{2}\omega(\tilde{p}^2 + \tilde{x}^2)$$

$$[\tilde{p}, \tilde{x}] = -i\hbar$$

Raising and lowering operators:

$$a = (\tilde{x} + i\tilde{p})/\sqrt{2\hbar}$$

$$a^\dagger = (\tilde{x} - i\tilde{p})/\sqrt{2\hbar}$$

Hamiltonian and commutation relations:

$$H = \hbar\omega\left(a^\dagger a + \frac{1}{2}\right)$$

$$[a, a^\dagger] = 1$$

The commutation relations,

$$[H, a^\dagger] = \hbar\omega a^\dagger$$

$$[H, a] = -\hbar\omega a$$

imply that there is a ladder of states,

$$Ha^\dagger \mid E\rangle = (E + \hbar\omega)a^\dagger \mid E\rangle$$

$$Ha \mid E\rangle = (E - \hbar\omega)a \mid E\rangle$$

This ladder will continue down to negative energies (which it can't since the Hamiltonian is manifestly positive deπ finite) unless there is an $E_0 \geq 0$ such that

$$a \mid E_0\rangle = 0$$

Such a state has $E_0 = \hbar\omega/2$.

We label the states by their a^{y}a eigenvalues. We have a complete set of H eigenstates, $|n\rangle$, such that

$$H \mid n\rangle = \hbar\omega\left(n + \frac{1}{2}\right) \mid n\rangle$$

and $(a^\dagger)^n|0\rangle \propto |n\rangle$. To get the normalization, we write $a^\dagger \mid n\rangle = c_n \mid n+1\rangle$. Then,

$$|c_n|^2 = \langle n|aa^\dagger|n\rangle$$
$$= n+1$$

Hence,

$$a^\dagger \,|\, n\rangle = \sqrt{n+1}\,|\, n+1\rangle$$
$$a\,|\, n\rangle = \sqrt{n}\,|\, n-1\rangle$$

DOUBLE WELL

$$H = -\frac{\hbar^2}{2m}\frac{\partial^2}{\partial x^2} + V(x)$$

where

$$V(x) = \begin{cases} \infty & \text{if } |x| > 2a+2b \\ 0 & \text{if } b < |x| < a+b \\ V_0 & \text{if } |x| < b \end{cases}$$

Symmetrical solutions:

$$\psi(x) = \begin{cases} A\cos k'x & \text{if } |x| < b \\ \cos(k|x| - \phi) & \text{if } b < |x| < a+b \end{cases}$$

with

$$k' = \sqrt{k^2 - \frac{2mV_0}{\hbar^2}}$$

The allowed k's are determined by the condition that $\psi(a + b) = 0$:

$$\phi = \left(n + \frac{1}{2}\right)\pi - k(a+b)$$

the continuity of $\psi(x)$ at $|x| = b$:

$$A = \frac{\cos(kb - \phi)}{\cos k'b}$$

and the continuity of $\psi'(x)$ at $|x| = b$:

$$k\tan\left(\left(n + \frac{1}{2}\right)\pi - ka\right) = k'\tan k'b$$

If k' is imaginary, cos → cosh and tan → i tanh in the above equations.
Antisymmetrical solutions:

$$\psi(x) = \begin{cases} A\sin k'x & \text{if } |x| < b \\ \text{sgn}(x)\cos(k|x| - \phi & \text{if } b < |x| < a+b \end{cases}$$

The allowed k's are now determined by

$$\phi = \left(n + \frac{1}{2}\right)\pi - k(a+b)$$

$$A = \frac{\cos(kb - \phi)}{\sin k'b}$$

$$k \tan\left(\left(n + \frac{1}{2}\right)\pi - ka\right) = -k' \cot k'b$$

Suppose we have n wells? Sequences of eigenstates, classified according to their eigenvalues under translations between the wells.

SPIN

The electron carries spin-1/2. The spin is described by a state in the Hilbert space:

$$\alpha|+\rangle + \beta|-\rangle$$

spanned by the basis vectors $|\pm\rangle$. Spin operators:

$$S_x = \frac{1}{2}\begin{pmatrix} 0 & 1 \\ 1 & 0 \end{pmatrix}$$

$$S_y = \frac{1}{2}\begin{pmatrix} 0 & -i \\ i & 0 \end{pmatrix}$$

$$S_z = \frac{1}{2}\begin{pmatrix} 1 & 0 \\ 0 & -1 \end{pmatrix}$$

Coupling to an external magnetic field:

$$H_{\text{int}} = -g\mu_B \vec{s} \cdot \vec{B}$$

States of a spin in a magnetic field in the $\hat{z}$ direction:

$$H|+\rangle = -\frac{g}{2}\mu_B|+\rangle$$

$$H|-\rangle = \frac{g}{2}\mu_B|-\rangle$$

2.8MANY-PARTICLE HILBERT SPACES: BOSONS, FERMIONS

When we have a system with many particles, we must now specify the states of all of the particles. If we have two distinguishable particles whose Hilbert spaces are

$$|i, 1\rangle$$

spanned by the bases

$$|\alpha, 2\rangle$$

and

Then the two-particle Hilbert space is spanned by the set:

$$|i, 1; \alpha, 2\rangle \equiv |i, 1\rangle \otimes |\alpha, 2\rangle$$

Suppose that the two single-particle Hilbert spaces are identical, e.g. the two particles are in the same box. Then the two-particle Hilbert space is:

$$|i,j\rangle \equiv |i,1\rangle \otimes |j,2\rangle$$

If the particles are identical, however, we must be more careful. $|i,j\rangle$ and $|j,i\rangle$ must be physically the same state, i.e.

$$|i,j\rangle = e^{i\alpha}\,|j,i\rangle$$

Applying this relation twice implies that

$$|i,j\rangle = e^{2i\alpha}\,|i,j\rangle$$

so $e^{i\pi} = \pm 1$. The former corresponds to bosons, while the latter corresponds to fermions. The two-particle Hilbert spaces of bosons and fermions are respectively spanned by:

$$|i,j\rangle + |j,i\rangle$$

and

$$|i,j\rangle - |j,i\rangle$$

The n-particle Hilbert spaces of bosons and fermions are respectively spanned by:

$$\sum_{\pi} |i_{\pi(1)},\ldots,i_{\pi(n)}\rangle$$

and

$$\sum_{\pi} (-1)^{\pi}\,|i_{\pi(1)},\ldots,i_{\pi(n)}\rangle$$

In position space, this means that a bosonic wavefunction must be completely symmetric:

$$\psi(x_1,\ldots,x_i,\ldots,x_j,\ldots,x_n) = \psi(x_1,\ldots,x_j,\ldots,x_i,\ldots,x_n)$$

while a fermionic wavefunction must be completely antisymmetric:

$$\psi(x_1,\ldots,x_i,\ldots,x_j,\ldots,x_n) = -\psi(x_1,\ldots,x_j,\ldots,x_i,\ldots,x_n)$$

STATISTICAL MECHANICS

MICROCANONICAL, CANONICAL, ENSEMBLES

In statistical mechanics, we deal with a situation in which even the quantum state of the system is unknown. The expectation value of an observable must be averaged over:

$$\langle O\rangle = \sum_i w_i \langle i|O|i\rangle$$

where the states $|i\rangle$ form an orthonormal basis of H and w_i is the probability of being in state $|i\rangle$. The w_i's must satisfy $\Sigma w_i = 1$. The expectation value can be written in a basis-independent form:

$$\langle O\rangle = \mathrm{Tr}\{\rho O\}$$

where ρ is the density matrix. In the above example, ρ = . The condition, wi = 1, i.e. that the probabilities add to 1, is:

$$\mathrm{Tr}\ \{\rho\} = 1$$

We usually deal with one of three ensembles: the microcanonical emsemble, the canonical ensemble, or the ensemble. In the microcanonical ensemble, we assume that our system is isolated, so the energy is fixed to be E, but all states with energy E are taken with equal probability:

$$\rho = C\ \delta(H - E)$$

C is a normalization constant which is determined by (3.3). The entropy is given by,

$$S = -\ln C$$

In other words,

$$S(E) = \ln\ (\#\text{ of states with energy } E)$$

Inverse temperature, $\beta = 1/(k_B T)$:

$$\beta \equiv \left(\frac{\partial S}{\partial E}\right)_V$$

Pressure, *P*:

$$\frac{P}{k_B T} \equiv \left(\frac{\partial S}{\partial V}\right)_E$$

where V is the volume.

First law of thermodynamics:

$$dS = \frac{\partial S}{\partial E} dE + \frac{\partial S}{\partial V} dV$$

$$dE = k_B T dS - P dV$$

:

$$F = E - k_B TS$$

Differential relation:

$$dF = -k_B S\ dT - P dV$$

or,

$$S = -\frac{1}{k_B}\left(\frac{\partial F}{\partial T}\right)_V$$

$$P = -\left(\frac{\partial F}{\partial T}\right)_T$$

while

$$E = F + k_B TS$$

$$= F - T\left(\frac{\partial F}{\partial T}\right)_V$$

$$= -T^2 \frac{\partial}{\partial T} \frac{F}{Tm}$$

In the canonical ensemble, we assume that our system is in contact with a heat reservoir so that the temperature is constant. Then,

$$\rho = Ce^{-\beta H}$$

It is useful to drop the normalization constant, C, and work density matrix so that we can define the partition function:

$$Z = Tr\ \{\rho\}$$

or,

$$Z = \sum_a e^{-\beta E_a}$$

The average energy is:

$$E = \frac{1}{Z}\sum_a E_a e^{-\beta E_a}$$

$$= -\frac{\partial}{\partial\beta}\ln Z$$

$$= -k_B T^2 \frac{\partial}{\partial T}\ln Z$$

Hence,

$$F = -k_B T \ln Z$$

The chemical potential, μ, is defined by

$$\mu = \frac{\partial F}{\partial N}$$

where N is the particle number.

In the ensemble, the system is in contact with a reservoir of heat and particles. Thus, the temperature and chemical potential are held fixed and

$$\rho = Ce^{-\beta(H-\mu N)}$$

We can again work with an unnormalized density matrix and construct the partition function:

$$Z = \sum_{N,a} e^{-\beta(Ea-\mu N)}$$

The average number is:

$$N = -k_B T \frac{\partial}{\partial\mu}\ln Z$$

while the average energy is:

$$E = -\frac{\partial}{\partial\beta}\ln Z + \mu k_B T \frac{\partial}{\partial\mu}\ln Z$$

BOSE-EINSTEIN AND PLANCK DISTRIBUTIONS

Bose-Einstein Statistics

For a system of free bosons, the partition function

$$Z = \sum_{Ea,N} e^{-\beta(Ea - \mu N)}$$

can be rewritten in terms of the single-particle eigenstates and the single-particle energies $\in_i$:

$$E_a = n_0 \in_0 + n_1 \in_1 + \ldots$$

$$Z = \sum_{\{n_i\}} e^{-\beta\left(\sum_i n_i \in_i - \mu \sum_i n_i\right)}$$

$$= \prod_i \left(\sum_{n_i} e^{-\beta(n_i \in_i - \mu n_i)} \right)$$

$$= \prod_i \frac{1}{1 - e^{-\beta(\in_i - \mu)}}$$

$$\langle n_i \rangle = \frac{1}{e^{\beta(\in_i - \mu)} - 1}$$

The chemical potential is chosen so that

$$N = \sum_i \langle n_i \rangle$$

$$= \sum_i \frac{1}{e^{\beta(\in_i - \mu)} - 1}$$

The energy is given by

$$E = \sum_i \langle n_i \rangle \in_i$$

$$= \sum_i \frac{1}{e^{\beta(\in_i - \mu)} - 1}$$

N is increased by increasing μ ($\mu \le 0$ always). Bose-Einstein condensation occurs when

$$N > \sum_{i \neq 0} \langle n_i \rangle$$

In such a case, $\langle n_0 \rangle$ must become large. This occurs when $\mu = 0$.

The Planck Distribution

Suppose N is not fixed, but is arbitrary, e.g. the numbers of photons and neutrinos are not fixed. Then there is no Lagrange multiplier μ and

$$\langle n_i \rangle = \frac{1}{e^{\beta \epsilon i} - 1}$$

Consider photons (two polarizations) in a cavity of side L with $\epsilon_k = \hbar ck$ $= \hbar ck$ and

$$k = \frac{2\pi}{L}(m_x, m_y, m_z)$$

$$E = 2\sum_{m_x, m_y, m_z} \omega_{m_x, m_y, m_z} \langle n_{m_x, m_y, m_z} \rangle$$

We can take the thermodynamic limit, $L \to \infty$, and convert the sum into an integral. Since the allowed $\vec{k}$'s are $\frac{2\pi}{L}(m_x, m_y, m_z)$, the $\vec{k}$-space volume per allowed $\vec{k}$ is $(2\pi)^3/L^3$. Hence, we can take the inπ finite -volume limit by making the replacement:

$$\sum_k f(\vec{k}) = \frac{1}{(\Delta\vec{k})^3} \sum_k f(\vec{k})(\Delta\vec{k})^3$$

$$= \frac{L^3}{(2\pi)^3} \int d^3\vec{k} f(\vec{k})$$

Hence,

$$E = 2V \int_0^{k_{\max}} \frac{d^3k}{(2\pi)^3} \frac{\hbar\omega k}{e^{\beta\hbar\omega k} - 1}$$

$$= 2V \int_0^{k_{\max}} \frac{d^3k}{(2\pi)^3} \frac{\hbar ck}{e^{\beta\hbar ck} - 1}$$

$$= \frac{Vk_B^4}{\pi^2(\hbar c)^3} T^4 \int_0^{\beta\hbar ck_{\max}} \frac{x^3 dx}{e^x - 1}$$

For $\beta\hbar ck_{\max} \gg 1$,

$$E = \frac{Vk_B^4}{\pi^2(\hbar c)^3} T^4 \int_0^{\infty} \frac{x^3 dx}{e^x - 1}$$

and

$$C_V = \frac{4Vk_B^4}{\pi^2(\hbar c)^3} T^3 \int_0^{\infty} \frac{x^3 dx}{e^x - 1}$$

For $\beta\hbar ck_{\max} \ll 1$,

$$E = \frac{Vk_{\max}^3}{3\pi^2} k_B$$

and

$$C_V = \frac{Vk_{\max}^3 k_B}{3\pi^2}$$

FERMI-DIRAC DISTRIBUTION

For a system of free fermions, the partition function

$$Z = \sum_{E_a, N} e^{-\beta(Ea - \mu N)}$$

can again be rewritten in terms of the single-particle eigenstatesand the single-particle energies $\in_i$:

$$E_a = n_0 \in_0 + n_1 \in_1 + ...$$

but now

$$n_i = 0, 1$$

so that

$$Z = \sum_{\{n_i\}} e^{-\beta\left(\sum_i n_i \in_i - \mu \sum_i n_i\right)}$$

$$= \prod_i \left(\sum_{n_i=0}^{1} e^{-\beta(n_i \in_i - \mu n_i)} \right)$$

$$= \prod_i \left(1 + e^{-\beta(\in_i - \mu)}\right)$$

$$\langle n_i \rangle = \frac{1}{e^{\beta(\in_i + \mu)} + 1}$$

The chemical potential is chosen so that

$$N = \sum_i \frac{1}{e^{\beta(\in_i + \mu)} + 1}$$

The energy is given by

$$E = \sum_i \frac{\in_i}{e^{\beta(\in_i + \mu)} + 1}$$

THERMODYNAMICS OF THE

Free electron gas in a box of side L:

$$\in_k = \frac{\hbar^2 k^2}{2m}$$

with

$$k = \frac{2\pi}{L}(m_x, m_y, m_z)$$

Then, taking into account the 2 spin states,

$$E = 2 \sum_{m_x, m_y, m_z} \in_{m_x, m_y, m_z} \left\langle n_{m_x, m_y, m_z} \right\rangle$$

$$= 2V \int_0^{k_{\max}} \frac{d^3 k}{(2\pi)^3} \frac{\frac{\hbar^2 k^2}{2m}}{e^{\beta\left(\frac{\hbar^2 k^2}{2m} - \mu\right)} + 1}$$

$$N = 2V \int_0^{k_{\max}} \frac{d^3 k}{(2\pi)^3} \frac{1}{e^{\beta\left(\frac{\hbar^2 k^2}{2m} - \mu\right)} + 1}$$

At $T = 0$,

$$\frac{1}{e^{\beta\left(\frac{\hbar^2 k^2}{2m} - \mu\right)} + 1} = \theta\left(\mu - \frac{\hbar^2 k^2}{2m}\right)$$

All states with energies less than μ are filled; all states with higher energies are empty.

We write

$$k_F = \frac{\sqrt{2m\mu_{T=0}}}{\hbar}, \qquad \in_F = \mu_{T=0}$$

$$\frac{N}{V} = 2\int_0^{k_F} \frac{d^3 k}{(2\pi)^3} = \frac{k_F^3}{3\pi^2}$$

$$\frac{E}{V} = 2\int_0^{k_F} \frac{d^3 k}{(2\pi)^3} = \frac{\hbar^2 k^2}{2m}$$

$$= \frac{1}{\pi^2} \frac{\hbar^2 k_F^5}{10m}$$

$$= \frac{3}{5} \frac{N}{V} \in_F$$

$$2\int \frac{d^3 k}{(2\pi)^3} = \frac{m^{\frac{3}{2}} 2^{\frac{1}{2}}}{\pi^2 \hbar^2} \int d\in \in^{\frac{1}{2}}$$

For $k_B T \gg e_F$,

$$\frac{N}{V} = \frac{m^{\frac{3}{2}} 2^{\frac{1}{2}}}{\pi^2 \hbar^3} \int_0^\infty d\in \in^{\frac{1}{2}} \frac{1}{e^{\beta(\in-\mu)}+1}$$

$$= \frac{m^{\frac{3}{2}} 2^{\frac{1}{2}}}{\pi^2 \hbar^3} \int_0^\infty d\in \in^{\frac{1}{2}} + \frac{m^{\frac{3}{2}} 2^{\frac{1}{2}}}{\pi^2 \hbar^3} \int_0^\mu d\in \in^{\frac{1}{2}} \left(\frac{1}{e^{\beta(\in-\mu)}+1} - 1 \right) +$$

$$\frac{m^{\frac{3}{2}} 2^{\frac{1}{2}}}{\pi^2 \hbar^3} \int_\mu^\infty d\in \in^{\frac{1}{2}} \frac{1}{e^{\beta(\in-\mu)}+1}$$

$$= \frac{(2m)^{\frac{3}{2}}}{3\pi^2 \hbar^3} \mu^{\frac{3}{2}} - \frac{m^{\frac{3}{2}} 2^{\frac{1}{2}}}{\pi^2 \hbar^3} \int_0^\mu d\in \in^{\frac{1}{2}} \frac{1}{e^{-\beta(\in-\mu)}+1} + \frac{m^{\frac{3}{2}} 2^{\frac{1}{2}}}{\pi^2 \hbar^3} \int_\mu^\infty d\in \in^{\frac{1}{2}} \frac{1}{e^{\beta(\in-\mu)}+1}$$

$$= \frac{(2m)^{\frac{3}{2}}}{3\pi^2 \hbar^3} \mu^{\frac{3}{2}} - \frac{m^{\frac{3}{2}} 2^{\frac{1}{2}}}{\pi^2 \hbar^3} \int_0^\infty \frac{k_B T dx}{e^x+1} ((\mu + k_B T x)^{\frac{1}{2}} + O(e^{-\beta\mu})$$

$$= \frac{(2m)^{\frac{3}{2}}}{3\pi^2 \hbar^3} \mu^{\frac{3}{2}} + \frac{(2m)^{\frac{3}{2}}}{\pi^2 \hbar^3} \int_0^\infty \frac{k_B T dx}{e^x+1} ((\mu + k_B T x)^{\frac{1}{2}} - (\mu + k_B T x)^{\frac{1}{2}} + O(e^{-\beta\mu})$$

$$= \frac{(2m)^{\frac{3}{2}}}{3\pi^2 \hbar^3} \mu^{\frac{3}{2}} + \frac{(2m)^{\frac{3}{2}}}{\pi^2 \hbar^3} \sum_{n=1}^\infty (k_B T)^{2n} \left[\mu^{\frac{3}{2}-2n} \frac{1}{(2n-1)!} \frac{\Gamma\left(\frac{3}{2}\right)}{\Gamma\left(\frac{5}{2}-2n\right)} \int_0^\infty dx \frac{x^{2n-1}}{e^x+1} \right]$$

$$= \frac{(2m)^{\frac{3}{2}}}{3\pi^2 \hbar^3} \mu^{\frac{3}{2}} \left[1 + \frac{3}{2} \left(\frac{k_B T}{\mu} \right)^2 I_1 + O(T^4) \right]$$

with

$$I_k = \int_0^\infty dx \frac{x^k}{e^x+1}$$

We will only need

$$I_1 = \frac{\pi^2}{12}$$

Hence,

$$(\in_F)^{\frac{3}{2}} = \mu^{\frac{3}{2}}\left[1+\frac{3}{2}\left(\frac{k_BT}{\mu}\right)^2 I_1 + O(T^4)\right]$$

To lowest order in T, this gives:

$$\mu = \in_F\left(1-\left(\frac{k_BT}{\in_F}\right)^2 I_1 + O(T^4)\right)$$

$$=\in_F\left(1-\frac{\pi^2}{12}\left(\frac{k_BT}{\in_F}\right)^2 + O(T^4)\right)$$

$$\frac{N}{V} = \frac{m^{\frac{3}{2}}2^{\frac{1}{2}}}{\pi^2\hbar^3}\int_0^\infty d\in \in^{\frac{3}{2}}\frac{1}{e^{\beta(\in-\mu)}+1}$$

$$=\frac{m^{\frac{3}{2}}2^{\frac{1}{2}}}{\pi^2\hbar^3}\int_0^\infty d\in \in^{\frac{3}{2}} + \frac{m^{\frac{3}{2}}2^{\frac{1}{2}}}{\pi^2\hbar^3}\int_0^\mu d\in \in^{\frac{3}{2}}\left(\frac{1}{e^{\beta(\in-\mu)}+1}-1\right)+$$

$$\frac{m^{\frac{3}{2}}2^{\frac{1}{2}}}{\pi^2\hbar^3}\int_\mu^\infty d\in \in^{\frac{1}{2}}\frac{1}{e^{\beta(\in-\mu)}+1}$$

$$=\frac{(2m)^{\frac{3}{2}}}{5\pi^2\hbar^3}\mu^{\frac{5}{2}} - \frac{m^{\frac{3}{2}}2^{\frac{1}{2}}}{\pi^2\hbar^3}\int_0^\mu d\in \in^{\frac{3}{2}}\frac{1}{e^{-\beta(\in-\mu)}+1} + \frac{m^{\frac{3}{2}}2^{\frac{1}{2}}}{\pi^2\hbar^3}\int_\mu^\infty d\in \in^{\frac{3}{2}}\frac{1}{e^{\beta(\in-\mu)}+1}$$

$$=\frac{(2m)^{\frac{3}{2}}}{5\pi^2\hbar^3}\mu^{\frac{5}{2}} - \frac{m^{\frac{3}{2}}2^{\frac{1}{2}}}{\pi^2\hbar^3}\int_0^\mu \frac{k_BTdx}{e^x+1}\left((\mu+k_BTx)^{\frac{3}{2}} - (\mu-k_BTx)^{\frac{3}{2}}\right) + O(e^{-\beta\mu})$$

$$=\frac{(2m)^{\frac{3}{2}}}{5\pi^2\hbar^3}\mu^{\frac{5}{2}} - \frac{(2m)^{\frac{3}{2}}}{\pi^2\hbar^3}\sum_{n=2}^\infty (k_BT)^{2n}\left[\mu^{\frac{5}{2}-2n}\frac{1}{(2n-1)!}\frac{\Gamma\left(\frac{5}{2}\right)}{\Gamma\left(\frac{7}{2}-2n\right)}\int_0^\infty dx\frac{x^{2n-1}}{e^x+1}\right]$$

$$=\frac{(2m)^{\frac{3}{2}}}{5\pi^2\hbar^3}\mu^{\frac{5}{2}}\left[1+\frac{15}{2}\left(\frac{k_BT}{\mu}\right)^2 I_1 + O(T^4)\right]$$

$$=\frac{3}{5}\frac{N}{V}\in_F\left(1+\frac{5\pi^2}{12}\left(\frac{k_BT}{\in_F}\right)^2 + O(T^4)\right)$$

Hence, the specific heat of a gas of free fermions is:

$$C_V = \frac{\pi^2}{2} N k_B \frac{k_B T}{\epsilon_F}$$

Note that this can be written in the more general form:

$$C_V = (\text{const.}) \cdot k_B \cdot g(\epsilon_F) k_B T$$

The number of electrons which are thermally excited above the ground state is $\sim g(\epsilon_F)\, k_B T$; each such electron contributes energy $\sim k_B T$ and, hence, gives a specific heat contribution of k_B. Electrons give such a contribution to the specific heat of a metal.

ISING MODEL, MEAN FIELD THEORY, PHASES

Consider a model of spins on a lattice in a magnetic field:

$$H = -g\mu_B B \sum_i S_i^z \equiv 2h \sum_i S_i^z$$

with $S_i^z = \pm 1/2$. The partition function for such a system is:

$$Z = \left(2\cosh \frac{h}{k_B T} \right)^N$$

The average magnetization is:

$$S_i^z = \frac{1}{2} \tanh \frac{h}{k_B T}$$

The susceptibility, χ, is defined by

$$\chi = \left(\frac{\partial}{\partial h} \sum_i S_i^z \right)_{h=0}$$

For free spins on a lattice,

$$\chi = \frac{1}{2} N \frac{1}{k_B T}$$

A susceptibility which is inversely proportional to temperature is called a Curie suc-septibility. In problem set 3, you will show that the susceptibility is much smaller for a system of electrons.

Now consider a model of spins on a lattice such that each spin interacts with its neighbors according to:

$$H = -\frac{1}{2} \sum_{\langle i,j \rangle} J S_i^z S_j^z$$

This Hamiltonian has a symmetry

$$S_i^z \to -S_i^z$$

For $k_B T \gg J$, the interaction between the spins will not be important and the susceptibility will be of the Curie form. For $k_B T < J$, however, the behaviour will be much different. We can understand this qualitatively using mean field theory.

Let us approximate the interaction of each spin with its neighbors by an interaction with a mean-field, h:

$$H = -\sum_i h S_i^z$$

with h given by

$$h = \sum_i J\langle S_i^z \rangle = Jz\langle S_i^z \rangle$$

where z is the coordination number. In this field, the partition function is just $2\cosh\frac{h}{k_B T}$ and

$$\langle S^z \rangle = \tanh\frac{h}{k_B T}$$

Using the self-consistency condition, this is:

$$\langle S^z \rangle = \tanh\frac{Jz\langle S^z \rangle}{k_B T}$$

For $k_B T < Jz$, this has non-zero solutions, $S^z \neq 0$ which break the symmetry $S_i^z \to S_i^z$. In this phase, there is a spontaneous magnetization. For $k_B T > Jz$, there is only the solution $S^z = 0$. In this phase the symmetry is unbroken and the is no spontaneous magnetization. At $k_B T = Jz$, there is a critical point at which a phase transition occurs.

Chapter 11

Broken Translational Invariance

SIMPLE ENERGETICS OF SOLIDS

Why do solids form? The Hamiltonian of the electrons and ions is:

$$H = \sum_i \frac{p_i^2}{2m_e} + \sum_a \frac{P_a^2}{2M} + \sum_{i>j} \frac{e^2}{|r_i - r_j|} + \sum_{a>b} \frac{Z^2 e^2}{|R_a - R_b|} - \sum_{i,a} \frac{Ze^2}{|r_i - R_b|}$$

It is invariant under the symmetry, $\vec{r}_i \to \vec{r}_i + \vec{a}, \vec{R}_a \to \vec{R}_a + \vec{a}$. However, the energy can usually be minimized by forming a crystal. At low enough temperature, this will win out over any possible entropy gain in a competing state, so crystallization will occur. Why is the crystalline state energetically favorable? This depends on the type of crystal. Different types of crystals minimize different terms in the Hamiltonian. In molecular and ionic crystals, the potential energy is minimized. In a molecular crystal, such as solid nitrogen, there is a van der Waals attraction between molecules caused by polarization of one by the other. The van der Waals attraction is balanced by short-range repulsion, thereby leading to a crystalline ground state. In an ionic crystal, such as NaCl, the electrostatic energy of the ions is minimized (one must be careful to take into account charge neutrality, without which the electrostatic energy diverges).

In covalent and metallic crystals, crystallization is driven by the minimization of the electronic kinetic energy. In a metal, such as sodium, the kinetic energy of the electrons is lowered by their ability to move throughout the metal. In a covalent solid, such as diamond, the same is true. The kinetic energy gain is high enough that such a bond can even occur between just two molecules (as in organic chemistry). The energetic gain of a solid is called the cohesive energy.

QUANTUM MECHANICS OF A LINEAR CHAIN

As a toy model of a solid, let us consider a linear chain of masses m connected by springs with spring constant B. Suppose that the equilibrium spacing between the masses is a. The equilibrium positions define a $1D$ lattice. The lattice 'vectors', $\vec{R}_j$, are defined by:

$$\vec{R}_j = ja$$

They connect the origin to all points of the lattice. If $\vec{R}$ and $\vec{R}'$ are lattice vectors, then $\vec{R}+\vec{R}'$ are also lattice vectors. A set of basis vectors is a minimal set of vectors which generate the full set of lattice vectors by taking linear combinations of the basis vectors. In our 1D lattice, a is the basis vector.

Let u_i be the displacement of the ith mass from its equilibrium position and let pi be the corresponding momentum. Let us assume that there are N masses, and let's impose a periodic boundary condition, $u_i = u_{i+N}$. The Hamiltonian for such a system is:

$$H = \frac{1}{2m}\sum_i p_i^2 + \frac{1}{2}B\sum_i (u_i - u_{i+1})^2$$

Let us use the Fourier transform representation:

$$u_j = \frac{1}{\sqrt{N}}\sum_k u_k e^{ikja}$$

$$p_j = \frac{1}{\sqrt{N}}\sum_k p_k e^{ikja}$$

As a result of the periodic boundary condition, the allowed k's are:

$$k = \frac{2\pi n}{Na}$$

We can invert (4.4):

$$\frac{1}{\sqrt{N}}\sum_j u_k e^{ik'ja} = \frac{1}{\sqrt{N}}\sum_k\sum_j u_k e^{i(k-k')ja}$$

$$\frac{1}{\sqrt{N}}\sum_j u_j e^{ik'ja} = u_{k'}$$

Note that $u_k^\dagger = u\_k, p_k^\dagger = p\_k$ since $u_j^\dagger = u_j, p_j^\dagger = p_j$. They satisfy the commutation relations:

$$[p_k, u_{k'}] = \frac{1}{N}\sum_{j,j'} e^{ikja} e^{ik'j'a}[p_j, u_{j'}]$$

$$= \frac{1}{N}\sum_{j,j'} e^{ikja} e^{ik'j'a} - i\hbar\delta_{jj'}$$

$$= -i\hbar\frac{1}{N}\sum_j e^{i(k-k')ja} = -i\hbar\delta_{kk'}$$

Hence, p_k and $u_{k'}$ commute unless $k = k'$.

The displacements described by u_k are the same as those described by $u_{k+\frac{2\pi m}{a}}$ for a any integer n:

$$u_{k+\frac{2\pi m}{a}} = \frac{1}{\sqrt{N}}\sum_j u_j e^{-i\left(k+\frac{2\pi m}{a}\right)ja}$$

$$= \frac{1}{\sqrt{N}}\sum_j u_j e^{-ikja}$$

$$= u_k$$

Hence,

$$k \equiv k + \frac{2\pi m}{a}$$

Hence, we can restrict attention to $n = 0, 1, ...; N - 1$ in (4.5). The Hamiltonian can be rewritten:

$$H = \sum_k \frac{1}{2m} p_k p_{-k} + 2B\left(\sin + \frac{ka}{2}\right)^2 u_k u_{-k}$$

Recalling the solution of the problem, we define:

$$a_k = \frac{1}{\sqrt{2\hbar}}\left[\left(4mB\left(\sin\frac{ka}{2}\right)^2\right)^{\frac{1}{4}} u_k + \frac{i}{\left(4mB\left(\sin\frac{ka}{2}\right)^2\right)^{\frac{1}{4}}} p_k\right]$$

$$a_k^\dagger = \frac{1}{\sqrt{2\hbar}}\left[\left(4mB\left(\sin\frac{ka}{2}\right)^2\right)^{\frac{1}{4}} u_{-k} - \frac{i}{\left(4mB\left(\sin\frac{ka}{2}\right)^2\right)^{\frac{1}{4}}} p_{-k}\right]$$

(Recall that $u_k^\dagger = u_{-k}, p_k^\dagger = p_{-k}$.) which satisfy:

$$\left[a_k, a_k^\dagger\right] = 1$$

Then:

$$H = \sum_k \hbar\omega_k\left(a_k^\dagger a_k + \frac{1}{2}\right)$$

with

$$\omega_k = 2\left(\frac{B}{m}\right)^{\frac{1}{2}}\left|\sin\frac{ka}{2}\right|$$

Hence, the linear chain is equivalent to a system of N independent s. Its thermodynamics can be described by the Planck distribution.

The operators $a^{\dagger}$, a_k are said to create and annihilate phonons. We say that a state $|\psi\rangle$

$$a_k^{\dagger}a_k\,|\psi\rangle = N_k\,|\psi\rangle$$

with has N_k phonons of momentum k. Phonons are the quanta of lattice vibrations, analogous to photons, which are the quanta of oscillations of the electromagnetic eld.

Observe that, as $k \to 0$, $\omega_k \to 0$:

$$\omega_{k\to 0} = \left(\frac{Ba}{m/a}\right)^{\frac{1}{2}} k$$

$$= \left(\frac{Ba}{\rho}\right)^{\frac{1}{2}} k$$

The physical reason for this is simple: an oscillation with $k = 0$ is a uniform translation of the linear chain, which costs no energy.

Note that the reason for this is that the Hamiltonian is invariant under translations $u_i \to u_i + \lambda$. However, the ground state is not: the masses are located at the points $x_j = j_a$. Translational invariance has been spontaneously broken. Of course, it could just as well be broken with $x_j = ja + \lambda$ and, for this reason,$\omega_k \to 0$ as $k \to 0$. An oscillatory mode of this type is called a Goldstone mode.

Consider now the case in which the masses are not equivalent. Let the masses alternate between m and M. As we will see, the phonon spectra will be more complicated. Let a be the distance between one m and the next m. The Hamiltonian is:

$$H = \sum\left(\frac{1}{2m}p_{1,i}^2 + \frac{1}{2m}p_{2,i}^2 + \frac{1}{2}B(u_{1,i} - u_{2,i})^2 + \frac{1}{2}B(u_{2,i} - u_{1,i+1})^2\right)$$

The equations of motion are:

$$m\frac{d^2}{dt^2}u_{1,i} = -B[(u_{1,i} - u_{2,i}) - (u_{2,i-1} - u_{1,i})]$$

$$M\frac{d^2}{dt^2}u_{2,i} = -B[(u_{2,i} - u_{1,i}) + (u_{2,i} - u_{1,i+1})]$$

Going again to the Fourier representation, $\alpha = 1; 2$

$$u_{\alpha,j} = \frac{1}{\sqrt{N}} \sum_k u_{\alpha,k} e^{ikja}$$

Where the allowed k's are:

$$k = \frac{2\pi}{N} n$$

if there are $2N$ masses. As before,

$$u_{\alpha,k} = u_{\alpha,k+\frac{2\pi n}{a}}$$

$$\begin{pmatrix} m\dfrac{d^2}{dt^2} & 0 \\ 0 & M\dfrac{d^2}{dt^2} \end{pmatrix} \begin{pmatrix} u_{1,k} \\ u_{2,k} \end{pmatrix} = \begin{pmatrix} 2B & -B(1+e^{ika}) \\ -B(1+e^{-ika}) & 2B \end{pmatrix} \begin{pmatrix} u_{1,k} \\ u_{2,k} \end{pmatrix}$$

Fourier transforming in time:

$$\begin{pmatrix} -m\omega_k^2 & 0 \\ 0 & -Mm\omega_k^2 \end{pmatrix} \begin{pmatrix} u_{1,k} \\ u_{2,k} \end{pmatrix} = \begin{pmatrix} 2B & -B(1+e^{ika}) \\ -B(1+e^{-ika}) & 2B \end{pmatrix} \begin{pmatrix} u_{1,k} \\ u_{2,k} \end{pmatrix}$$

This eigenvalue equation has the solutions:

$$\omega^2 = B\left[\frac{1}{M} + \frac{1}{m} \pm \sqrt{\left(\frac{1}{M} + \frac{1}{m}\right)^2 - \frac{4}{nM}\left(\sin\frac{ka}{2}\right)^2}\,\right]$$

Observe that

$$\omega_{k\to 0}^- = \left(\frac{Ba}{2(m+M)/a}\right)^{\frac{1}{2}} k$$

This is the acoustic branch of the phonon spectrum in which m and M move in phase. As $k \to 0$, this is a translation, so $\omega_k^- \to 0$. Acoustic phonons are responsible for sound. Also note that

$$\omega_{k=\pi/a}^- = \left(\frac{2B}{M}\right)^{\frac{1}{2}}$$

Meanwhile, ω^+ is the optical branch of the spectrum (these phonons scatter light), in which m and M move in opposite directions.

$$\omega_{k\to 0}^+ = \sqrt{2B\left(\frac{1}{m} + \frac{1}{M}\right)}$$

$$\omega_{k=\pi/a}^+ = \left(\frac{2B}{M}\right)^{\frac{1}{2}}$$

so there is a gap in the spectrum of width

$$\omega_{\text{gap}} = \left(\frac{2B}{m}\right)^{\frac{1}{2}} - \left(\frac{2B}{M}\right)^{\frac{1}{2}}$$

Note that if we take $m = M$, we recover the previous results, with $a \to a/2$.

This is an example of what is called a lattice with a basis. Not every site on the chain is equivalent. We can think of the chain of 2N masses as a lattice with N sites.

Each lattice site has a two-site basis: one of these sites has a mass m and the other has a mass M. Sodium Chloride is a simple subic lattice with a two-site basis: the sodium ions are at the vertices of an FCC lattice and the chlorine ions are displaced from them.

STATISTICAL MECHNICS OF A LINEAR CHAIN

Let us return to the case of a linear chain of masses m separated by springs of force constant B, at equilibrium distance a. The excitations of this system are phonons which can have momenta $k \in [-\pi/a, \pi/a]$ (since $kk \equiv k + \frac{2\pi m}{a}$), corresponding to energies

$$\hbar\omega_k = 2\left(\frac{B}{m}\right)^{\frac{1}{2}}\left|\sin\frac{ka}{2}\right|$$

Phonons are bosons whose number is not conserved, so they obey the Planck distribution. Hence, the energy of a linear chain at finite temperature is given by:

$$E = \sum_k \frac{\hbar\omega_k}{e^{\beta\hbar\omega_k} - 1}$$

$$= L\int_{-\frac{\pi}{a}}^{\frac{\pi}{a}} \frac{dk}{(2\pi)} \frac{\hbar\omega_k}{e^{\beta\hbar\omega_k} - 1}$$

Changing variables from k to ω,

$$E = 2\cdot\frac{L}{2\pi}\int_0^{\sqrt{4B/m}} \frac{2}{a} \frac{d\omega}{\sqrt{\frac{4B}{m} - \omega^2}} \frac{\hbar\omega}{e^{\beta\hbar\omega} - 1}$$

$$= \frac{2N}{\pi}\int_0^{\sqrt{4B/m}} \frac{d\omega}{\sqrt{\frac{4B}{m} - \omega^2}} \frac{\hbar\omega}{e^{\beta\hbar\omega} - 1}$$

$$= \frac{2N}{\pi} \frac{(k_B T)^2}{\hbar} \int_0^{\beta\hbar\sqrt{4B/m}} \frac{1}{\sqrt{\frac{4B}{m} - \left(\frac{x}{\beta\hbar}\right)}} \frac{x\,dx}{e^x - 1}$$

In the limit $k_B T \ll \hbar\sqrt{4B/m}$, we can take the upper limit of integration to ∞ and drop the x-dependent term in the square root in the denominator of the integrand:

$$E = \frac{N}{\pi}\sqrt{\frac{m}{4B}}\frac{(k_B T)^2}{\hbar}\int_0^\infty \frac{x\,dx}{e^x - 1}$$

Hence, $C_v \sim T$ at low temperatures.

In the limit $k_B T \gg \hbar\sqrt{4B/m}$, we can approximate $e^x \approx 1 + x$:

$$E = \frac{2N}{\pi}\frac{(k_B T)^2}{\hbar}\int_0^{\beta\hbar\sqrt{4B/m}} \frac{dx}{\sqrt{\frac{4B}{m} - \left(\frac{x}{\beta\hbar}\right)^2}}$$

$$= Nk_B$$

In the intermediate temperature regime, a more careful analysis is necessary. In particular, note that the density of states, $1/\sqrt{\frac{4B}{m} - \omega^2}$ diverges at $\omega = \sqrt{4B/m}$; this is an example of a van Hove singularity. If we had alternating masses on springs, then the expression for the energy would have two integrals, one over the acoustic modes and one over the optical modes.

LESSONS FROM THE 1*D* CHAIN

In the course of our analysis of the 1*D* chain, we developed the following strategy, which we will apply to crystals more generally in subsequent sections.

- Expand the Hamiltonian to Quadratic Order
- Fourier transform the Hamiltonian into momentum space
- Identify the Brillouin zone (range of distinct $\vec{k}$s)
- Rewrite the Hamiltonian in terms of creation and annihilation operators
- Obtain the Spectrum
- Compute the Density of States
- Use the Planck distribution to obtain the thermodynamics of the vibrational modes of the crystal.

RECIPROCAL LATTICE, BRILLOUIN ZONES, CRYSTAL MOMENTUM

Note that, in the above, momenta were only defined up to $\frac{2\pi n}{a}$. The momenta $\frac{2\pi n}{a}$ form a lattice in k-space, called the reciprocal lattice. This is true of any function which, like the ionic discplacements, is a function defined at the lattice sites. For such a function, $f(\vec{R})$, defined on an arbitrary lattice, the Fourier transform

$$\tilde{f}(\vec{k}) = \sum_R e^{i\vec{k}\cdot\vec{R}} f(\vec{R})$$

satisfies

$$\tilde{f}(\vec{k}) = \tilde{f}(\vec{k} + \vec{G})$$

if $\vec{G}$ is in the set of reciprocal lattice vectors, defined by:~

$$e^{i\vec{G}\cdot\vec{R}} = 1, \text{ for all } \vec{R}$$

The reciprocal lattice vectors also form a lattice since the sum of two reciprocal lattice vectors is also a reciprocal lattice vector. This lattice is called the reciprocal lattice or dual lattice.

In the analysis of the linear chain, we restricted momenta to $|k| < \pi/a$ to avoid double-counting of degrees of freedom. This restricted region in k-space is an example of a Brillouin zone. All of k-space can be obtained by translating the Brillouin zone through all possible reciprocal lattice vectors.

We could have chosen our Brillouin zone differently by taking $0 < k < 2\pi/a$. Physically, there is no difference; the choice is a matter of convenience. What we need is a set of points in k space such that no two of these points are connected by a reciprocal lattice vector and such that all of k space can be obatained by translating the Brillouin zone through all possible reciprocal lattice vectors. We could even choose a Brillouin zone which is not connected, e.g. $0 < k < \pi/a$. or $3\pi/a < k < 4\pi/a$.

Later, we will consider solids with a more complicated lattice structure than our linear chain. Once again, phonon spectra will be defined in the Brillouin zone. Since $\tilde{f}(\vec{k}) = \tilde{f}(\vec{k} + \vec{G})$, the phonon modes outside of the Brillouin zone are not physically distinct from those inside. One way of defining the Brillouin zone for an arbitrary lattice is to take all points in k space which are closer to the origin than to any other point of the reciprocal lattice. Such a choice of Brillouin zone is also called the Wigner-Seitz cell of the reciprocal lattice. We will discuss this in some detail later but, for now, let us consider the case of a cubic lattice. The lattice vectors of a cubic lattice of side a are:

$$\vec{R}_{n_1,n_2,n_3} = a(n_1\hat{x} + n_2\hat{y} + n_3\hat{z})$$

The reciprocal lattice vectors are:

$$\vec{G}_{m_1,m_2,m_3} = \frac{2\pi}{a}(m_1\hat{x} + m_2\hat{y} + m_3\hat{z})$$

The reciprocal lattice vectors also form a cubic lattice. Theπ first Brillouin zone (Wigner-Seitz cell of the reciprocal lattice) is given by the cube of side $\frac{2\pi}{a}$ centered at the origin. The volume of this cube is related to the density according to:

$$\int_{B.Z.} \frac{d^3k}{(2\pi)^3} = \frac{1}{a^3} = \frac{N_{\text{ions}}}{V}$$

As we have noted before, the ground state (and the Hamiltonian) of a crystal is invariant under the discrete group of translations through all lattice vectors. Whereas full translational invariance leads to momentum conservation, lattice translational symmetry leads to the conservation of crystal momentum – momentum up to a reciprocal lattice vector.

For instance, in a collision between phonons, the difference between the incoming and outgoing phonon momenta can be any reciprocal lattice vector, G. Physically, one may think of the missing momentum as being taken by the lattice as a whole. This concept will also be important when we condsider the problem of electrons moving in the background of a lattice of ions.

PHONONS: CONTINUUM ELASTIC THEORY

Consider the lattice of ions in a solid. Suppose the equilibrium positions of the ions are the sites $\vec{R}_i$. Let us describe small displacements from these sites by a displacement field $\vec{u}(\vec{R}_i)$. We will imagine that the crystal is just a system of masses connected by springs of equilibrium length *a*.

Before considering the details of the possible lattice structures of 2D and 3D crystals, let us consider the properties of a crystal at length scales which are much larger than the lattice spacing; this regime should be insensitive to many details of the lattice.

At length scales much longer than its lattice spacing, a crystalline solid can be modelled as an elastic medium. We replace $\vec{u}(\vec{R}_i)$ by $\vec{u}(\vec{r})$ (i.e. we replace the lattice vectors, $\vec{R}_i$, by a continuous variable, $\vec{r}$). Such an approximation is valid at ~ length scales much larger than the lattice spacing, a, or, equivalently, at wavevectors $q \ll 2\pi/a$.

In 1D, we can take the continuum limit of our model of masses and springs:

$$H = \frac{1}{2} m \sum_i \left(\frac{du_i}{dt} \right)^2 + \frac{1}{2} B \sum_i (u_i - u_{i+1})^2$$

$$= \frac{1}{2} \frac{m}{a} \sum_i a \left(\frac{du_i}{dt} \right)^2 + \frac{1}{2} Ba \sum_i \left(\frac{u_i - u_{i+1}}{a} \right)^2$$

$$\to \int dx \left(\frac{1}{2}\rho \left(\frac{du}{dt}\right)^2 + \frac{1}{2}\bar{B}\left(\frac{du}{dx}\right) \right)^2$$

where ρ is the mass density and $\bar{B}$ is the bulk modulus. The equation of motion,

$$\frac{d^2u}{dt^2} = \bar{B}\frac{d^2u}{dx^2}$$

has solutions

$$u(x,t) = \sum_k u_k e^{i(kx-\omega t)}$$

with

$$\omega = \sqrt{\frac{\bar{B}}{\rho}}k$$

The generalization to a $3D$ continuum elastic medium is:

$$\rho \partial_i^2 \vec{u} = (\mu + \lambda)\vec{\nabla}(\nabla \cdot \vec{u}) + \mu \nabla^2 \vec{u}$$

where ρ is the mass density of the solid and μ and λ are the Lame coefficients. Under a dilatation, $\vec{u}(\vec{r}) = a\vec{r}$, the change in the energy density of the elastic medium is $\alpha^2(\lambda + 2\mu/3)/2$; under a shear stress, $u_x = \pi y$, $u_y = u_z = 0$, it is $\alpha^2\mu/2$. In a crystal – which has only a discrete rotational symmetry – there may be more parameters than just μ and λ, depending on the symmetry of the lattice. In a crystal with cubic symmetry, for instance, there are, in general, three independent parameters. We will make life simple, however, and make the approximation of full rotational invariance.

The solutions are,

$$\vec{u}(\vec{r},t) = \in e^{i(\vec{k}\cdot\vec{r}-\omega t)}$$

where is $\vec{\in}$ a unit polarization vector, satisfy

$$-\rho\omega^2 \vec{\in} - (\mu + \lambda)\vec{k}(\vec{k}\cdot\vec{\in}) - \mu k^2 \vec{\in}$$

For longitudinally polarized waves, $\vec{k} = \vec{k}\cdot\vec{\in}$,

$$\omega_k^l = \pm\sqrt{\frac{2\mu+\lambda}{\rho}}k \equiv \pm v_l k$$

while transverse waves, $\vec{k}\cdot\vec{\in} = 0$ have

$$\omega_k^t = \pm\sqrt{\frac{\mu}{\rho}}k \equiv \pm v_s k$$

Above, we introduced the concept of the polarization of a phonon. In 3D, the displacements of the ions can be in any direction. The two directions perpendicular to $\vec{k}$ are called transverse. Displacements along $\vec{k}$ are called longitudinal.

The Hamiltonian of this system,

$$H = \int d^3r \left(\frac{1}{2}\rho(\dot{\vec{u}})^2 + \frac{1}{2}(\mu+\lambda)(\vec{\nabla}\cdot\vec{u})^2 - \frac{1}{2}\mu\vec{u}\cdot\nabla^2\vec{u} \right)$$

can be rewritten in terms of creation and annihilation operators,

$$a_{k,s} = \frac{1}{\sqrt{2\hbar}}\left[\sqrt{\rho\omega_{k,s}}\,\vec{\epsilon}_s\cdot\vec{u}_k + i\sqrt{\frac{\rho}{\omega_{k,s}}}\,\vec{\epsilon}_s\cdot\dot{\vec{u}}_k\right]$$

$$a^\dagger_{k,s} = \frac{1}{\sqrt{2\hbar}}\left[\sqrt{\rho\omega_{k,s}}\,\vec{\epsilon}_s\cdot\vec{u}_{-k} - i\sqrt{\frac{\rho}{\omega_{k,s}}}\,\vec{\epsilon}_s\cdot\dot{\vec{u}}_{-k}\right]$$

as

$$H = \sum_{k,s}\hbar\omega_{k,s}\left(a^\dagger_{k,s}a_{k,s} + \frac{1}{2}\right)$$

Inverting the above definitions, we can express the displacement $\vec{u}(r)$ in terms of the creation and annihilation operators:

$$\vec{u}(\vec{r}) = \sum_{k,s}\sqrt{\frac{\hbar}{2\rho V\omega^s_k}}\,\vec{\epsilon}_s\,(a_{\vec{k},s} + a^\dagger_{-\vec{k},s})e^{i\vec{k}.\vec{r}}$$

s = 1, 2, 3 corresponds to the longitudinal and two transverse polarizations. Acting with $\vec{u}_k$ either annihilates a phonon of momentum k or creates a phonon of momentum k.

The allowed $\vec{k}$ values are determined by the boundary conditions in a finite system. For periodic boundary conditions in a cubic system of size $V = L^3$, the allowed $\vec{k}$'s are $\frac{2\pi}{L}(n_1, n_2, n_3)$. Hence, the ~k-space volume per allowed $\vec{k}$ is $(2\pi)^3$/V. Hence, we can take the inπ finite -volume limit by making the replacement:

$$\sum_k f(\vec{k}) = \frac{1}{(\Delta\vec{k})^3}\sum_k f(\vec{k})(\Delta\vec{k})^3$$

$$= \frac{V}{(2\pi)^3}\int d^3\vec{k}\, f(\vec{k})$$

DEBYE THEORY

Since a solid can be modelled as a collection of independent oscillators, we canobtain the energy in thermal equilibrium using the Planck distribution function:

$$E = V\sum_s \int_{\text{B.Z.}} \frac{d^3k}{(2\pi)^3} \frac{\hbar\omega_s(k)}{e^{\beta\hbar\omega_s(k)}-1}$$

where $s = 1, 2, 3$ are the three polarizations of the phonons and the integral is over the Brillouin zone.

This can be rewritten in terms of the phonon density of states, $g(\omega)$ as:

$$E = V\int_0^\infty d\omega\, g(\omega) \frac{\hbar\omega}{e^{\beta\hbar\omega}-1}$$

where

$$g(\omega) = \sum_s \int_{\text{B.Z.}} \frac{d^3k}{(2\pi)^3} \delta(\omega - \omega_s(k))$$

The total number of states is given by:

$$\int_0^\infty d\omega\, g(\omega) = \int_0^\infty d\omega \sum_s \int_{\text{B.Z.}} \frac{d^3k}{(2\pi)^3} \delta(\omega - \omega_s(k))$$

$$= 3\int_{\text{B.Z}} \frac{d^3k}{(2\pi)^3}$$

$$= 3\frac{N_{\text{ions}}}{V}$$

The total number of normal modes is equal to the total number of ion degrees of freedom.

For a continuum elastic medium, there are two transverse modes with velocity v_t and one longitudinal mode with velocity v_l. In the limit that the lattice spacing is very small, $a \to 0$, we expect this theory to be valid. In this limit, the Brillouin zone is all of momentum space, so

$$g_{\text{CEM}}(\omega) = \int \frac{d^3k}{(2\pi)^3} (2\delta(\omega - v_t k)) + \delta(\omega - v_l k))$$

$$= \frac{1}{2\pi^2}\left(\frac{2}{v_t^3} + \frac{1}{v_l^3}\right)\omega^2$$

In a crystalline solid, this will be a reasonable approximation to $g(\omega)$ for $k_B T \ll hv_t/a$ where the only phonons present will be at low energies, far from the Brillouin zone boundary. At high temperatures, there will be thermally excited phonons near the Brillouin zone boundary, where the spectrum is definitely not linear, so we cannot use the continuum approximation. In particular, this $g(\omega)$ does not have a finite integral, which violates the condition that the integral should be the total number of degrees of freedom.

A simple approximation, due to Debye, is to replace the Brillouin zone by a sphere of radius k_D and assume that the spectrum is linear up to k_D. In other words, Debye assumed that:

$$g_D(\omega) = \begin{cases} \dfrac{3}{2\pi^2 v^3} & \text{if } \omega < \omega_D \\ 0 & \text{if } \omega > \omega_D \end{cases}$$

Here, we have assumed, for simplicity, that $v_l = v_t$ and we have written $\omega_D = v k_D$. ω_D is chosen so that

$$3\frac{N_{\text{ions}}}{V} = \int_0^\infty d\omega\, g(\omega)$$

$$= \int_0^{\omega_D} d\omega \frac{3}{2\pi^2 v^3}\omega^2$$

$$= \frac{\omega_D^2}{2\pi^2 v^3}$$

i.e.

$$\omega_D = (6\pi^2 v^3 N_{\text{ions}}/V)^{\frac{1}{3}}$$

With this choice,

$$E = V\int_0^{\omega_D} d\omega \frac{3}{2\pi^2 v^3}\omega^2 \frac{\hbar\omega}{e^{\beta\hbar\omega}-1}$$

$$= V\frac{3(k_BT)^4}{2\pi^2 v^2 \hbar^3}\int_0^{\beta\hbar\omega_D} dx \frac{x^3}{e^x - 1}$$

In the low temperature limit, $\beta\hbar\omega_D \to \infty$, we can take the upper limit of the integral to ∞ and:

$$E \approx V\frac{3(k_BT)^4}{2\pi^2 v^3 \hbar^3}\int_0^{\infty} dx \frac{x^3}{e^x - 1}$$

The specific heat is:

$$C_V \approx T^3\left[V\frac{12k_B^4}{2\pi^2 v^2 \hbar^3}\int_0^{\infty} dx \frac{x^3}{e^x - 1}\right]$$

The T^3 contribution to the specific heat of a solid is often the most important contribution to the measured specific heat.

For $T \to \infty$,

$$E \approx V\frac{3(k_BT)^4}{2\pi^2 v^2 \hbar^3}\int_0^{\beta\hbar\omega_D} dx\, x^2$$

$$= V\frac{\omega_D^2}{2\pi^2 v^3} k_BT$$

$$= 3N_{\text{ions}} k_BT$$

so

$$C_V \approx 3N_{\text{ions}}\, k_B T$$

The high-temperature specific heat is just $k_B/2$ times the number of degrees of freedom, as in classical statistical mechanics. At high-temperature, we were guaranteed the right result since the density of states was normalized to give the correct total number of degrees of freedom. At low-temperature, we obtain a qualitatively correct result since the spectrum is linear.

To obtain the exact result, we need to allow for longitudinal and transverse velocities which depend on the direction, $v_t(\hat{k}), v_l(\hat{k})$, since rotational invariance is not present.

Debye's formula interpolates between these well-understood limits.

We can define θ_D by $k_B\theta_D = \hbar\omega_D$. For lead, which is soft, $\theta_D \approx 88\text{K}$, while for diamond, which is hard, $\theta_D \approx 1280\text{K}$.

MORE REALISTIC PHONON SPECTRA: OPTICAL PHONONS, VAN

Although Debye's theory is reasonable, it clearly oversimplifies certain aspects of the physics. For instance, consider a crystal with a two-site basis. Half of the phonon modes will be optical modes. A crude approximation for the optical modes is an Einstein spectrum:

$$g_E(\omega) = \frac{N_{\text{ions}}}{2}\,\delta(\omega - \omega_E)$$

In such a case, the energy will be:

$$E = V\frac{3(k_B T)^4}{2\pi^2 v^3 \hbar^3}\int_0^{\beta\hbar\omega_{\max}} dx \frac{x^3}{e^x - 1} + V\frac{N_{\text{ions}}}{2}\frac{\hbar\omega_E}{e^{\beta\hbar\omega_E} - 1}$$

with $\omega_{\max}$ chosen so that

$$3\frac{N_{\text{ions}}}{2} = \frac{\omega_{\max}^3}{2\pi^2 v^3}$$

Another feature missed by Debye's approximation is the existence of singularities in the phonon density of states. Consider the spectrum of the linear chain:

$$\omega(k) = 2\left(\frac{B}{m}\right)^{\frac{1}{2}}\left|\sin\frac{ka}{2}\right|$$

The minimum of this spectrum is at $k = 0$. Here, the density of states is well described by Debye theory which, for a $1D$ chain predicts $g(\omega) \sim \text{const}$:. The maximum is at $k = \pi/a$. Near the maximum, Debye theory breaks down; the density of states is singular:

$$g(\omega) = \frac{2}{\pi a}\frac{1}{\sqrt{\omega_{\max}^2 - \omega^2}}$$

In 3D, the singularity will be milder, but still present. Consider a cubic lattice.

The spectrum can be expanded about a maximum as:

$$\omega(k) = \omega_{\max} - \alpha_x (k_x^{\max} - k_x)^2 - \alpha_y (k_y^{\max} - k_y)^2 - \alpha_z (k_z^{\max} - k_z)^2$$

Then (6 maxima; 1/2 of each ellipsoid is in the B.Z.)

$$G(\omega) \equiv \int_{\omega}^{\omega_{\max}} d\omega\, g(\omega)$$

$$= 6 \cdot \frac{1}{2} \cdot \frac{V}{(2\pi)^3} (\text{vol. of ellipsoid})$$

$$= 3 \frac{V}{(2\pi)^3} \frac{4}{3} \pi \frac{(\omega_{\max} - \omega)^{\frac{3}{2}}}{(\alpha_z, \alpha_y, \alpha_z)}$$

Differentiating:

$$g(\omega) = \frac{3V}{4\pi^2} \frac{(\omega_{\max} - \omega)^{\frac{1}{2}}}{(\alpha_z, \alpha_y, \alpha_z)^{\frac{1}{2}}}$$

In $2D$ and $3D$, there can also be saddle points, where $\vec{\nabla}_k \omega(k) = 0$, but the eigenvalues of the second derivative matrix have different signs. At a saddle point, the phonon spectrum again has a square root singularity. van Hove proved that every 3D phonon spectrum has at least one maximum and two saddle points (one with one negative eigenvalue, one with two negative eigenvalues).

To see why this might be true, draw the spectrum in the full k-space, repeating the Brillouin zone. Imagine drawing lines connecting the minima of the spectrum to the nearest neighboring minima (i.e. from each copy of the B.Z. to its neighbors). Imagine doing the same with the maxima. These lines intersect; at these intersections, we expect saddle points.

LATTICE STRUCTURES

Thus far, we have focussed on general properties of the vibrational physics of crystalline solids. Real crystals come in a variety of different lattice structures, to which we now turn our attention.

BRAVAIS LATTICES

Bravais lattices are the underlying structure of a crystal. A 3D Bravais lattice is defined by the set of vectors $\vec{R}$

$$\{\vec{R} \mid \vec{R} = n_1\vec{a}_1 + n_2\vec{a}_2 + n_3\vec{a}_3; ni \in Z\}$$

where the vectors $\vec{a}_i$ are the basis vectors of the Bravais lattice. (Do not confuse with a lattice with a basis.) Every point of a Bravais lattice is equivalent to every other point. In an elemental crystal, it is possible that the elemental ions are located at the vertices of a Bravais lattice. In general, a crystal structure will be a Bravais lattice with a basis.

The symmetry group of a Bravais lattice is the group of translations through the lattice vectors together with some discrete rotation group about (any) one of the lattice points. In the problem set you will show that this rotation group can only have 2-fold, 3-fold, 4-fold, and 6-fold rotation axes.

There are 5 different types of Bravais lattice in 2D: square, rectangular, hexagonal, oblique, and body-centered rectangular. There are 14 different types of Bravais lattices in 3D. We will content ourselves with listing the Bravais lattices and discussing some important examples.

Bravais lattices can be grouped according to their symmetries. All but one can be obtained by deforming the cubic lattices to lower the symmetry.

- *Cubic symmetry:* cubic, FCC, BCC
- *Tetragonal:* stretched in one direction, $a \times a \times c$; tetragonal, centered tetragonal
- *Orthorhombic:* sides of 3 different lengths $a \times b \times c$, at right angles to each other; orthorhombic, base-centered, face-centered, body-centered.
- *Monoclinic:* One face is a parallelogram, the other two are rectangular; monoclinic, centered monoclinic.
- *Triclinic:* All faces are parallelograms.
- *Trigonal:* Each face is an $a \times a$ rhombus.
- *Hexagonal:* 2D hexagonal lattices of side a, stacked directly above one another, with spacing c.

Examples:

- Simple cubic lattice: $\vec{a}_i = a\hat{x}_i$.
- Body-centered cubic (BCC) lattice: points of a cubic lattice, together with the centers of the cubes ≅ interpenetrating cubic lattices opset by 1=2 the body-diagonal.

$$\vec{a}_1 = a\hat{x}_1, \quad \vec{a}_2 = a\hat{x}_2, \quad \vec{a}_3 = \frac{a}{2}(\hat{x}_1 + \hat{x}_2 + \hat{x}_3)$$

  Examples: Ba, Li, Na, Fe, K, Tl

- Face-centered cubic (FCC) lattice: points of a cubic lattice, together with the centers of the sides of the cubes, ≅ interpenetrating cubic lattices opset by 1=2 a face-diagonal.

$$\vec{a}_1 = \frac{a}{2}(\hat{x}_2 + \hat{x}_3), \qquad \vec{a}_2 = \frac{a}{2}(\hat{x}_1 + \hat{x}_3), \qquad \vec{a}_3 = \frac{a}{2}(\hat{x}_1 + \hat{x}_2)$$

  Examples: Al, Au, Cu, Pb, Pt, Ca, Ce, Ar.

- *Hexagonal Lattice:* Parallel planes of triangular lattices.

$$\vec{a}_1 = a\hat{x}_1, \quad \vec{a}_2 = \frac{a}{2}\left(\hat{x}_1 + \sqrt{3}\hat{x}_2\right), \quad \vec{a}_3 = c\hat{x}_3.$$

Bravais lattices can be broken up into unit cells such that all of space can be recovered by translating a unit cell through all possible lattice vectors. A primitive unit cell is a unit cell of minimal volume. There are many possible choices of primitive unit cells. Given a basis, $\vec{a}_1, \vec{a}_2, \vec{a}_3$, a simple choice of unit cell is the region:

$$\{\vec{r} \mid \vec{r} = x_1\vec{a}_1 + x_2\vec{a}_2 + x_3\vec{a}_3; x_i \in [0,1]\}$$

The volume of this primitive unit cell and, thus, any primitive unit cell is:

$$\vec{a}_1 \cdot \vec{a}_2 \times \vec{a}_3$$

An alternate, symmetrical choice is the Wigner-Seitz cell: the set of all points which are closer to the origin than to any other point of the lattice. Examples:

Wigner-Seitz for square=square, hexagonal=hexagon (not parallelogram), oblique = distorted hexagon, BCC=octohedron with each vertex cut oπ to give an extra square face (A + M).

Reciprocal Lattices

If $\vec{a}_1, \vec{a}_2, \vec{a}_3$, span a Bravais lattice, then

$$\vec{b}_1 = 2\pi \frac{\vec{a}_2 \times \vec{a}_3}{\vec{a}_1 \cdot \vec{a}_2 \times \vec{a}_3}$$

$$\vec{b}_2 = 2\pi \frac{\vec{a}_3 \times \vec{a}_1}{\vec{a}_1 \cdot \vec{a}_2 \times \vec{a}_3}$$

$$\vec{b}_3 = 2\pi \frac{\vec{a}_1 \times \vec{a}_2}{\vec{a}_1 \cdot \vec{a}_2 \times \vec{a}_3}$$

span the reciprocal lattice, which is also a bravais lattice.

The reciprocal of the reciprocal lattice is the set of all vectors $\vec{r}$ satisfying $e^{i\vec{G}\cdot\vec{r}} = 1$ for any recprocal lattice vector $\vec{G}$, i.e. it is the original lattice.

As we discussed above, a simple cubic lattice spanned by

$$a\hat{x}_1, \quad a\hat{x}_2, \quad a\hat{x}_3$$

has the simple cubic reciprocal lattice spanned by:

$$\frac{2\pi}{a}\hat{x}_1, \quad \frac{2\pi}{a}\hat{x}_2, \quad \frac{2\pi}{a}\hat{x}_3$$

An FCC lattice spanned by:

$$\frac{a}{2}(\hat{x}_2 + \hat{x}_3), \quad \frac{a}{2}(\hat{x}_1 + \hat{x}_3), \quad \frac{a}{2}(\hat{x}_1 + \hat{x}_2)$$

has a BCC reciprocal lattice spanned by:

$$\frac{4\pi}{a}\frac{1}{2}(\hat{x}_2 + \hat{x}_3 - \hat{x}_1), \quad \frac{4\pi}{a}\frac{1}{2}(\hat{x}_1 + \hat{x}_3 - \hat{x}_2), \quad \frac{4\pi}{a}\frac{1}{2}(\hat{x}_1 + \hat{x}_2 - \hat{x}_3)$$

Conversely, a BCC lattice has an FCC reciprocal lattice.

The Wigner-Seitz primitive unit cell of the reciprocal lattice is theπ first Brillouin zone. In the problem set, you will show that the Brillouin zone has volume $(2\pi)^3/v$ if the volume of the unit cell of the original lattice is v. The first Brillouin zone is enclosed in the planes which are the perpendicular bisectors of the reciprocal lattice vectors. These planes are called Bragg planes for reasons which will become clear below.

Bravais Lattices with a Basis

Most crystalline solids are not Bravais lattices: not every ionic site is equivalent to every other. In a compound this is necessarily true; even in elemental crystals it is often the case that there are inequivalent sites in the crystal structure. These crystal structures are lattices with a basis. The classification of such structures is discused in Ashcroft and Mermin. Again, we will content ourselves with discussing some important examples.

- *Honeycomb Lattice (2D):* A triangular lattice with a two-site basis. The triangular lattice is spanned by:

$$\vec{a}_1 = \frac{a}{2}\left(\sqrt{3}\,\hat{x}_1 + 3\hat{x}_2\right), \quad \vec{a}_2 = a\sqrt{3}\,\hat{x}_1$$

- The two-site basis is:

$$0, \quad a\hat{x}_2$$

Example: Graphite

- *Diamond Lattice:* FCC lattice with a two-site basis: The two-site basis is:

$$0, \quad \frac{a}{4}(\hat{x}_1 + \hat{x}_2 + \hat{x}_3)$$

Example: Diamond, Si, Ge

- *Hexagonal Close-Packed (HCP):* Hexagonal lattice with a two-site basis:

$$0, \quad \frac{a}{2}\hat{x}_1 + \frac{a}{2\sqrt{3}}\hat{x}_2 + \frac{c}{2}\hat{x}_3$$

Examples: Be,Mg,Zn,...

- *Sodium Chloride:* Cubic lattice with Na and Cl at alternate sites ≅ FCC lattice with a two-site basis:

$$0, \quad \frac{a}{2}(\hat{x}_1 + \hat{x}_2 + \hat{x}_3)$$

Examples: NaCl, NaF, KCl

BRAGG SCATTERING

One way of experimentally probing a condensed matter system involves scattering a photon or neutron oπ the system and studying the energy and angular dependence of the resulting cross-section. Crystal structure experiments have usually been done with X-rays.

Let us first examine this problem intuitively and then in a more systematic fashion. Consider, first, elastic scattering of X-rays. Think of the X-rays as photons which can take different paths through the crystal. Consider the case in which $\vec{k}$ is the wave vector of an incoming photon and $\vec{k}'$ is the wave vector of an outgoing photon. Let us, furthermore, assume that the photon only scatters off one of the atoms in the crystal (the probability of multiple scattering is very low). This atom can be any one of the atoms in the crystal. These different scattring events will interfere constructively if the path lengths differ by an integer number of wavelengths. The extra path length for a scattering off an atom at $\vec{R}$, as compared to an atom at the origin is:

$$\vec{R}\cdot\vec{k}+(-\vec{R}\cdot\vec{k}')=\vec{R}\cdot(\vec{k}\cdot\vec{k}')$$

If this is an integral multiple of 2π for all lattice vectors $\vec{R}$, then scattering interferes constructively. By definition, this implies that $\vec{k}-\vec{k}'$ must be a reciprocal lattice vector. For elastic scatering, $|\vec{k}|=|\vec{k}'|$, so this implies that there is a reciprocal lattic vector $\vec{G}$ of magnitude

$$|\vec{G}|=2|\vec{k}|\sin\theta$$

where θ is the angle between the incoming and outgoing X-rays.

To rederive this result more formally, let us assume that our crystal is in thermal equilibrium at inverse temperature β and that photons interact with our crystal via the Hamiltonian H'. Suppose that photons of momentum $\vec{k}_i$, and energy ω_i are scattered by our system. The differential cross-section for the photons to be scattered into a solid angle $d\Omega$ centered about $\vec{k}_f$ and into the energy range between $\omega_f \pm d\omega$ is:

$$\frac{d^2\sigma}{d\Omega d\omega}=\sum_{m,n}\frac{k_f}{k_i}\left|\left\langle\vec{k}_f;m|H'|\vec{k}_i;n\right\rangle\right|^2 e^{-\beta E_n}\delta(\omega+E_n-E_m)$$

where $\omega=\omega_i-\omega_f$ and n and m label the initial and final states of our crystal. Let $\vec{q}-\vec{k}_f-\vec{k}_i$.

Let us assume that the interactions between the photon and the ions in our system is of the form:

$$H'=\sum_R U(\vec{x}-(\vec{R}+\vec{u}(\vec{R})))$$

Then

$$\begin{aligned}
\langle \vec{k}_f; m|H'|\vec{k}_i; n\rangle &= \int d^3\vec{x}\frac{1}{V}e^{i\vec{q}\cdot\vec{x}}\left\langle m\left|\sum_R U(\vec{x}-(\vec{R}+\vec{u}(\vec{R})))\right|n\right\rangle \\
&= \left\langle m\left|\frac{1}{V}\sum_R e^{-i\vec{q}\cdot(\vec{R}+\vec{u}(\vec{R}))}\right|n\right\rangle \\
&= \frac{1}{V}\sum_R\left[e^{-i\vec{q}\cdot\vec{R}}\left\langle m\left|e^{-i\vec{q}\cdot\vec{u}(\vec{R})}\right|n\right\rangle\right]\tilde{U}(\vec{q}) \\
&= \left[\frac{1}{V}\sum_R e^{-i\vec{q}\cdot\vec{R}}\right]\left\langle m\left|e^{-i\vec{q}\cdot\vec{u}(0)}\right|n\right\rangle\tilde{U}(\vec{q})
\end{aligned}$$

Let us consider,π first, the case of elastic scattering, in which the state of the crystal does not change. Then $|n\rangle = |m\rangle$, $|k_i| = |k_f| \equiv k$, $|q|\ 2k\sin\frac{\theta}{2}$, and:

$$\langle \vec{k}_f; n|H'|\vec{k}_i; n\rangle = \left[\frac{1}{V}\sum_R e^{-i\vec{q}\cdot\vec{R}}\right]\left\langle n\left|e^{-i\vec{q}\cdot\vec{u}(0)}\right|n\right\rangle\tilde{U}(\vec{q})$$

Let us focus on the sum over the lattice:

$$\frac{1}{V}\sum_R e^{-i\vec{q}\cdot\vec{R}} = \sum_{\text{R.L.V.G}}\delta_{\vec{q},\vec{G}}$$

The sum is 1 if $\vec{q}$ is a reciprocal lattice vector and vanishes otherwise. The scattering cross-section is given by:

$$\frac{d^2\sigma}{d\Omega d\omega} = \sum_n e^{-\beta E_n}\left|\left\langle n\left|e^{-i\vec{q}\cdot\vec{u}(0)}\right|n\right\rangle\right|^2\left|\tilde{U}(\vec{q})\right|^2\sum_{\text{R.L.V.G}}\delta_{\vec{q},\vec{G}}$$

In other words, the scattering cross-section is peaked when the photon is scattered through a reciprocal lattice vector $\vec{k}_f = \vec{k}_i + \vec{G}$. For elastic scattering, this requires

$$\left(\vec{k}_i\right)^2 = \left(\vec{k}_i + \vec{G}\right)^2$$

or,

$$\left(\vec{G}\right)^2 = -2\vec{k}_i\cdot\vec{G}$$

This is called the Bragg condition. It is satisπ ed when the endpoint of $\vec{k}$ is on a Bragg plane. When it is satisfied, Bragg scattering occurs.

When there is structure within the unit cell, as in a lattice with a basis, the formula is slightly more complicated. We can replace the photon-ion interaction by:

$$H' = \sum_R\sum_b U_b\left(\vec{x}-\left(\vec{R}+\vec{b}+\vec{u}\left(\vec{R}+\vec{b}\right)\right)\right)$$

Then,

$$\frac{1}{V}\sum_{R} e^{-i\vec{q}.\vec{R}\tilde{U}(\vec{q})}$$

is replaced by

$$\frac{1}{V}\sum_{R} f_q e^{-i\vec{q}.\vec{R}}$$

where

$$f_q = \sum_{b} U_b(\vec{x}) e| ^{-i\vec{q}.\vec{x}}$$

As a result of the structure factor, f_q, the scattering amplitude need not have a peak at every reciprocal lattic vector, $\vec{q}$.

Of course, the probability that the detector is set up at precisely the right angle to receive $\vec{k}_f = \vec{k}_i + \vec{G}$ is very low. Hence, these experiments are usually done with a powder so that there will be Bragg scattering whenever $2k\sin\frac{\theta}{2} = |G|$. By varying θ, a series of peaks are seen at, e.g. π/6, π/4, etc., from which the reciprocal lattice
vectors are reconstructed.

Since $|k| \sim |G| \sim (1\mathring{A})^{-1}$, the energy of the incoming photons is ~ħck~ 10^4eV which is definitely in the X-ray range.

Thus far, we have not looked closely at the factor:

$$\left|\left\langle m \right| e^{-i\vec{q}.\vec{u}(0)} \left| n \right\rangle\right|^2$$

This factor results from the vibration of the lattice due to phonons. In elastic scattering, the amplitude of the peak will be reduced by this factor since the probability of the ions forming a perfect lattice is less than 1. The inelastic amplitude will contain contributions from processes in which the incoming photon or neutron creates a phonon, thereby losing some energy. By measuring inelastic neutron scattering (for which the energy resolution is better than for X-rays), we can learn a great deal about the phonon spectrum.

Chapter 12

Electronic Bands

Thus far, we have ignored the dynamics of the elctrons and focussed on the ionic vibrations. However, the electrons are important for many properties of solids. In metals, the specific heat is actually $C_V = \gamma T + \alpha T^3$. The linear term is due to the electrons. Electrical conduction is almost always due to the electrons, so we will need to understand the dynamics of electrons in solids in order to compute, for instance, the conductivity $\sigma(T, \omega)$.

In order to do this, we will need to understand the quantum mechanics of electrons in the periodic potential due to the ions. Such an analysis will enable us to understand some broad features of the electronic properties of crystalline solids, such as the distinction between metals and insulators.

BLOCH'S THEOREM

Let us first neglect all interactions between the electrons and focus on the interactions between each electron and the ions. This may seem crazy since the inter-electron interaction isn't small, but let us make this approximation and proceed. At some level, we can say that we will include the electronic contribution to the potential in some average sense so that the electrons move in the potential created by the ions and by the average electron density (of course, we should actually do this self consistently). Later, we will see why this is sensible.

When the electrons do not interact with each other, the many-electron wavefunction can be constructed as a Slater determinant of single-electron wavefunctions.

Hence, we have reduced the problem to that of a single electron moving in a lattice of ions. The Hamiltonian for such a problem is:

$$H = -\frac{\hbar^2}{2m}\nabla^2 + \sum_R V\left(\vec{x} - \vec{R} - \vec{u}(\vec{R})\right)$$

expanding in powers of $\vec{u}(\vec{R})$,

$$H = -\frac{\hbar^2}{2m}\nabla^2 + \sum_R V(\vec{x} - \vec{R}) - \sum_R \vec{\nabla} V(\vec{x} - \vec{R}).\vec{u}(\vec{R}) + \ldots$$

The third term and the ... are electron-phonon interaction terms. They can be treated as a perturbation. W e w ill focus on the π first two terms, which describe an electron moving in a periodic potential. This highly simplified problem already contains much of the qualitative physics of a solid.

Let us begin by proving an important theorem due to Bloch.

Bloch's Theorem: If V $(\vec{r} + \vec{R})$ = V $(\vec{r})$ for all lattice vectors $\vec{R}$ of some given lattice, then for any solution of the Schrodinger equation in this potential,

$$-\frac{\hbar^2}{2m}\nabla^2\psi(\vec{r}) + V(\vec{r})\psi(\vec{r}) = E\psi(\vec{r})$$

there exists a $\vec{k}$ such that

$$\psi(\vec{r} + \vec{R}) = e^{i\vec{k}\cdot\vec{R}}\psi(\vec{r})$$

Proof: Consider the lattice translation operator T_R which acts according to

$$T_{R\chi}(\vec{r}) = \chi(\vec{r} + \vec{R})$$

Then

$$T_R H_\chi(\vec{r}) = HT_R\chi(\vec{r} + \vec{R})$$

i.e.$[T_R, H] = 0$. Hence, we can take our energy eigenstates to be eigenstates of T_R.

Hence, for any energy eigenstate $\psi(\vec{r})$,

$$T_{R\psi}(\vec{r}) = c(\vec{R})\psi(\vec{r})$$

The additivity of the translation group implies that,

$$c(\vec{R})c(\vec{R}) = c(\vec{R} + \vec{R}')$$

Hence, there is some k such that

$$c(\vec{R}) = c^{i\vec{k}\cdot\vec{R}}$$

Since $e^{i\vec{G}\cdot\vec{R}} = 1$ if $\vec{G}$ is a reciprocal lattice vector, we can always take ~k to be in the first Brillouin zone.

TIGHT-BINDING MODELS

Let's consider a very simple model of a 1D solid in which we imagine that the atomic nuclei lie along a chain of spacing a. Consider a single ion and focus on two of its electronic energy levels. In real systems, we will probably consider s and d orbitals, but this is not important here; in our toy model, these are simply two electronic states which are localized about the atomic nucleus.

We'll call them $|1\rangle$ and $|2\rangle$, with energies $\in_1^0$ and $\in_2^0$. Let's further imagine that the splitting $\in_2^0 - \in_1^0$ between these levels is large. Now, when we put this atom in the linear chain, there will be some overlap between these levels and the corresponding energy levels on neighboring atoms. We can model such a system by the Hamiltonian:

$$H = \sum_R \left(\epsilon_0^1 \,|R,1\rangle\langle R,|1 + \epsilon_0^2 |R,2\rangle\langle R,2| \right) - \sum_{R,R'\text{n.n}} \left(t_1|R,1\rangle\langle R',1| + t_2|R,2\rangle\langle R',2| \right)$$

We have assumed that t_1 is the amplitude for an electron at $|R, 1\rangle$ to hop to $|R', 1\rangle$, and similarly for t_2. For simplicity, we have ignored the possibility of hopping from $|R, 1\rangle$ to $|R', 2\rangle$, which is unimportant anyway when ϵ_0 is large. The eigenstates of this Hamiltonian are:

$$|k,1\rangle = \sum_R e^{ikR} |R,1\rangle$$

with energy

$$\epsilon_1(k) = \epsilon_1^0 - 2t_1 \cos ka$$

and

$$|k,2\rangle = \sum_R e^{ikR} |R,2\rangle$$

with energy

$$\epsilon_2(k) = \epsilon_2^0 - 2t_2 \cos ka$$

Note, first, that k lives in the first Brillouin zone since

$$|k,i\rangle \equiv \left| k + \frac{2\pi n}{a}, i \right\rangle$$

Now, observe that the two atomic energy levels have broadened into two energy bands. There is a band gap between these bands of magnitude $\epsilon_2^0 - \epsilon_1^0 - 2t_1 2t_2$. This is a characteristic feature of electronic states in a periodic potential: the states break up into bands with energy gaps separating the bands.

How many states are there in each band? As we discussed in the context of phonons, there are as many allowed k's in the Brillouin zone as there are ions in the crystal. Let's repeat the argument. The Brillouin zone has k-space extent $2\pi/a$. In a finite-size system of length L with periodic boundary conditions, allowed k's are of the form $2\pi n/L$ where n is an integer. Hence, there are $L/a = N_{\text{ions}}$ allowed k's in the Brillouin zone. (This argument generalizes to arbitrary lattices in arbitrary dimension.)

Hence, there are as many states as lattice sites. Each state can be filled by one up-spin electon and one down-spin electron.

Hence, if the atom is monovalent – i.e. if there is one electron per site – then, in the ground state, the lower band, $|k, 1\rangle$ is half-filled and the upper band is empty. The Fermi energy is at ϵ_1^0. The Fermi momentum (or, more properly, Fermi crystal momentum) is at $\pm\pi/2a$. At low temperature, the fact that there is a gap far away from the Fermi momentum is unimportant, and the Fermi sea will behave just like the Fermi sea of a free Fermi gas. In particular, there is no energy gap in the many-electron spectrum since we can always excite an electron from a filled state just below the Fermi surface to one of the unfilled states just above the gap.

For instance, the electronic contribution to the specific heat will be $C_V \sim T$. The difference is that the density of states will be different from that of a free Fermi gas. In situations such as this, when a band is partially filled, the crystal is (almost always) a metal.

If there are two electrons per lattice site, then the lower band isϕilled and the upper band is empty in the ground state. In such a case, there is an energy gap $E_g = \in_2^0 - \in_1^0 - 2t_1 - 2t_2$ between the ground state and the lowest excited state which necessarily involves exciting an electron from the lower band to the upper band. Crystals of this type, which have no partiallyϕilled bands, are insulators. The electronic contribution to the specific heat will be suppressed by a factor of $e^{-Eg/T}$.

Note that the above tight-binding model can be generalized to arbitrary dimension of lattice. For instance, a cubic lattice with one orbital per site has tight-binding spectrum:

$$\in(k) = -2t(\cos k_x a + \cos k_y a + \cos k_z a)$$

Again, if there is one electron per site, the band will be half-filled (and metallic); if there are two electrons per site the band will be filled (and insulating).

The model which we have just examined is grossly oversimplified but can, never-theless, be justified, to an extent. Let us reconsider our lattice of atoms.

Electronic orbitals of an isolated atom:

$$\varphi_n(\vec{r})$$

with energies $\in_n$:

$$-\frac{\hbar^2}{2m}\nabla^2\varphi_n(\vec{r}) + V(\vec{r})\varphi_n(\vec{r}) = \in_n \varphi_n(\vec{r})$$

We now want to solve:

$$-\frac{\hbar^2}{2m}\nabla^2\psi_n(\vec{r}) + \sum_R V(\vec{r}+\vec{R})\psi_n(\vec{r}) = \in(k)\psi_n(\vec{r})$$

Let's try the ansatz:

$$\psi_k(\vec{r}) = \sum_{R,n} c_n e^{i\vec{k}\cdot\vec{R}}\varphi_n(\vec{r}+\vec{R})$$

which satisfies Bloch's theorem. Substituting into SchrodingeΔs equation and taking the matrix element with φ_m, we get:

$$\int d^3r\,\varphi_m^*(\vec{r})\left(-\frac{\hbar^2}{2m}\nabla^2+\sum\nolimits_{R'}V(r'+\vec{R}')\right)\sum\nolimits_{R,n}c_n e^{i\vec{k}\cdot\vec{R}}\varphi_n(\vec{r}+\vec{R}) = \in(k)\int d^3r\,\varphi_m^*(\vec{r})\sum\nolimits_{R,n}c_n e^{i\vec{k}\cdot\vec{R}}\varphi_n(\vec{r}+\vec{R})$$

Let's write

$$-\frac{\hbar^2}{2m}\nabla^2+\sum_R V(\vec{r}+\vec{R}') = -\frac{\hbar^2}{2m}\nabla^2+V(\vec{r}+\vec{R}')\sum_{R'\neq R}V(\vec{r}+\vec{R}')$$

$$= H_{\text{at},R}+\Delta V_R(r)$$

Then, we have

$$\int d^3r\varphi_m^*(\vec{r})\sum_{R,n}c_n e^{i\vec{k}\cdot\vec{R}}(H_{\text{at},R}+\Delta V_R(r))\,\varphi_n(\vec{r}+\vec{R}) = \in(k)$$

$$\int d^3r\varphi_m^*(\vec{r})\sum_{R,n}c_n e^{i\vec{k}\cdot\vec{R}}\varphi_n(\vec{r}+\vec{R})$$

Let's write

$$-\frac{\hbar^2}{2m}\nabla^2+\sum_{R'}V(\vec{r}+\vec{R}') = -\frac{\hbar^2}{2m}\nabla^2+V(\vec{r}+\vec{R}')+\sum_{R'\neq R}V(\vec{r}+\vec{R}')$$

$$= H_{\text{at},R}+\Delta V_R(r)$$

Then, we have

$$\int d^3r\varphi_m^*(\vec{r})\sum_{R,n}c_n e^{i\vec{k}\cdot\vec{R}}(H_{\text{at},R}+\Delta V_R(r))\,\varphi_n(\vec{r}+\vec{R}) = \in(k)$$

$$\int d^3r\varphi_m^*(\vec{r})\sum_{R,n}c_n e^{i\vec{k}\cdot\vec{R}}\,\varphi_n(\vec{r}+\vec{R})$$

$$\sum_{R,n}c_n\in_n e^{i\vec{k}\cdot\vec{R}}\int d^3r\varphi_m^*(\vec{r})\varphi_n(\vec{r}+\vec{R})+\sum_{R,n}c_n e^{i\vec{k}\cdot\vec{R}}$$

$$\int d^3r\varphi_m^*(\vec{r})\Delta V_R(r)\,\varphi_n(\vec{r}+\vec{R}) = \in(k)c_m+\in(k)$$

$$\sum_{R\neq 0,n}e^{i\vec{k}\cdot\vec{R}}c_n\int d^3r\varphi_m^*(\vec{r})\varphi_n(\vec{r}+\vec{R})$$

$$c_m\in_m+\sum\nolimits_{R\neq 0,n}c_n\in_n e^{e\vec{k}\cdot\vec{R}}\int d^3r\varphi_m^*(\vec{r})\varphi_n(\vec{r}+\vec{R})+$$

$$\sum\nolimits_{R,n}c_n e^{e\vec{k}\cdot\vec{R}}\int d^3r\varphi_m^*(\vec{r})[\Delta V_R(r)]\varphi_n(\vec{r}+\vec{R}) =$$

$$\in(k)c_m+\sum\nolimits_{R\neq 0,n}e^{e\vec{k}\cdot\vec{R}}c_n\int d^3r\varphi_m^*(\vec{r})\varphi_n(\vec{r}+\vec{R})$$

Writing:

$$\alpha_{mn}(\vec{R}) = \int d^3r\varphi_m^*(\vec{r})\varphi_n(\vec{r}+\vec{R})$$

$$\gamma_{mn}(\vec{R}) = -\int d^3r\varphi_m^*(\vec{r})[\Delta V_R(r)]\,\varphi_n(\vec{r}+\vec{R})$$

We have:

$$c_m(\in_m - \in(k)) + \sum_{R\neq 0,n} c_n(\in_n - \in(k))\, e^{i\vec{k}\cdot\vec{R}}\alpha_{mn}(\vec{R}) = \sum_{R,n} c_n\, e^{i\vec{k}\cdot\vec{R}}\gamma_{mn}(\vec{R})$$

Both $\alpha_{mn}(\vec{R})$ and $_{mn}(\vec{R})$ are exponentially small, $\sim e^{-R/a_0}$. In particular, $\alpha_{mn}(\vec{R})$ and $\gamma_{mn}(\vec{R})$ are much larger for nearest neighbors than for any other sites, so let's neglect the other matrix elements and write $\alpha_{mn} = \alpha_{mn}(\vec{R}_{\text{n.n.}})$, $\gamma_{mn} = \gamma_{mn}(\vec{R}_{\text{n.n}})$, $v_{\text{mn}} = \gamma_{mn}(0)$. In problem 2 of problem set 7, so may make these approximations. Suppose that we make the approximation that the lth orbital is well separated in energy from the others. Then we can neglect $\alpha_{ln}(\vec{R})$ and $\gamma_{ln}(\vec{R})$ for $n \neq l$. We write $\beta = v_{ll}$. Focusing on the $m = l$ equation, we have:

$$(\in_l - \in(k) + \alpha_{ll}(\in_l - \in(k))\sum_{R_{\text{n.n}}} e^{i\vec{k}\cdot\vec{R}} = \beta + \gamma_{ll}\sum_{R_{\text{n.n}}} e^{i\vec{k}\cdot\vec{R}}$$

Hence,

$$\in(k) = \in(k) = \in_l - \frac{\beta + \gamma_{ll}\sum_{R_{\text{n.n.}}} e^{i\vec{k}\cdot\vec{R}}}{1 + \alpha_{ll}\sum_{R_{\text{n.n.}}} e^{i\vec{k}\cdot\vec{R}}}$$

If we neglect the π's and retain only the γ's, then we recover the result of our phenomenological model. For instance, for the cubic lattice, we have:

$$\in(k) = [\in_l - \beta] - 2\gamma_{ll}\,[\cos k_x a + \cos k_y a + \cos k_z a]$$

Tight-binding models give electronic wavefunctions which are a coherent superposition of localized atomic orbitals. Such wavefunctions have very small amplitude in the interstitial regions between the ions.

Such models are valid, as we have seen, when there is very little overlap between atomic wavefunctions on neighboring atoms. In other words, a tight-binding model will be valid when the size of an atomic orbital is smaller than the interatomic distance, i.e. $a_0 \ll R$. In the case of core electrons, e.g. $1s$, $2s$, $2p$, this is the case. However, this is often not the case for valence electrons, e.g. 3s electrons. Nevertheless, the tight-binding method is a simple method which gives many qualitative features of electronic bands. In the study of high-T_c superconductivity, it has proven useful for this reason.

THE δ-

Let us now consider another simple toy-model of a solid, a 1D array of δ-functions:

$$\left(-\frac{\hbar^2}{2m}\frac{d^2}{dx^2}+V\sum_{n=-\infty}^{\infty}\delta(x-na)\right)\psi(x)=E\psi(x)$$

Between the peaks of the δ functions, $\psi(x)$ must be a superposition of the plane waves e^{iqx} and e^{-iqx} with energy $E(q) = \hbar^2q^2/2m$. Between $x = 0$ and $x = a$,

$$\psi(x) = e^{iqx+i\alpha} + e^{-iqx-i\pi}$$

with α complex. According to Bloch's theorem,

$$\psi(x + a) = e^{ika}\,\psi(x)$$

Hence, in the region between $x = a$ and $x = 2a$,

$$\psi(x)=e^{ika}\left(e^{iq(x-a)+ia}+e^{-iq(x-a)-i\alpha}\right)$$

Note that k which determines the transformation property under a translation $x \to x + a$ is not the same as q, which is the 'local' momentum of the electron, which determines the energy. Continuity at $x = a$ implies that

$$\cos(qa + \alpha) = e^{ika}\cos\alpha$$

or,

$$\tan\alpha=\frac{\cos qa-e^{ika}}{\sin qa}$$

Integrating Schrodinge's equation from $x=a-\epsilon$ to $x=a+\epsilon$, we have,

$$\sin(qa+\alpha)-e^{ika}\sin\alpha=\frac{2mV}{\hbar^2q}e^{ika}\cos\alpha$$

or,

$$\tan\alpha=\frac{\dfrac{2mV}{\hbar^2q}e^{ika}-\sin qa}{\cos qa-e^{ika}}$$

Combining these equations,

$$e^{2ika}-2\left(\cos qa+\frac{mV}{\hbar^2q}\sin qa\right)e^{ika}+1=0$$

The sum of the two roots is cos *ka*:

$$\cos ka=\frac{\cos(qa-\delta)}{\cos\delta}$$

where

$$\tan\delta=\frac{mV}{\hbar^2q}$$

For each $k\in\left[-\frac{\pi}{a},\frac{\pi}{a}\right]$, there are infinitely many roots q of this equation, $q_n(k)$. The energy spectrum of the nth band is:

$$E_n(k) = \frac{\hbar^2}{2m}[q_n(k)]^2$$

$\pm k$ have the same root $q_n(k) = q_n(-k)$. Not all q's are allowed. For instance, the values $qa - \delta = n\pi$ are not allowed. These regions are the energy gaps between bands. Consider, for instance, $k = \pi/a$.

$$\cos(qa - \pi) = \cos\delta$$

This has the solutions

$$qa = \pi, \pi + 2\delta$$

For V small, the latter solution occurs at $qa = \pi + \dfrac{2mV_a}{\pi\hbar^2}$. The energy gap is:

$$E_2\left(\frac{\pi}{a}\right) - E_1\left(\frac{\pi}{a}\right) \approx E\left(\frac{\pi}{a} + \frac{2mV_a}{\pi\hbar^2}\right) - E\left(\frac{\pi}{a}\right)$$

$$\approx 2V/a$$

NEARLY FREE ELECTRON APPROXIMATION

According to Bloch's theorem, electronic wavefunctions can be expanded as:

$$\psi(x) = \sum_G c_{k-G} e^{i(\vec{k}-\vec{G})\cdot\vec{x}}$$

In the nearly free electron approximation, we assume that electronic wavefunctions are given by the superposition of a small number of plane waves. This approximation is valid, for instance, when the periodic potential is weak and contains a limited number of reciprocal lattice vectors.

Let's see how this works. Schrodinge's equation in momentum space reads:

$$\left(\frac{\hbar^2 k^2}{2m} - \in(\vec{k})\right) + c_k + \sum_G c_{k-G} V_G = 0$$

Second-order perturbation theory tells us that (let's assume that $V_k = 0$)

$$\in(\vec{k}) = \in_0(\vec{k}) + \sum_{G\neq 0} \frac{|V_G|^2}{\in_0(\vec{k}) - \in_0(\vec{k}-\vec{G})}$$

where

$$\in_0(\vec{k}) = \frac{\hbar^2 k^2}{2m}$$

Perturbation theory will be valid so long as the second term is small, i.e. so long as

$$|V_G| \ll \in_0(\vec{k}) - \in_0(\vec{k}-\vec{G})$$

For generic $\vec{k}$, this will be valid if V_G is small. The correction to the energy will be $O\left(|V_G|^2\right)$. However, no matter how small V_G is, perturbation theory fails for degenerate states,

$$\frac{\hbar^2 k^2}{2m} = \frac{\hbar^2 (k-G)^2}{2m}$$

or, when the Bragg condition is satisfied,

$$G^2 = 2\vec{k}\cdot\vec{G}$$

In other words, perturbation theory fails when $\vec{k}$ is near a Brillouin zone boundary. Suppose that V_G is very small so that we can neglect it away from the Brillouin zone boundaries. Near a zone boundary, we can focus on the reciprocal lattice vector which it bisects, $\vec{G}$ and ignore $V_{G'}$ for $\vec{G} \neq \vec{G}'$. We keep only c_k and c_{k-G}, where $\in_0(k) \approx \in_0(k - G)$. We can thereby reduce Schrodinge's equation to a 2 × 2 equation:

$$\left(\in_0(\vec{k}) - \in(\vec{k})\right)c_k + c_{k-G}V_G = 0$$

$$\left(\in_0(\vec{k}-\vec{G}) - \in(\vec{k}-\vec{G})\right)c_{k-G} + c_k V_G^* = 0$$

$V_{G'}$ for $\vec{G} \neq \vec{G}'$ can be handled by perturbation theory and, therefore, neglected in the small $V_{G'}$ limit. In this approximation, the eigenvalues are:

$$\in_\pm(\vec{k}) = \frac{1}{2}\left[\in_0(\vec{k}) + \in_0(\vec{k}-\vec{G}) \pm \sqrt{(\in_0(\vec{k}) - \in_0(\vec{k}-\vec{G}))^2 + 4|V_G|^2}\right]$$

At the zone boundary, the bands have been split by

$$\in_+(\vec{k}) - \in_-(\vec{k}) = 2|V_G|$$

The effects of $V_{G'}$ for $\vec{G} \neq \vec{G}'$ are now handled perturbatively.

To summarize, the nearly free electron approximation gives energy bands which are essentially free electron bands away from the Brillouin zone boundaries; near the Brillouin zone boundaries, where the electronic crystal momenta satisfy the Bragg condition, gaps are opened.

Though intuitively appealing, the nearly free electron approximation is not very reasonable for real solids. Since

$$V_G \approx \frac{4\pi Ze^2}{G^2} \sim 13.6\text{eV}$$

while $\pi_F \sim 10$eV,

$$|V_G| \sim \in_0(\vec{k}) - \in_0(\vec{k}-\vec{G})$$

and the nearly free electron approximation is not valid.

SOME GENERAL PROPERTIES OF ELECTRONIC BAND STRUCTURE

Much, much more can be said about electronic band structure. There are many approximate methods of obtaining energy spectra for more realistic potentials. We will content ourselves with two observations. Band Overlap. In 2D and 3D, bands can overlap in energy. As a result, both the first and second bands can be partiallyϕilled when there are two electrons per site. Consider, for instance, a weak periodic potential of rectangular symmetry:

$$V(x,y) = V_x \cos\frac{2\pi}{a_1}x + V_y \cos\frac{2\pi}{a_2}x$$

with $V_{x,y}$ very small and $a_1 \gg a_2$. Using the nearly free electron approximation, we have a spectrum which is essentially a free-electron parabola, with small gaps opening at the zone boundary.

Since the Brillouin zone is much shorter in the k_x direction, the Fermi sea will cross the zone boundary in this direction, but not in the k_y-direction. Hence, there will be empty states in the first Brillouin zone, near $(0, \pm\pi a_2)$ and occupied states in the second Brillouin zone, near $(\pm\pi a_1, 0)$.

This is a general feature of 2D and 3D bands. As a result, a solid can be metallic even when it has two electrons per unit cell. Van . A second feature of electronic energy spectra is the existence of van . They are singularities in the electronic density of states, $g(\in)$

$$\int d\in g(\in)\, f(\in) = \int \frac{d^3k}{(2\pi)^3} f(\in(k))$$

They occur for precisely the same reason as in the case of phonon spectra – as a result of the lattice periodicity.

Consider the case of a tight-binding model on the square lattice with nearestneighbor hopping only.

$$\in(k) = -2t(\cos k_x a + \cos k_y a)$$

$$\nabla_k \in(k) = 2ta(\sin k_x a + \sin k_y a)$$

The density of states is given by:

$$g(\in) = 2\int \frac{d^2k}{(2\pi)^2} \delta(\in - \in(k))$$

Let's change variables in the integral on the right to E and S which is the arc length around an equal energy contour $\in\ = \in(k)$:

$$g(\in) = \frac{1}{2\pi^2} \int dS \frac{dE}{|\vec{\nabla}_k \in(k)|} \delta(\in - E)$$

$$= \frac{1}{2\pi^2} \int dS \frac{dE}{|\vec{\nabla}_k \in(k)|}$$

The denominator on the right-hand-side vanishes at the minimum of the band, $\vec{k}\ = (0, 0)$, the maxima $\vec{k}\ = (\pm\pi/a, \pm\pi/a)$ and the saddle points

$\vec{k} = (\pm\pi/a, 0), (0, \pm\pi/a)$. At the latter points, the density of states will have divergent slope.

THE FERMI SURFACE

The Fermi surface is defined by

$$\in_n (k) = \mu$$

By the Pauli principle, it is the surface in the Brillouin zones which separates the occupied states, $\in_n(k) < \mu$, inside the Fermi surface from the unoccupied states $\in_n (k) > \mu$ outside the Fermi surface. All low-energy electronic excitations involve holes just below the Fermi surface or electrons just above it. Metals have a Fermi surface and, therefore, low-energy excitations. Insulators have no Fermi surface: μ lies in a band gap, so there is no solution to (5.64).

In the low-density limit the Fermi surface is approximately circular (in 2D) or spherical (in 3D). Consider the 2D tight-binding model

$$\in(k) = -2t\,(\cos k_x a + \cos k_y a)$$

For $\vec{k} \to 0$,

$$\in(k) \approx -4t + ta^2(k_x^2 + k_x^2)$$

Hence, for $\mu + 4t \ll t$, the Fermi surface is given by the circle:

$$k_x^2 + k_x^2 + \frac{\mu + 4t}{ta^2}$$

Similarly, in the nearly free electron approximation,

$$\in(\vec{k}) = \frac{\hbar^2 k^2}{2m} + \sum_{G \neq 0} \frac{|V_G|^2}{\dfrac{\hbar^2 k^2}{2m} - \dfrac{\hbar^2 (k-G)^2}{2m}}$$

For $\pi \to 0$ and V_G small, we can neglect the scond term, and, as case, the Fermi surface is given by

$$k = \frac{1}{\hbar}\sqrt{2m\mu}$$

in the free electron

Away from the bottom of a band, however, the Fermi surface can look quite different. In the tight-binding model, for instance, for $\mu = 0$, the Fermi surface is the diamond $k_x \pm k_y = \pm\pi/a$.

The chemical potential at zero temperature is usually called the Fermi energy, $\in_F$. The key measure of the number of low-lying states which are available to an electronic system is the density of states at the Fermi energy, $g(\in_F)$. When $g(\in_F)$ is large, the C_V, σ, etc. are large; when $g(\in_F)$ is small, these quantities are small.

METALS, INSULATORS, AND SEMICONDUCTORS

Earlier we saw that, in order to compute the vibrational properties of a solid, we needed to determine the phonon spectra of the crystal. A characteristic feature of these phonon spectra is that there is always an acoustic mode with $\omega(k) \sim k$ for k small. This mode is responsible for carrying sound in a solid, and it always gives a $C_V^{ph} \sim T^3$ contribution to the specific heat.

In order to compute the electronic properties of a solid, we must similarly determine the electronic spectra. If we ignore the interactions between electrons, the electronic spectra are determined by the single-electron energy levels in the periodic potential due to the ions.

These energy spectra break up into bands. When there is a partially filled band, there are low energy excitations, and the solid is a metal. There will be a $C_V^{el} \sim T$ electronic contribution to the specific heat, as in *a*.

When all bands are either filled or completely empty, there is a gap between the many-electron ground state and then first excited state; the solid is an insulator and there is a negligible contribution to the low-temperature specific heat. Let us recall how this works. Once we have determined the electronic band structure, $\in_n(k)$, we can determine the electronic density-of-states:

$$g(\in) = 2\sum_n \int_{\text{B.Z.}} \frac{d^2k}{(2\pi)^2} \delta(\in - \in_n(k))$$

With the density-of-states in hand, we can compute the thermodynamics. In the limit $k_B T \ll \in F$,

$$\frac{N}{V} = \int_0^\infty d\in g(\in) \frac{1}{e^{\beta(e-\mu)}+1}$$

$$= \int_0^\mu d\in g(\in) + \int_0^\mu d\in g(\in)\left(\frac{1}{e^{\beta(e-\mu)}+1} - 1\right) + \int_\mu^\infty d\in g(\in)\frac{1}{e^{\beta(\in-\mu)}+1}$$

$$= \int_0^{\in_F} d\in g(\in) + \int_{\in_F}^\mu d\in g(\in) - \int_0^\mu d\in g(\in)\frac{1}{e^{-\beta(e-\mu)}+1} +$$

$$\int_\mu^\infty d\in g(\in)\frac{1}{e^{\beta(\in-\mu)}+1}$$

$$\approx \frac{N}{V} + (\mu - \in_F)\, g(\in_F) + \int_0^\infty \frac{k_B T\, dx}{e^x+1} + (g(\mu + k_B Tx) -$$

$$g(\mu - k_B Tx)) + O(e^{-\beta\mu})$$

$$= \frac{N}{V} + (\mu - \in_F)\, g(\in_F) + \sum_{n=1}^\infty \frac{(k_B T)^{n+1}}{n!} g^{(n)}(\mu) \int_0^\infty dx \frac{x^{2n-1}}{e^x+1}$$

$$\approx \frac{N}{V} + (\mu - \in_F)\, g(\in_F) + (k_B T)^2\, g'(\in_F)\, I_1$$

with

$$I_k = \int_0^\infty dx \frac{x^k}{e^x + 1}$$

We will only need

$$I_1 = \frac{\pi^2}{6}$$

Hence, to lowest order in T,

$$(\mu - \in_F) g(\in_F) \approx -(k_B T)^2\, g'(\in_F) I_1$$

Meanwhile,

$$\frac{N}{V} = \int_0^\infty d\in \in g(\in) \frac{1}{e^{\beta(e-\mu)} + 1}$$

$$= \int_0^\mu d\in \in g(\in) + \int_0^\mu d\in \in g(\in)\left(\frac{1}{e^{\beta(e-\mu)} + 1} - 1\right) + \int_\mu^\infty d\in \in g(\in) \frac{1}{e^{\beta(\in-\mu)} + 1}$$

$$\approx \frac{N}{V} + (\mu - \in_F) \in_F\, g(\in_F) - \int_0^\mu d\in \in g(\in) \int_0^\mu d\in \in g(\in) \frac{1}{e^{-\beta(e-\mu)} + 1} +$$

$$\int_\mu^\infty d\in \in g(\in) \frac{1}{e^{\beta(\in-\mu)} + 1}$$

$$= \frac{N}{V} + (\mu - \in_F) \in_F\, g(\in_F) + \int_0^\infty \frac{k_B T\, dx}{e^x + 1} + ((\mu + k_B T x)\, g(\mu - k_B T x)) -$$

$$(\mu - k_B T x)\, g(\mu - k_B T x) + O(e^{-\beta\mu})$$

$$= \frac{E_0}{V} + (\mu - \in_F) \in_F\, g(\in_F) + (k_B T)^2 [g(\in_F) + \in_F\, g'(\in_F)] I_1$$

Substituting (5.74) into theπ nal line of (5.75), we have:

$$\frac{E}{V} = \frac{E_0}{V} + (k_B T)^2\, g(\in_F)\, I_1$$

Hence, the electronic contribution to the low-temperature specific heat of a crystalline solid is:

$$\frac{C_V}{V} = \frac{\pi^2}{3} + k_B^2 T g(\in_F)$$

In a metal, the Fermi energy lies in some band; hence $g(\in_F)$ is non-zero. In an insulator, all bands are either completely full or completely empty. Hence, the Fermi energy lies between two bands, and $g(\in_F) = 0$.

Each band contains twice as many single-electron levels (the factor of 2 comes from the spin) as there are lattice sites in the solid. Hence, an insulator must have an even number of electrons per unit cell. A metal will result if there is an odd number of electrons per unit cell (unless the electron-electron interactions, which we have neglected, are strong); as a result of band overlap, a metal can also result if there is an even number of electrons per unit cell.

A semiconductor is an insulator with a small band gap. A good insulator will have a band gap of $E_g \sim 4$eV. At room temperature, the number of electrons which will be excited out of the highestfilled band and into the lowest empty band will be $\sim e^{-E_g/2k^B T} \sim 10^{-35}$ which is negligible.

Hence, theϕilled and empty bands will remain filled and empty despite thermal excitation. A semiconductor can have a band gap of order $E_g \sim 0.25$ 1eV. As a result, the thermal excitation of electrons can be as high as $\sim e^{-E_g/2k^B T} \sim 10^{-2}$. Hence, there will be a small number of carriers excited into the empty band, and some conduction can occur. Doping a semiconductor with impurities can enhance this.

The basic property of a metal is that it conducts electricity. Some insight into electrical conduction can be gained from the classical equations of motion of a electron, i.e. Drude theory:

$$\frac{d}{dt}\vec{r} = \frac{1}{m}\vec{p}$$

$$\frac{d}{dt}\vec{p} = -eE(\vec{r},t) - \frac{e}{m}\vec{p} \times B(\vec{r},t)$$

If we continue to treat the electric and magnetic field s classically, but treat the electrons in a periodic potential quantum mechanically, this is replaced by:

$$\frac{d}{dt}\vec{r} = v_n(k) = \frac{1}{\hbar}\vec{\nabla}_k \in_n (k)$$

$$\hbar\frac{d}{dt}\vec{k} = -eE(\vec{r},t) - ev_n(k) \times B(\vec{r},t) - \frac{\hbar}{\tau}\vec{k}$$

Theπ nal term in the second equation is the scattering rate. It is caused by effects which we have neglected in our analysis thus far: impurities, phonons, and electronelectron interactions. Without these effects, electrons would accelerate forever in a constant electric field, and the conductivity would be inπ finite. As a result of scattering, σ is finite. Hence, a finite electric field leads to aπ finite current:

$$\vec{j} = \sum_n \int \frac{d^3k}{(2\pi)^3} \frac{1}{\hbar}\vec{\nabla}_k \in_n (k)$$

Filled bands give zero contribution to the current since they vanish by integration by parts. Since an insulator has onlyϕilled or empty bands, it cannot carry current. Hence, it is not characterized by its conductivity but, instead, by its dielectric constant, $\in$.

ELECTRONS IN A MAGNETIC FIELD: LANDAU BANDS

In 1879, E.H. Hall performed an experiment designed to determine the sign of the current-carrying particles in metals. If we suppose that these particles have charge *e* (with a sign to be determined) and mass m, the classical equations of motion of charged particles in an electric field, $E = E_x\hat{x} + E_y\hat{y}$, and a magnetic field, $B = B\hat{z}$ are:

$$\frac{dp_x}{dt} = eE_x - \omega_c p_y - p_x / \tau$$

$$\frac{dp_y}{dt} = eE_y - \omega_c p_x - p_y / \tau$$

where $\omega_c = eB/m$ and τ is a relaxation rate determined by collisions with impurities, other electrons, etc. These are the equations which we would expect for free particles. In a crystalline solid, the momentum $\vec{p}$ must be replaced by the crystal momentum and the velocity of an electron is no longer $\vec{p}/m$, but is, instead,

$$\vec{v}(p) = \vec{\nabla}_p \in (p)$$

We won't worry about these subtleties for now. In the systems which we will be considering, the electron density will be very small. Hence, the electrons will be close to the bottom of the band, where we can approximate:

$$\in (k) = \in_0 + \frac{\hbar^2 k^2}{2m_b} + \cdots$$

where m_b is called the band mass. For instance, in the square lattice nearest-neighbour tight-binding model,

$$\in (k) = -2t(\cos k_x a + \cos k_y a)$$

$$\approx -4t + ta^2k^2 + \cdots$$

Hence,

$$m_b = \frac{\hbar^2}{2ta^2}$$

In GaAs, $m_b \approx 0{:}07 m_e$. Once we replace the mass of the electron by the band mass, we can approximate our electrons by free electrons.

Let us, following Hall, place a wire along the $\hat{x}$ direction in the above magnetic elds and run a current, j_x, through it. In the steady state, $dp_x/dt\ /dp_y\ /dt/j_y = 0$, we must have $E_x = \dfrac{m}{ne^2 r} j_x$ and

$$E_y = -\frac{B}{ne} j_x = \frac{-e}{|e|} \frac{h}{e^2} \frac{\Phi / \Phi_0}{N} j_x$$

where n and N are the density and number of electrons in the wire, Φ is the magnetic ux penetrating the wire, and $\Phi_0 = h/e$ is the ux quantum. Hence, the sign of the charge carriers can be determined from a measurement of the transverse voltage in a magnetic field. Furthermore, according to equation, the density of charge carriers Figure ρ_{xx} and ρ_{xy} vs. magnetic field, B, in the quantum Hall regime. *A* number of integer and fractional plateaus can be clearly seen. This data was taken at Princeton on a GaAs-AlGaAs heterostructure.

i.e. electrons – can be determined from the slope of the $\rho_{xy} = E_y/j_x$ vs B. At high temperatures, this is roughly what is observed.

In the quantum Hall regime, namely at low-temperatures and high magnetic field s, very different behaviour is found in two-dimensional electron systems. ρ_{xy} passes through a series of plateaus, $\rho_{xy} = \frac{1}{\nu}\frac{h}{e^2}$, where ν is a rational number, at which ρ_{xx} vanishes. The quantization is accurate to a few parts in 10^8, making this one of the most precise measurements of theπ ne structure constant, $\alpha = \frac{e^2}{\hbar c}$, and, in fact, one of the highest precision experiments of any kind.

Some insight into this phenomenon can be gained by considering the quantum mechanics of a single electron in a magnetic field. Let us suppose that the electron's motion is planar and that the magnetic field is perpendicular to the plane. For now, we will assume that the electron is spin-polarized by the magnetic field and ignore the spin degree of freedom. The Hamiltonian,

$$H = \frac{1}{2m}(-i\hbar\nabla + eA)^2$$

takes the form of a Hamiltonian in the gauge $A_x = -By$, $A_y = 0$. (Here, and in what follows, we will take $e = |e|$; the charge of the electron is $-e$.) If we write the wavefunction $\phi(x, y) = e^{ik_x x}\,\phi(y)$, then:

$$H\psi\left[\frac{1}{2m}(eBy + \hbar k_x)^2 + \frac{1}{2m}(-i\hbar\partial_y)^2\right]\phi(y)e^{ik_x x}$$

The energy levels $E_n = \left(n + \frac{1}{2m}\hbar\omega_c\right)$, called Landau levels, are highly degenerate because the energy is independent of k. To analyse this degeneracy, let us consider a system of size $L_x \times L_y$. If we assume periodic boundary conditions, then the allowed k_x values are $2\pi n/L_x$ for integer n. The wavefunctions are centered at $y = \hbar k/(eB)$, i.e. they have spacing $y_n - y_{n-1} = h/(eBL_x)$.

The number of these which will fit in L_y is $eBL_xL_y/h = BA/\Phi_0$. In other words, there are as many degenerate states in a Landau level as there are ux quanta.

It is often more convenient to work in symmetric gauge, $A = \frac{1}{2}B\times r$ Writing $z = x + iy$, we have:

$$H = \frac{\hbar^2}{m}\left[-2\left(\partial - \frac{\overline{z}}{4\ell_0^2}\right)\left(\overline{\partial} - \frac{z}{4\ell_0^2}\right) + \frac{1}{2\ell_0^2}\right]$$

with (unnormalized) energy eigenfunctions:

$$\psi_{n,m}(z,\overline{z}) = z^m L_n^m(z,\overline{z})\, e^{-\frac{|z|^2}{4\ell_0^2}}$$

at energies $E_n = (n + \frac{1}{2})\hbar\omega_c$, where $L_n^m(z,\overline{z})$ are the Laguerre polynomials and $\sqrt{\hbar/(eB)}$ is the magnetic length.

Let's concentrate on the lowest Landau level, $n = 0$. The wavefunctions in the lowest Landau level,

$$\psi_{n=0,m}(z,\overline{z}) = z^m e^{-\frac{|z|^2}{4\ell_0^2}}$$

are analytic functions of z multiplied by a Gaussian factor. The general lowest Landau level wavefunction can be written:

$$\psi_{n=0,m}(z,\overline{z}) = f(z) e^{-\frac{|z|^2}{4\ell_0^2}}$$

The state $\psi_{n=0,m}$ is concentrated on a narrow ring about the origin at radius $r_m = \ell_0\sqrt{2(m+1)}$. Suppose the electron is conπ ned to a disc in the plane of area A. Then the highest m for which $\psi_{\mathrm{n=0,m}}$ lies within the disc is given by $A = \pi r_{m\max}$, or, simply, $m_{\max} + 1 = \Phi/\Phi_0$, where $\Phi = BA$ is the total ux. Hence, we see that in the thermodynamic limit, there are Φ/Φ_0 degenerate single-electron states in the lowest Landau level of a two-dimensional electron system penetrated by a uniform magnetic flux Φ. The higher Landau levels have the same degeneracy. Higher Landau levels can, at a qualitative level, be thought of as copies of the lowest Landau level. The detailed structure of states in higher Landau levels is different, however.

Let us now imagine that we have not one, but many, electrons and let us ignore the interactions between these electrons. To completely fill p Landau levels, we need $N_e = p(\Phi/\Phi_0)$ electrons. Lorentz invariance tells us that if

$$n = p\frac{e^2}{h}B$$

then

$$j_x = p\frac{e^2}{h}E_y$$

i.e.

$$\sigma_{xy} = p\frac{e}{h}$$

The same result can be found by inverting the semi-classical resistivity matrix, and substituting this electron number.

Suppose that we fix the chemical potential, μ. As the magnetic field is varied, the energies of the Landau levels will shift relative to the chemical potential. However, so long as the chemical potential lies between two Landau levels, an integer number of Landau levels will beϕilled, and we expect toπ nd the quantized Hall conductance.

These simple considerations neglected two factors which are crucial to the observation of the quantum , namely the effects of impurities and inter-electron interactions. The integer quantum occurs in the regime in which impurities dominate; in the fractional quantum , interactions dominate.

The density of states in a pure system. So long as the chemical potential lies between Landau levels, a quantized conductance is observed. (b) Hypothetical density of states in a system with impurities. The Landau levels are broadened into bands and some of the states are localized. The shaded regions denote extended states. (c) As we mention later, numerical studies indicate that the extended state(s) occur only at the centre of the band.

The Integer Quantum

Let us model the effects of impurities by a random potential in which non-interacting electrons move. Clearly, such a potential will break the degeneracy of the different states in a Landau level.

More worrisome, still, is the possibility that some of the states might be localized by the random potential and therefore unable to carry any current at all. As a result of impurities, the Landau levels are broadened into bands and some of the states are localized.

Hence, we would be led to naively expect that the Hall conductance is less than $\frac{e^2}{h}p$ when p Landau levels areϕilled. In fact, this conclusion, though intuitive, is completely wrong. In a very instructive calculation (at least from a pedagogical standpoint), Prange analyzed the exactly solvable model of electrons in the lowest Landau level interacting with a single δ-function impurity. In this case, a single localized state, which carries no current, is formed. The current carried by each of the extended states is increased so as to exactly compensate for the localized state, and the conductance remains at the quantized value, $\sigma_{xy} = \frac{e^2}{h}$.

This calculation gives an important hint of the robustness of the quantization, but cannot be easily generalized to the physically relevant situation in which there is a random distribution of impurities. To understand

(a) The Corbino annular geometry. (b) Hypothetical distribution of energy levels as a function of radial distance. The quantization of the Hall conductance in this more general setting, we will turn to the beautiful arguments of Laughlin (and their refinement by Halperin), which relate it to gauge invariance.

Let us consider a two-dimensional electron gas confined to an annulus such that all of the impurities are confined to a smaller annulus. Since, as an experimental fact, the quantum is independent of the shape of the sample, we can choose any geometry that we like. This one, the Corbino geometry, is particularly convenient.

Outside the impurity region, there will simply be a Landau level, with energies that are pushed up at the edges of the sample by the walls (or a smooth conπ ning potential). In the impurity region, the Landau level will broaden into a band. Let us suppose that the chemical potential, μ, is above the lowest Landau level, $\mu > \hbar\omega_c/2$.

Then the only states at the chemical potential are at the inner and outer edges of the annulus and, possibly, in the impurity region. Let us further assume that the states at the chemical potential in the impurity region – if there are any – are all localized.

Now, let us slowly thread a time-dependent ux $\Phi(t)$ through the centre of the annulus. Locally, the associated vector potential is pure gauge. Hence, localized states, which do not wind around the annulus, are completely unaffected by the ux. Only extended states can be affected by the ux.

When an integer number of ux quanta thread the annulus, $\Phi(t) = p\Phi_0$, the ux can be gauged away everywhere in the annulus. As a result, the Hamiltonian in the annulus is gauge equivalent to the zero- ux Hamiltonian. Then, according to the adiabatic theorem, the system will be in some eigenstate of the $\Phi(t) = 0$ Hamiltonian.

In other words, the single-electron states will be unchanged. The only possible difference will be in the occupancies of the extended states near the chemical potential. Localized states are unaffected by the ux; states far from the chemical potential will be unable to make transitions to unoccupied states because the excitation energies associated with a slowly-varying ux will be too small. Hence, the only states that will be affected are the gapless states at the inner and outer edges. Since, by construction, these states are unaffected by impurities, we know how they are affected by the ux: each ux quantum removes an electron from the inner edge and adds an electron to the outer edge. Then, $\int I\,dt = e$ and $\int V\,dt = \int \frac{d\Phi}{dt} = h/e$, so:

$$I = \frac{e^2}{h} V,$$

Clearly, the key assumption is that there are no extended states at the chemical potential in the impurity region. If there were – and there probably are in samples that are too dirty to exhibit the quantum – then the above arguments break down. Numerical studies indicate that, so long as the strength of the impurity potential is small compared to $\hbar\omega_c$, extended states exist only at the centre of the Landau band. Hence, if the chemical potential is above the centre of the band, the conditions of our discussion are satisfied.

The other crucial assumption, emphasized by Halperin, is that there are gapless states at the edges of the system. In the special setup which we assumed, this was guaranteed because there were no impurities at the edges. In the integer quantum , these gapless states are a one-dimensional chiral Fermi liquid. Impurities are not expected to affect this because there can be no backscattering in a totally chiral system. More general arguments, which we will mention in the context of the fractional quantum , relate the existence of gapless edge excitations to gauge invariance.

One might, at first, be left with the uneasy feeling that these gauge invariance arguments are somehow too 'slick.'

To allay these worries, consider the annulus with a wedge cut out, which is topologically equivalent to a rectangle. In such a case, some of the Hall current will be carried by the edge states at the two cuts (i.e. the edges which run radially at fixed azimuthal angle). However, probes which measure the Hall voltage between the two cuts will effectively couple these two edges leading, once again, to annular topology.

Laughlin's argument for exact quantization will apply to the fractional quantum if we can show that the clean system has a gap. Then, we can argue that for an annular setup similar to the above there are no extended states at the chemical potential except at the edge. Then, if threading q ux quanta removes p electrons from the inner edge and adds p to the outer edge, as we would expect at $\nu = p/q$, we would have $\sigma_{xy} = \frac{p}{q}\frac{e^2}{h}$.

Index

U

V

W

X

Z